APPLIED SPATIAL ECOLOGY AND GISCIENCE

APPLIED SPATIAL ECOLOGY AND GISCIENCE

Foundations, Methods, Case Studies, and Lab Exercises

BRYAN C. PIJANOWSKI

PURDUE UNIVERSITY PRESS
WEST LAFAYETTE, INDIANA

Cataloging-in-Publication Data on file at the Library Congress.

978-1-62671-385-7 (cloth)

Cover images: Drone view over summer rural landscape, Ukraine: Panama7/iStock via Getty Images Plus; topographic map vector abstract background: Mindea/iStock via Getty Images Plus; close-up on a European Adonis blue butterfly, Polyommates bellargus on a white Clematis vitable flower: Henk Wallays/iStock via Getty Images Plus; space station deploys solar panels over planet Earth: 3DSculptor/iStock via Getty Images Plus; sompass with vintage map: AlexStar/iStock via Getty Images Plus; futuristic Holographic Topographic Environment: LV4260/iStock via Getty Images Plus, Earth seen from space showing lightmap over Asia: GettyImages-2164197505/iStock via Getty Images Plus.

To Dawn, Alex, and Benjamin

CONTENTS

PREFACE

This book is the culmination of over 20 years of teaching this topic to juniors in the College of Agriculture at Purdue University. The content of the course came mostly from my funded projects over the last 20+ years that helped me to advance areas in landscape ecology, biodiversity conservation, and applications of geographic information systems (GIS). The broad subjects, spatial ecology and its associated technical underpinnings of geographic information systems, the science of geography, and remote sensing of the environment, have formed the basis of my long-term research spanning 30 years.

Much of the research that I have conducted over these past decades has relied heavily on GIS. I was one of early converts to GIS, having been given the opportunity to learn Arc/INFO in 1991 while I was still a graduate student at Michigan State University. I was introduced to GIS by Professor Stuart Gage in the Department of Entomology at Michigan State University. Gage was one of the early innovators of applying GIS to study insect populations. He eventually hired me as a postdoc after I obtained my PhD in 1991 and there I learned to use GIS to map and analyze populations of the insect forest pest, the spongy moth. This was also the early days of the USDA's Slow the Spread project and so I was given the opportunity to serve on various national technical committees of this impressive large-scale project. From there, I used my knowledge of how to apply spatial technologies of modeling insect population expansion to urban systems. In the late 1990s, I worked with a variety of faculty and students at Michigan State University and the University of Michigan to develop a Land Transformation Model that coupled GIS and artificial neutral networks. That work eventually led to more transdisciplinary projects, such as the development of integrated climate-land change-food security models in East Africa with a focus on areas around Mt. Kenya. Subsequent funding helped to support work to also do integrated land change and hydrological and stream biota modeling with hydrologists and stream biologists that our team called the MegaModel project. This MegaModel project, focused in the Muskegon River Watershed, eventually transitioned into a community engagement project that involved researchers and extension professionals at Purdue University, Michigan State University, and the University of Michigan. This project, called the Tipping Point Project, focused on delivering spatial information from existing databases and also from models developed by the core research team for use in planning across the Great Lakes Basin.

This brief personal history, I hope, provides the users of this book with an appreciation for the specific applications that are described in the Exercises. Indeed, many of the Exercises leverage the data sets (i.e., case studies) that my colleagues and I developed as part of these funded projects. Indeed, I hope that the users can take advantage of not only learning about spatial ecology and GISci but also perhaps use the data to further the research and knowledge that already exists in these regions.

The book is organized around two separate course components: Foundations and Exercises. The Foundations Section contains those topics that I have presented in lecture format. This section is set up to present important spatial ecology topics in each week of a typical 15-week semester of instructional time. The 15 labs of the Exercises Section are listed in the order that I have presented these topics to students for over 20 years. I've attempted, in these computer-based labs, to have 2- to 3-hour sessions per lab with time outside of lab devoted to writing up the exercises. I have taught this course as both an undergraduate course (i.e., junior-level course, with one 2-credit sophomore course prior to it that focuses on data management and analysis, many sessions with GIS) and a graduate-level course. I would argue that this course could be offered as a first GIS course for students in their undergraduate training. This is very likely an appropriate first GIS course for graduate students. In the Exercises portion of this book, I offer two levels of assignments: one set for undergraduate students (I call these Intermediate Students) and the other for graduate students (Advanced Students). Intermediate students are often asked to answer 3–5 questions that require maps or figures and then write up a brief description of the outcomes of the work tasks. The Advanced Student assignments often have a few more tasks required and then a more comprehensive writeup following the style of a lab report (which is summarized in a Foundations chapter (Chapter 15).

The data for use in these exercises are a combination of data from my projects (and thus represent CC0 distributed/use rules) and from government and nonprofit organizations that are distributed as one of several forms of Creative Commons data sources. At Purdue, these data "packages" are stored in the Course Management System (CMS) and students are able to download these data and associated files for use. The Exercises Section content has been tested in my undergraduate and graduate courses for the last 15 years.

Motivation to write this book came from the College of Agriculture's Dean's Office and Purdue University's Book Initiative. They encourage faculty to take their course notes and transition these to a book. Over the years, I have continued to develop course notes and a variety of instructional materials that made the development of this book straightforward. Indeed, I used the copious notes, a variety of which I consulted during these years of teaching the subject, and superb textbooks such as Turner's *Landscape Ecology in Theory and Practice* (2001), Lomolino et al.'s *Biogeography: Biodiversity in Space and Time* (2017), and *Principles of Conservation Biology* by Groom et al. (2005) to write both sections. The book that I present here is a synthesis of core concepts that are exemplified in the lab exercises that come mostly from research case studies.

This book was written while I taught the undergraduate versions of the lectures and labs and while I was teaching two versions of the graduate course, both online (one for on-campus students and another for students in an online version of the course that is part of a MS certificate in Spatial Data Sciences).

While these chapters were written, the undergraduate students were offered the opportunity to review the Foundations chapters. Over 120 individual reviews of these chapters were made. Comments and scores were requested for: (1) clarity, depth, and interest of the narratives; (2) relevance and clarity of illustrations; (3) balance of theory and examples to illustrate concepts and terms; (4) cross referencing; and (5) overall appeal of each chapter. The enrollment for the undergraduate course is typically around 45 students and for the graduate version of the course, around 15. Lab sessions have been smaller for the undergraduate course, averaging around 15 per course.

The Exercises Section of the book relies heavily on Esri ArcGIS Pro and its Spatial Analyst Extension and several of the Microsoft Office applications (Excel, PowerPoint, and Word). A few of the labs require the use of FRAGSTATS, a popular categorical map spatial pattern statistics package developed for use in landscape ecology research and natural resources management.

I would like to thank the following Purdue University students who reviewed and commented on earlier drafts of chapters: Mackenna Bowling, Taylore Thompson, Sam Lillie, Anne Skinner, Teague McClury, Maddie MacPhee, Tyler Vanderveer, Beca Applemann, Caleb Barnett, Taryn Steed, and Ian Gabbert. I would also like to thank the graduate students who have TA'd for this course. These include Amelie Davis, Kim Robinson, Jim Plourde, David Savage, Ruth Bowers-Sword, Bowen Li, and Adebola Adeniji.

Additional information concerning the book can be found on the author's Purdue faculty website.

FOUNDATIONS

This section describes the foundations of spatial ecology and GIScience. Spatial ecology is composed of four intertwined subjects: landscape ecology, biogeography, conservation science, and GIScience. The chapters on landscape ecology focus on specific themes such as landscapes and land cover, landscapes and water resources, landscapes and organisms, landscapes and food production, and landscapes and the senses. The second foundational area is biogeography—patterns of species distributions in space and time that tend to be larger in extent than landscape ecology. The foundational material in biogeography covers global energy budgets, biomes and ecoregions, and the insular theory of biogeography. Conservation science is dispersed across landscape ecology and biogeographical background information and frame these two other spatial ecologies as an application focus. The foundations of GIScience describe its unique history, spatial data models, and common spatial statistical approaches developed by this community with a chapter on cartography and color theory to aid in developing strong mapmaking skills. This section on foundations ends with a chapter summarizing the best practices of scientific writing for use in developing traditional lab reports and scientific papers. Each chapter also includes a set of chapter aims and learning objectives at the beginning along with some suggested discussion questions at the end. Terms in bold denote the foundational vocabulary of these four subjects.

1. OVERVIEW OF SPATIAL ECOLOGY

Highlights

Aims. This chapter provides a very high-level review of the major topics covered in spatial ecology and associated technologies. Its purpose is to help frame the book.

Learning objectives. Students should be able to:

- Compare and contrast landscape ecology, biogeography, and conservation biology in terms of their main themes and scales of focus
- Describe the key themes for each of the major topics of spatial ecology and GIScience
- Relate how various spatial technologies support these three spatial ecologies

1.1 INTRODUCTION

Spatial ecology includes the areas of study referred to as landscape ecology, biogeography, and conservation science. The field of spatial ecology focuses on patterns and processes affecting the distribution of organisms across habitats (Tilman & Kareiva, 1997). **Landscape ecology** (Turner et al., 2001) considers linkages between spatial patterns and ecological processes that occur at the extent of a forest stand to several hundred square kilometers. In the United States, researchers with the US Forest Service were instrumental in its early development and application to natural resource management. Today, the principles of landscape ecology continue to shape the management of public terrestrial and aquatic natural resources across nearly all federal and state agencies in the United States. The intellectual roots of landscape ecology can be traced back to European research in plant ecology and geography in the 1920s. In the United States, the need to manage large areas of land managed by several federal agencies like the US Bureau of Land Management (BLM), the US Forest Service (USFS), and the US National Park Service (NPS) led to funding a lot of theoretical and applied research starting in the early 1980s. A key question asked by a landscape ecologist might be: What forest patch size is necessary to sustain a particular bird species that I manage at my protected area?

Biogeography, on the other hand, explores the factors that influence geographic variation in organisms, from genes to communities, at regional to global scales (Lomolino et al., 2017). Biogeography may also consider extinct, as well as extant, biological species or communities, as understanding the evolutionary history of organisms is a key feature of this discipline. Thus, in general, biogeography considers spatial extents and time extents much greater than those of landscape ecology. The principles of biogeography are likely very important for understanding the management of species that are closely related but differ in their abundances and geographic distributions. A typical question of biogeographers might be: Why are there so many species in the tropics compared to other areas of the world?

Conservation science (*in sensu* Soulé, 1985; Kareiva & Marvier, 2012) focuses on the preservation of specific species to communities (i.e., multiple species assemblages) of organisms. At the core of conservation science is the notion that the management of animals requires an understanding of the dependency of animal survival on abiotic (e.g., soils, water) and biotic (e.g., plants, microbes, other animal species) environmental conditions as well as how socioeconomics and policy are integrated into conservation practices. Thus, both landscape ecology and conservation science also consider how human activities (e.g., building roads) impact ecosystem structure and function.

What links these spatial sciences is the way they combine spatial technologies with ecological theory. In particular, GIS and remote sensing—imagery from drones, aircraft, and satellites—are routinely integrated with landscape ecology, biogeography, and conservation science using general geographic principles that are framed by geographic information science or **GIScience**. For the purposes of this textbook, conservation science is considered a cross-cutting application-oriented theme in spatial ecology whereas GIScience serves as the technical cross-cutting theme.

1.2 KEY PRINCIPLES AND TERMS OF LANDSCAPE ECOLOGY

A. CONCEPTUAL MODELS AND KEY TERMS OF LANDSCAPES

Landscape ecologists focus on both landscapes and seascapes and consider areas that contain several key features, and these features have been described in relation to several conceptual landscape models that are presented here. The first is that landscapes are mosaics of patches, corridors that are arranged in a matrix—this has been referred to as the *patch-corridor-matrix model* (Figure 1.1). Here, a *patch* is a distinct ecological unit

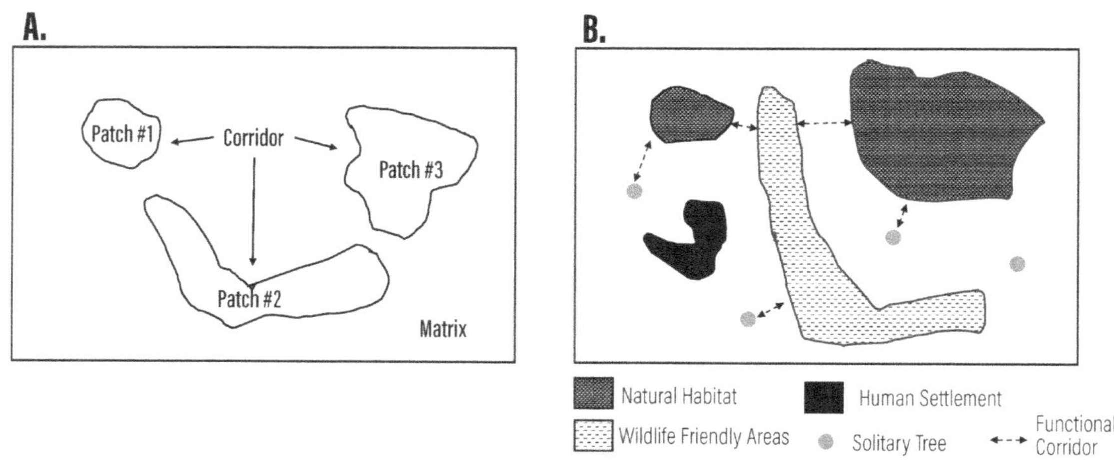

FIGURE 1.1. Two popular conceptualizations of a landscape including (A) the patch-corridor-matrix model and (B) the functional heterogeneity landscape model.

that has measurements of size, shape, and edge characteristics. Often, natural resource management concerns itself with all of these, as some species of birds prefer edge habitats (e.g., gray catbird) while others prefer deep interior locations (e.g., ovenbird), so the size of a forest patch is important. The contrast in *edges* can be sharp (e.g., a forest that borders an agricultural field) or soft (e.g., a forest that transitions to shrubs and then to a wetland). The *shape* of a patch will also dictate the amount of edge that a patch will have. It should be intuitive to you that the least amount of edge of a patch is a patch shape that is a circle. Likewise, "hot dog," or elongated, patch shapes contain large amounts of edge for any given patch size. Indeed, a key measurement made by landscape ecologists is the ratio of patch size to edge length.

The second popular conceptual model is that of the *functional heterogeneity landscape model* (Figure 1.1B). The functional heterogeneity model considers how patches and corridors of habitats function to support a variety of important life history attributes of animals (i.e., to support areas for breeding, feeding, predator avoidance, shelter). This model of a landscape considers how the spatial arrangement of these functional patches is connected in space. Functional habitat connections for wildlife, for example, may be close enough by distance (e.g., see the dashed arrows in the figure) so that wildlife can move between these dispersed habitat patches, creating a functional set of corridors.

Several key terms are used in landscape ecology (Table 1.1). It is best to know and use these specific terms rather than the more general term "scale," as scale can refer to any of the more specific terms. For example, is it best to use the term "resolution" rather than "scale of" to describe the size of a pixel used in grid or raster modeling of landscapes.

B. KEY THEMES OF LANDSCAPE ECOLOGY

Several key themes in landscape ecology are relevant to spatial ecology and GIScience. These themes (following Pijanowski, 2024) are:

TABLE 1.1. *Key terms used in landscape ecology (adapted from Turner et al., 2001)*

TERM	DEFINITION
Scale	Resolution, extent, or magnitude of any dimension of either space, time, or intensity.
Extent	Area under consideration that is defined by height and width but also may include depth. Units are often in square meters or square kilometers and in some cases, volumetric units, especially for seascapes. In GIScience parlance, a grid will represent the boundaries (i.e., extent) of the area of interest.
Resolution	Pixel size or grain size. This term is often used in spatial ecology and GIS to refer to the size of a cell that exists within a grid, image, or raster. In time, it can represent seconds, minutes, hours, days, years, decades, etc.
Intensity	Magnitude of a measure. Examples are stream flow rates, precipitation, or the magnitude of saturation in a band in a remote sensing image from space.
Level	All life on Earth can be viewed as hierarchically arranged. The dimensions of space, time, and magnitude all have levels of organization that need to be considered.
Threshold	Value of a measurement in space or time that results in a change of state (e.g., healthy changed to degraded).

Theme 1: Landscapes are heterogeneous.

One of the features of landscapes is that all are composed of land surface characteristics that are either natural or human dominated. The description of the surface of the land is called *land cover*. The broadest land cover types include forest, wetland, grassland, open water, rock, bare soil, and ice for natural surfaces, and agriculture and urban for human-dominated land surfaces. *Land use* is often used either in combination (e.g., as *land cover/use*) or separately to describe the use of the land for human purposes reflecting either economic or cultural purposing of the land.

Theme 2: Pattern and process are inextricably linked.

Landscape ecologists consider that the patterns occurring in a landscape (e.g., the arrangement of forest patches) influence processes (e.g., animal movement) that are critical to the landscape. There are many examples of the coupling of patterns and processes. The topography (e.g., pattern of elevation across the landscape) will influence water flow and eventually creates surface water bodies and potentially events like floods.

Landscape ecologists also argue that processes that occur in the landscape (e.g., the flow of water) will create patterns (e.g., the spatial arrangement of lakes and rivers). Another example would be that the rate of photosynthesis occurring at a place will influence the variety and abundances of tree species distributed across the landscape.

Theme 3: Human activities are part of the landscape.

The activities of humans (e.g., creating human settlements and roads, harvesting trees in forests, extracting soil for minerals) are nearly ubiquitous features of landscapes. Seascapes/aquatic systems (e.g., lakes and rivers, coral reefs) are also exposed to a variety of human activities that change their dynamics. For example, many surface bodies of the world have been changed due to nutrient runoff from agricultural fields.

Theme 4: Landscapes are dynamic.

Every location on Earth is undergoing change. Many of these changes result from natural influences (e.g., changes in climate) or alterations of the landscape by humans. Landscape change can be a slow process (e.g., tree species composition changing over hundreds to thousands of years) or fast (e.g., a forest stand being completely harvested). Some of the most significant changes over the past century have been urbanization as human settlement increases and agricultural expansion, which is needed to increase the human food supply.

Theme 5: Scaling is the critical lens for landscape ecology and natural resource management.

Life on Earth is hierarchically structured. We can consider the smallest level of biological organization to be molecules (e.g., amino acids), with the largest being the biosphere (i.e., all life on Earth). Within the spatial organization of life, landscape ecologists might consider spatial scaling to be forest stands (i.e., the smallest management unit), then compartments (e.g., collections of similar forest stands) and the full management unit (e.g., the protected area that is being managed). Society organizes itself using spatially arranged hierarchical units; the smallest are often referred to as municipalities, which are then grouped into townships, with townships grouped into counties, counties into states, and then states/provinces into a single nation.

Likewise, organisms are organized into the following hierarchical units: individuals, populations (of the same species), and then communities (the grouping of all species). One characteristic of most ecological hierarchies is emergent properties. As one moves up the hierarchical levels of organization, new, valuable properties emerge. Populations have measures of abundance and density, which are not measures of individuals. Likewise, communities possess the properties of richness (e.g., the number of species) and diversity (e.g., a measure of species abundance and species variety) that are not present at the lower levels of organization.

One of the general rules of thumb for landscape ecological analysis is to consider at least three spatial extents for the analysis with the focal extent having one level above and one below it to ensure that all considerations are studied. For example, if a natural resource manager is focusing on an endangered species of bird, the manager might consider tracking body measurements such as body weight (e.g., at the individual level), population abundances (e.g., at the population level), and avian diversity (e.g., at the community level), the latter of which might consider predator-prey relationships key to the survival of the endangered species under management focus.

1.3 KEY QUESTIONS ADDRESSED BY BIOGEOGRAPHERS

A. BRIEF HISTORY OF BIOGEOGRAPHY

Biogeography can also trace its historical roots back to the naturalists of the 18th and 19th centuries, including Charles Darwin, Alfred Russel Wallace, Alexander von Humboldt, Thomas Hooker, and Charles Lyell. Many of these naturalists traveled the world to learn and write about broad patterns of plant and animal life. Many inspirational stories about their adventures made their work very popular among the educated class in Europe and North America. Biogeography advanced in many areas of theory in the early 20th century with the work of Margaret Bryan Davis, Robert MacArthur, E. O. Wilson, and Robert Whittaker. These biogeographers formulated broad conceptual frameworks like the island theory of biogeography, the species-area curve relationship, the concept of an ecological niche, and the importance of climate parameters on ecological measures of biodiversity, carbon sequestration, and broad life history strategies (e.g., migration, number of offspring per season, and foraging strategies). These concepts will be covered in greater detail in subsequent chapters.

Some standard questions asked by biogeographers are listed in Table 1.2. Note that these questions were first posed by Alfred Russel Wallace in the early 19th century and nearly all are still relevant today.

B. KEY THEMES OF BIOGEOGRAPHY

There are several key themes in biogeography that are relevant to spatial ecology and GIScience. These themes (following Pijanowski, 2024 and also from Lomolino et al., 2017) are:

Theme 1: Classifying regions by their biotas.

Biogeographers are interested in developing approaches that classify areas based on the composition of plants, animals, and major dimensions of climate (e.g., average temperature and precipitation). Biogeographers have long viewed the planet's major regions as biomes. Biomes are typically described in broad terms of major plant types, which include forests, grasslands, deserts, and tundra; some of these are further subdivided. The nine major biomes and climate properties are shown in Figure 1.2. Note that areas with warm, wet

TABLE 1.2. *Common types of questions posed by biogeographers as originally suggested by Alfred Russel Wallace. Extracted from Lomolino et al. (2017).*

Why is a species or higher taxonomic (genus, family, order, and so on) confined to its present range?
Why are the animals and plants of large, isolated regions, such as Australia, New Caledonia and Madagascar, so distinctive?
Why are some groups of closely related species confined to the same region, and others found on opposite sides of the world?
Why are there so many [more] different species in the tropics than at temperate or arctic latitudes? What enables a species to live where it does and what prevents it from colonizing other areas?
What roles do climate, topography, and interactions with other organisms, play in limiting the distribution of a species?
What are the species' closest relatives and where are they found? Where did its ancestors live?
How are isolated oceanic islands colonized and why are there nearly always fewer species on islands than in the same kinds of habitats on continents?
How do different kinds of organisms replace each other as we go up a mountain, or move from a rocky shore to a sandy beach nearby?
How did a species come to be confined to its present range?
How have historical events, such as continental drift, Pleistocene glaciations and recent climatic change, shaped the species distributions?

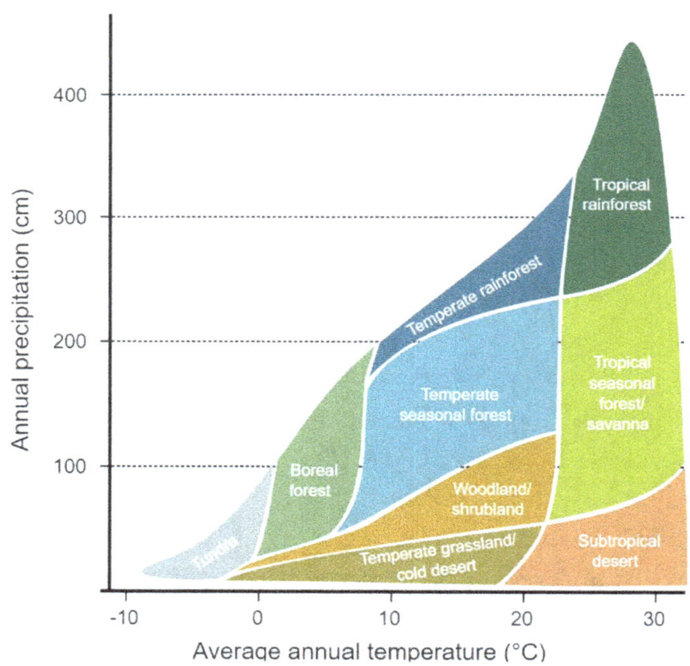

FIGURE 1.2. Major life zones as they are influenced by annual climate measures of air temperature and precipitation. For the Public Domain, CCo distribution rights. Diagram by Navarras.

climates support tropical rainforests and as areas become drier, forests become deciduous (e.g., trees lose their leaves annually).

Theme 2: Explaining differences in abundances and species richness among geographic regions.

It has been well known for centuries that areas near the equator support a greater number of plant and animal species and that as one moves toward the poles, the diversity of life decreases. Several latitudinal gradient trends, broadly referred to as the latitude diversity gradient (LDG) pattern, have also been observed for species abundances, biomass, size, and productivity (Brown, 2014; Field et al., 1998; Huston, 1994; Mittlebach et al., 2007; Wright, 1983). Attempts to explain the underlying mechanisms that give rise to these global trends have occupied biogeographers for centuries.

Theme 3: Explaining differences between closely related species, especially traits like morphology, behavior, physiology, genetics, and demography.

Another key theme has been examining patterns of similarity in traits between closely related species in attempts to understand the evolutionary mechanisms that give rise to differences through speciation. Intense debates have occurred over the last several decades on such topics as the mode of speciation (e.g., allopatric versus sympatric), the importance of barriers in creating species (called vicariance biogeography), and the role of ecological factors in the ways that broad multispecies assemblages are formed.

Theme 4: Studying how dispersal patterns relate to the geographic distribution of a species or community.

Perhaps one of the oldest and deepest set of discussions occurring among biogeographers is about the importance and impact of dispersal on shaping the composition of communities across space. Proposed by Darwin (1859) and his contemporaries (e.g., Wallace, 1876) as one of the most fundamental features of all life (all life on Earth disperses), the study of the dispersal and then the fragmentation of large populations by barriers is a cornerstone theme in biogeography.

1.4 KEY GOAL AND THEMES OF CONSERVATION SCIENCE

A. GOALS OF CONSERVATION BIOLOGY AND CONSERVATION SCIENCE

Modern conservation biology focuses on the preservation of species, often ones that are endangered, with the ultimate goal of protecting all biodiversity on Earth. Conservation biology has emerged to be a very cross-cutting, applied science that attempts to integrate ecological knowledge with natural resource management practices. Conservation science, presented formally by Kareiva and Marvier (2012), focuses on conservation practices in coupled natural-human systems. Doing so means that conservation needs to consider social and cultural features of local people, their needs for natural resources, and also how management of biological resources should be placed within the human system.

B. KEY THEMES OF CONSERVATION SCIENCE

The main themes of conservation science (see Kareiva & Marvier, 2012) are:

Theme 1: Conservation science is a synthetic approach to natural resource management.

Conservation biology combines habitat management, genetics, biodiversity assessments, principles of landscape ecology and biogeography, along with monitoring approaches and technologies, and spatial technologies with the goal "to provide principles and tools for preserving biological diversity" (Soulé, 1985; and also Kareiva & Marvier, 2012).

Theme 2: Conservation biology is one of the crisis disciplines in science.

Soule (1985) also argues that conservation biology belongs to one of several scientific fields that are crisis driven, the others being public health, cures for critical diseases (e.g., cancer), hazards mitigation (e.g., floods), environmental toxicology, and planetary health. Other crisis-driven disciplines that have been suggested in more recent literature have been food insecurity, sustainability of urban systems, and biosecurity (i.e., preventing the engineering and release of harmful organisms).

Theme 3: Management of habitat ultimately improves animal biodiversity.

There is also a recognition that, based on research in landscape ecology and biodiversity, there is a close relationship between habitat (e.g., vegetation structure and species composition) and animal diversity. Thus, a key principle in conservation biology is the management of habitats to support species richness and abundance of animals at landscape to larger scales.

Theme 4: Conservation science is inherently a spatial science.

Although rarely explicitly stated, conservation biology is inherently a spatial science. As part of a nearly half-century-old debate referred to as a SLOSS (single large or several small) debate on the spatial configuration of protected areas, the recognition that conservation biology needs to consider spatial properties of biodiversity was part of the discussions regarding general rules of protected area designs and how they impact goals of conservation (Soulé & Simberloff, 1986).

1.5 COMPARISONS OF SPATIAL ECOLOGIES

Although a bit oversimplified, the differences in the focus of spatial and temporal extents for landscape ecologists and biogeographers are shown in Figure 1.3. Note that landscape ecological foci are at the smaller/shortest spatial and temporal extents compared to those of biogeography. There is likely some overlap (shown with the intersection of transparent patterns for each). It is also possible for landscape ecological approaches to be applied to problems that are long-term (e.g., millions of years, also called mega-annum or Ma) and large spatial extents (e.g., continental).

1.6 SPATIAL TECHNOLOGIES AND GISCIENCE

A. WHAT ARE SPATIAL TECHNOLOGIES?

Providing the instrumental foundations for all of these spatial ecologies are a variety of technologies that form the basis of GIScience (geographic information science, sometimes referred to informally as GIScience).

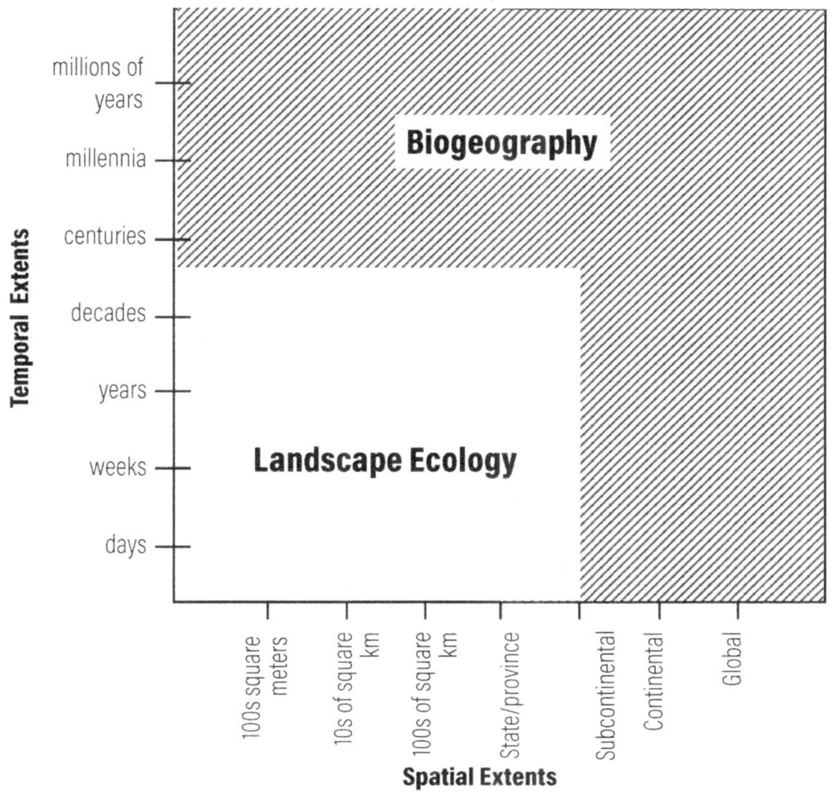

FIGURE 1.3. Spatial and temporal extents for landscape ecology and biogeographical applications. The solid gray area represents landscape ecology and the hashed lines represent biogeography. The area of overlap is shown as a transparency of the gray areas over the hashed areas. Source: Bryan C. Pijanowski.

GIScience is the study of space and how spatial information represents processes and objects distributed in space. GIScience is the scientific discipline that provides both the theoretical and methodological foundations and methodologies underpinning GIS technology (Goodchild, 1992). In a broad sense, which we cover in more detail in Chapter 11, GIScience includes the following: methods for developing spatial data models, software tools organized around theories of space, cartographic principles, remote sensing technologies (e.g., space based, airborne, in situ, wearables, etc.), monitoring techniques, and principles of geography. GIS, a subset of GIScience, stands for **geographic information systems**, the software systems that allow users to create, store, manage, analyze, integrate, and display spatial information. **Spatial technologies** form the fundamental instrumental infrastructure by which GIScience operates. These technologies include instruments (e.g., sensors for imagery and global positioning systems or GPS handheld units) that collect, store, and transmit spatial data.

B. KEY THEMES OF SPATIAL TECHNOLOGIES AND GISCIENCE FOR SPATIAL ECOLOGIES

The advancement of GIScience and associated spatial technologies can be organized under the following themes:

Theme 1: How can spatial data be used to quantify patterns and processes in landscapes and assess large-scale biogeographic trends?

Spatial data comes in a variety of formats (see Chapter 12 on spatial data models), and these represent a feature of an area that is changing. Mapping these dynamics is a key goal of experts who use spatial technologies like remote sensing and GIS (Masser, 2005; O'Sullivan & Perry, 2013).

Theme 2: Which approaches can integrate disparate sources of spatial data to characterize ecological change?

Data that are necessary to understand changes in space come in a diverse array of formats (e.g., lines, point locations, remote sensing imagery, and tabular). How can these data be integrated and then used to analyze these changes (Campbell & Wynne, 2011)?

Theme 3: In what ways can the delivery of spatial data online provide decision support for natural resource management?

One of the most important advances in the last few decades has been the development and distribution of spatial data, either through data portals like Esri's LiveAtlas or through the development of online decision support systems (DSS) that provide users around the world with access to spatial data and associated visualization and analysis tools, all online.

Theme 4: What spatial models and geostatistical tools can be used to quantify spatial and temporal variability in space?

Across all of the ecological sciences, various types of models (see Exercise 4) and statistical approaches have been developed and then applied to help understand complex dynamics in ecosystems. A subset of these have been developed for use in a spatial context (Anselin, 1995). This book will introduce the concepts of models and the application of neutral landscape models in Exercise 4 and the use of geostatistical tools in Exercise 7.

1.7 SUMMARY

Traditionally, there have been two major areas of ecology that are explicitly spatially oriented—landscape ecology and biogeography. A third, conservation biology, is inherently spatially oriented although it has not been explicitly so. Major themes of these spatial ecologies have focused on ecosystem health and the spatial distribution of organisms. It is not surprising, then, that the ecological sciences rely on spatial tools, which with major theories and methodological frameworks related to space—called GIScience (or geographic information science)—form an important foundation of spatial ecology.

1.8 DISCUSSION QUESTIONS

1. What are the key themes for each of the spatial ecologies, and how do the themes of spatial technologies support these? Select two and expand on this based on what you know about spatial technologies.
2. Select an endangered species and use this as an example for how principles of landscape ecology, biogeography, conservation biology, and spatial technologies and GIScience can address the challenges of managing this species.

2. LANDSCAPES AND LAND COVER

Highlights

Aims. Standards have been developed to classify the surface of the Earth. These standards reflect the broad categories of land cover types that are commonly found around the world. These include land cover classes such as forests, grasslands, and bare soil. There are also several ways to characterize the properties of the land surface, each of which are summarized here.

Learning objectives. Students should be able to:

- Describe the LCCS classification system developed by Anderson et al. in 1976
- Compare and contrast land cover features such as albedo, greenness, land surface temperature
- Compare and contrast some similar features of land cover that include canopy cover, fractional cover, and leaf area index

2.1 INTRODUCTION

The term **land cover** is used to describe "the physical material on the surface of the Earth—such as vegetation, water, bare ground, or artificial structures—mapped through the interpretation of remotely sensed data and categorized according to the NLCD classification system" (Anderson et al., 1976). Major land cover classes include developed, barren land, forests, shrub, grassland, crops, pasture, wetlands, perennial ice/snow, and open water. There have been several reports that summarize the percentages of land cover by each class for the world; one of the most often used is that by the FAO (2014) from the Global Land Cover SHARE Project (Table 2.1).

Over time, nearly all landscapes on Earth undergo change, either naturally or by transformation brought about by humans. One of the foundational papers in the scholarship of land cover science is that of Lambin, Geist, and Lepers (2003). They used a variety of methods to determine forest cover in the last 8,000 years and reported that this cover class changed from about 50% of global land cover to about 30% today. Agriculture expansion in the last 300 years was 4.5 to 5 times what it was in the 1700s, and there was a 50% increase in the 20th century alone. They also determined that in the same 300-year period (1700–2000), there was a seven-fold increase in the amount of land in pasture (i.e., land used to raise livestock) and that over half of savanna, steppe, and grasslands land cover was lost, mostly due to conversion to pastureland and cropland.

Recent (Li et al., 2023) models of historical changes in land cover in the continental United States (CONUS) have shown that cropland, pastureland, and urban land started to increase in the early 1800s, comprising about less than 1% of the postsettlement landscape and rising to a little more than 25% of the land area. It should be noted that the total area included in this analysis, which combined US population census data by

TABLE 2.1. *Summary of the percentages of land cover by class for all areas of the Earth, excluding Antarctica (from Latham et al., 2014)*

LAND COVER CLASS	APPROXIMATE PERCENTAGE OF GLOBAL LAND AREA
Artificial Surfaces	~0.6%
Cropland	~12.6%
Grassland/Shrubs/Herbaceous/Sparse Vegetation	~31.5%
Tree-Covered Area	~27.7%
Bare Soil (Bare Ground)	~15.2%
Snow & Glaciers	~9.7%
Water Bodies & Mangroves (combined)	~2.7%
Total	~100%

county and known natural land cover areas, changed considerably over the years as more states and more data became available for the model (Li et al., 2023). The major point here is that cropland and pastureland have replaced natural areas of shrubland, grassland, and forest over time. In addition, the increase in the amount of urban land added each year ranges from about 0.4% per year, expressed as percent gain of the original footprint. In terms of total amount, it is estimated that about 2.6% of the land surface in the CONUS region in 2020 is urban, up from 2.0% in 1985.

2.2 LAND COVER CLASSIFICATION SYSTEMS (LCCS)

Some land cover types are almost always in the category of "gain" in areal amount. These include urban and agriculture. As a result, the other types of land cover experience loss. Most often these are natural areas such as forests and wetlands. Many kinds of questions are asked by natural resource analysts. These include: How fast are forests being lost? How many forest patches exist in time 1 versus time 2? How much forest is lost in riparian areas (say, within a 300 m buffer of a river or stream)? What are the most frequent types of land cover transitions that are important for my management of these areas?

There have been many **land cover classification systems (LCCSs)** used by natural resource managers and geographers. LCCSs (Di Gregorio & Jansen, 1998) are systems of describing the physical properties of the land surface using a set of classifiers applied in a hierarchical system that creates clearly separate classes. It is important to develop classifiers that create land cover classes that are useful for the specific natural resource management problem at hand but also generic enough to be applied to a variety of problems. Most land cover classifiers use inputs from remote sensing imagery, like Landsat, Sentinel-1/2, or MODIS, or are developed by heads-up digitizing of aerial photographs from planes or from drones.

One of the most universally used LCCSs, proposed by Anderson et al. (1976), used, as input, satellite-based Landsat imagery. This is sometimes called the Anderson Land Cover Classification System. This LCCS was developed by a US Interagency Steering Committee on Land Use Information and Classification, formed in 1971, which was also part of a larger US initiative called Earth Resources Observation Satellite or EROS. The committee was composed of scientists, economists, and engineers from the United States Geological

TABLE 2.2. ***Level I and Level II land cover classes illustrating the hierarchical nature of the system and the standard nomenclature. Modified from Anderson et al., 1976.***

LEVEL I CLASSES	LEVEL II CLASSES
1 Urban or Built-Up Land	11 Residential
	12 Commercial and Services
	13 Industrial
	14 Transportation, Communications, and Utilities
	15 Industrial and Commercial Complexes
	16 Mixed Urban or Built-Up Land
	17 Other Urban or Built-Up Land
2 Agricultural Land	21 Cropland and Pasture
	22 Orchards, Groves, Vineyards, Nurseries, and Ornamental Horticultural Areas
	23 Confined Feeding Operations
	24 Other Agricultural Land
3 Rangeland	31 Herbaceous Rangeland
	32 Shrub and Brush Rangeland
	33 Mixed Rangeland
4 Forest Land	41 Deciduous Forest Land
	42 Evergreen Forest Land
	43 Mixed Forest Land
5 Water	51 Streams and Canals
	52 Lakes
	53 Reservoirs
	54 Bays and Estuaries
6 Wetland	61 Forested Wetland
	62 Nonforested Wetland
7 Barren Land	71 Dry Salt Flats
	72 Beaches
	73 Sandy Areas other than Beaches
	74 Bare Exposed Rock
	75 Strip Mines, Quarries, and Gravel Pits
	76 Transitional Areas
	77 Mixed Barren Land
8 Tundra	81 Shrub and Brush Tundra
	82 Herbaceous Tundra
	83 Bare Ground Tundra
	84 Wet Tundra
	85 Mixed Tundra
9 Ice	91 Perennial Snowfields
	92 Glaciers

Survey, NASA, and the USDA Soils Conservation Service (SCS), along with members from the Association of American Geographers (AAG) and the International Union of Geographers (Anderson et al., 1976). The timing of this committee was driven by the imminent launch of the first Landsat satellite land cover mapping program, which occurred in 1972 (i.e., eventually called Landsat 1). The objective of the interagency committee was to create a globally useful LCCS that would map land resources (e.g., areas of forest, crop cover, and urban) through time and at resolutions that could be useful for natural resource management. Prior to this time, mapping of land surface was done using costly, labor-intensive methods (e.g., heads-up digitizing) based on aerial imagery, coordinated mostly at the state level.

The first level classification of land cover by Anderson et al. (1976) is shown in Table 2.2. Note that nine major classes are listed in Level I. For each of the nine classes, there are between two and seven subclasses. Also note the nomenclature. Urban at Level I is a 1 but Residential at Level I is an 11. The classification system can be further extended to Level III (Table 2.3). Note how these classes provide very detailed information that begins to align more with economic or demographic characteristics (e.g., land use) than land cover.

Every class should also be accompanied by a description of the characteristics of that class. The complete description of Urban and Built-Up Areas (Level I class) from Anderson et al. (1976) is provided in Table 2.4 so that you can understand how these details are provided to users of the data.

TABLE 2.3. *Illustration for how Level III classes are organized under a Level II class.*

LEVEL II CLASSES	LEVEL III CLASSES
11. Residential	111. Single-Family Units
	112. Multi-Family Units
	113. Group Quarters
	114. Residential Hotels
	115. Mobile Home Parks
	116. Transient Lodging
	117. Other

TABLE 2.4. *Full description of the Urban Built-Up class. Taken directly from p. 19 of Anderson et al. (1976).*

URBAN OR BUILT-UP LAND CLASS DESCRIPTION
Urban or Built-Up Land is comprised of areas of intensive use with much of the land covered by structures. Included in this category are cities, towns, villages, strip developments along highways, transportation, power, and communications facilities, and areas such as those occupied by mills, shopping centers, industrial and commercial complexes, and institutions that may, in some instances, be isolated from urban areas. As development progresses, land having less intensive or nonconforming use may be located in the midst of Urban or Built-Up areas and will generally be included in this category. Agricultural land, forest, wetland, or water areas on the fringe of Urban or Built-Up areas will not be included except where they are surrounded and dominated by urban development. The Urban or Built-Up category takes precedence over others when the criteria for more than one category are met. For example, residential areas that have sufficient tree cover to meet Forest Land criteria will be placed in the Residential category.

2.3 IMPERVIOUS SURFACES

Within the broad land cover class of urban areas there is another land cover classification system that quantifies how impervious the surface is. According to the United States Geological Survey (USGS), impervious surfaces are "the built surfaces through which water cannot infiltrate the underlying soil," including roads, rooftops, driveways, and parking lots—used as a measure of urbanization and hydrologic alteration (Xian & Homer, 2010; Yang et al., 2003). Over the years, there have been two major ways to classify impervious surfaces. The first is to report on the fractional cover of a 30 m x 30 m pixel that was impervious to water seepage into the ground. In the 1990s through the mid-2010s, the USGS released versions of these maps that placed them in broad categories of high (> 80%), medium (50–79%), and low intensity (20–49%) and a developed, open space category (< 20%). These have been created as a companion dataset to the USGS's Land Use Cover Database (NLCD) (Homer et al., 2004; Xian & Homer, 2010).

2.4 FUNCTIONAL CHARACTERISTICS OF LAND COVER

Land cover is more than just a categorical measure such as land cover class. Other features of the land cover have ecological significance and are thus part of the assessment of the spatial distribution of vegetation cover around the world.

A. LEAF AREA

Leaf area is the surface area that plant leaves occupy. It is essentially a measure of vegetation density. Ecologists have long used **leaf area index (LAI)** as a measure of vegetation density (Monsi & Saeki, 1953); its equation is:

$$LAI = \text{leaf area}/\text{ground area}$$

Typical ranges for LAI are:

Bare soil/urban: 0–0.1
Grasslands/shrublands: 0.5–2
Crops (in peak season): 3–6
Forests (temperate/tropical): 5–10+

Measures of LAI are used in climate-land interaction modeling, modeling of carbon sequestration, estimating the complexity of vegetation structure to assess habitat suitability for animals, and estimating evapotranspiration and water balance during times of drought. Many of the space-based remote sensing platforms can be used to generate LAI; these include MODIS, Sentinel-2, and Landsat (see Exercise 2 for a review of space-based remote sensing technologies).

B. CANOPY HEIGHT

Canopy height is simply the distance from the ground to the top of the vegetation. It is the most direct measure of vegetation form and, combined with a 2D pixel, its additional dimension provides ecologists with information about a voxel (e.g., a 3D cube). Greater canopy heights means that the habitat is likely to support more subhabitats and thus more species of animals. Many ecologists (see Potapov et al., 2023) have found strong positive relationships between canopy height and animal species richness (i.e., the number of species). Canopy heights may also reflect stand age (of forests).

C. ABOVE GROUND BIOMASS (AGB)

A very important measure of land cover is the amount of **above ground biomass (AGB)**—the total mass of woody, leaf material, and other living organic matter (e.g., fungi) of an area. It is typically expressed as a dry weight, in megagrams, of all biomasses per hectare. Ecologists (e.g., Chave et al., 2014) have found that there is an allometric measure (one variable scales to the power of the second) between above ground biomass (AGB) and canopy height (height) as:

$$\text{AGB} \propto \text{height}^{b}$$

where b is typically between 2 to 3.

D. GROSS PRIMARY PRODUCTION

One of the most widely used measures of plant productivity is **gross primary productivity (GPP)**. GPP (Chapin et al., 2011) is the total amount of carbon, in grams, that is fixed by primary producers (mostly plants) through photosynthesis per unit area (typically square meters) and per unit time (typically a day or year). GPP is also related to net primary production (NPP) by the equation:

$$\text{GPP} = \text{NPP} + R_a$$

where R_a is autotropic (i.e., plant) respiration. There are several ways to measure GPP and NPP, and for the latter, the most often are through remote sensing data that is combined with models. Exercise 12 will introduce the use of the MODIS satellite sensor to analyze large-scale maps of NPP for the US CONUS.

E. GREENING

A spectral (i.e., light reflectance-based) set of measures are used by ecologists to quantify the amount of vegetation greenness (e.g., Tucker, 1979). These measures also quantify photosynthetic activity and water stress of the plants. Two indices are commonly used, NDVI (**normalized difference vegetation index**) and EVI (**enhanced vegetation index**). The latter is preferred in tropical ecosystems. Both NDVI and EVI range from -1 to +1.0 with values > 0.0 for vegetated areas. Ecologists often use a time series of NDVI or EVI to develop what are called phenology curves; these "greenness" curves quantify the seasonal changes in the greening up, peak greening and photosynthesis, senescence, and dormancy of plants, and therefore the trends in photosynthesis over time and across space.

F. ALBEDO

Albedo is a surface property that controls how land and climate interact. A nontechnical definition of **albedo** is that it is the proportion of sunlight energy that bounces off the land surface and into the sky (IPCC, 2021). Albedo values range from 0.0 to 1.0. A low albedo means that energy is absorbed and not reflected toward space, which is very characteristic of dark surfaces such as deep ocean water or dense, dark green forests. High albedo is characteristic of highly reflective surfaces such as snow and ice where energy is mostly reflected. It is also sometimes important to distinguish between white-sky and black-sky albedo. **Black-sky albedo** measures are reported for clear, sunny conditions and assume direct beam sun. **White-sky albedo** are measures of energy fluxes between the land and atmosphere during cloudy conditions that promote very diffuse light. Several NASA (i.e., the MCD43A3 product from MODIS) and ESA satellite remote sensing platforms report albedo measures, making it possible to compare land covers across regions or across time for the same landscape.

G. FRACTIONAL CANOPY COVER (FCC)

This is another simple measure—the fraction of the area that is covered by vegetation (Daubenmire, 1959). These values range from 0 to 100% (or sometimes 0.0 to 1.0). Several remote sensing platforms produce these measures (Jasinski et al., 2023). Slight variations in FCC measure are used. One is **fractional vegetation cover** or **FVC** (proportions of the ground covered by woody or herbaceous cover); **fractional tree cover** (**FTC**), which is often used in areas where trees are scattered across large areas of grassland (e.g., Miombo Woodlands); or **canopy closure**, which is typically measured on the ground and is a measure of the percentage of the sky that is seen through a fisheye camera (Rich, 1990).

H. RUGOSITY

A very spatially explicit measure of land cover is the variation in canopy height within a pixel or small area, which is termed **rugosity**, a technical term for roughness or textured surfaces. These are often derived from LiDAR measures. High rugosity is characteristic of old growth forests where there are trees of multiple ages and heights. A standard way to calculate rugosity is to create a TIN (triangular irregular network; see Chapter 12 on spatial data models) of surfaces that follow the surface of the canopy) and divide that by the 2D area that is comprised of the sampled space. Thus, rugosity, R, is:

$$R = A_{TIN}/A_{2D}$$

where values of perfectly flat surfaces, such as a lake, street, or crop, are 1.0. Values that are > 1.0 have complex surfaces with the greatest being in old growth forests where the ages of the trees vary a lot and/or there is high tree mortality and tree gaps are common.

I. SURFACE TEMPERATURE

Another feature of land cover is its **surface temperature**. Measured typically from space with sensors that quantify thermal infrared radiation, **land surface temperature (LST)** varies across space and is a function of latitude, altitude, and land cover type. LST is often reported in either Kelvin (K) or Celsius (°C). Low LSTs are typically found in areas of dense vegetation, high LSTs in areas of bare ground and/or a lot of impervious (e.g., concrete) surface. Grassland and croplands will often have moderate LSTs. High LSTs can contribute to greater air temperatures, as reflected in the urban heat island effect (Zhou et al., 2022). Several of the space-based remote sensing platforms, such as MODIS, Landsat, and Sentinel-2, collect this information, and special algorithms are used to create LST products at, generally, coarse resolution (100 m x 100 m to 1 km x 1 km).

2.5 DIGITAL ELEVATION

Although not directly a functional characteristic of land cover, the elevation of a location greatly influences land cover classes and the functional (e.g., LAI) characteristics of a location. Elevation feature influences include how soils are formed, energy balances, hydrological function, and habitat heterogeneity. Aggregated as terrain analysis (Wilson & Gallant, 2000), the land surface as a function of height above sea level, slope, and aspect are all important characteristics of the surface of the landscape. Spatial ecologists use a digital representation of elevation referred to as DEM (digital elevation model). DEMs are close but not perfect representations of the height, slope, and aspect of the terrain and are collected by special sensors on space-based, aerial-based, or ground-based sensors. Elevation features include:

1. Height—typically reported as meters above sea level (masl) where sea level is reported as the mean.
2. Slope—the amount of change in elevation across an area.
3. Aspect—the compass direction. In many cases, aspect influences the amount of daily radiation that directly hits a location, thereby influencing moisture and land covers.
4. Topographic position—the location of any point that is relative to others surrounding it.
5. Roughness—Similar to canopy roughness, the roughness of the terrain (at ground level) is an important measure of a landscape that can influence variation in the spatial arrangement of land covers.

2.6 SUMMARY

Land cover is a key feature of landscapes. The features can be characterized as categorical classes that have been placed into globally relevant terrestrial categories organized hierarchically. In addition to the categorical classifications of land cover, functional measures have been developed by ecologists over the years, first as field-based measurements, which then, in recent decades, transitioned to aerial- and space-based measurements. Many of these functional characteristics of land cover are available from weekly to monthly schedules with historical data dating back to as far as the mid-1970s.

2.7 DISCUSSION QUESTIONS

1. When considering a map of land covers and their distribution across space, how might you quantify this spatial variability given the hierarchical nature of the data? For example, consider the variability in urban use across a region. How might you consider this variability in Level I and Level II classes? Discuss your answer in light of how these Level II classes "interact with each other" in space.
2. Describe how each of the functional characteristics of land cover could affect the distribution of animals in terms of species numbers (i.e., species richness) or body sizes.
3. What might be missing from the list above that summarizes the functional characteristics of land cover?

3. LANDSCAPES AND SPATIAL PATTERNS

Highlights

Aims. This chapter provides a critical step-by-step introduction to the methods for quantifying spatial patterns, especially in categorical maps (e.g., land cover maps). The basics of these concepts (e.g., neighborhood rules) are used by many spatial statistical tools and in spatial statistics in GIS software.

Learning objectives. Students should be able to:

- Determine the number of patches based on 4- and 8-cell neighborhood rules
- Read and interpret an adjacency table for land cover maps
- Interpret output for patch, class, and landscape spatial statistical tools
- Describe fundamental patch statistics like edge, area, and shape

3.1 INTRODUCTION

Landscape ecologists and biogeographers have long focused on ways to quantify spatial patterns in landscapes and regions. This chapter will review some of the approaches used by these ecologists for categorical raster data models.

3.2 KEY PRINCIPLES AND TERMS FOR SPATIAL PATTERN

A. ORGANIZATIONAL SCALES OF ANALYSIS

In the raster spatial data model environment, two sets of spatial pattern metrics have been developed by landscape ecologists: those at the pixel level and those above the pixel level. Let us consider these sequentially. These levels of spatial patterns in categorical grid maps are summarized in Table 3.1.

B. ADJACENCY RULES: ANALYSIS AND DEFINING A PATCH

For all raster data models, spatial patterns are assessed first based on **adjacency** or "what's next to me" rules. Let us consider a pixel, such as the ones in Figure 3.1, that is coded as gold. Note that we can consider the nearest neighbor rules that only count cells that have full sides in common. The cell in Figure 3.1A has four adjacent cells and so using a **4-cell rule**, we could only count the cells to the top and bottom and to the right and left. If we count the cells positioned diagonally, then the number of cells (Figure 3.1B) that are adjacent to the center pixel are eight; this is referred to as the **8-cell rule**. For practical considerations, as a natural resource manager, you would consider counting diagonals if an animal that you are managing could easily move through the landscape into diagonal locations. In some rare cases, you might consider counting those cells that allow

TABLE 3.1. ***Different levels of spatial pattern analysis for a grid spatial data model***

AT THE PIXEL SCALE

1. Adjacency (what is next to me)
2. Contrast (e.g., how different neighbors are)

ABOVE THE PIXEL SCALE

3. Patch (cells of same type next to one another as a group)
4. Class (e.g., all urban patches)
5. Landscape (all patches of all types)

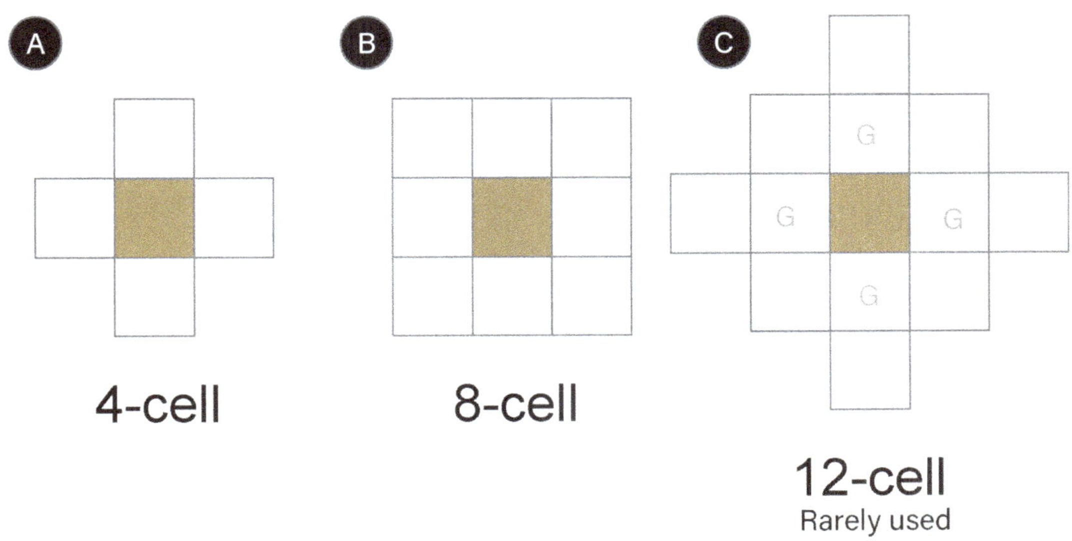

FIGURE 3.1. The three most common adjacency rules used by landscape ecologists to determine spatial patterns. These include counting cells that are adjacent only if they share a full side (A); if one considers full-side cells and those that are diagonally positioned (B); and for those cells where a full-side cell can be counted as a gap (e.g., an animal can traverse through these cells), then one can count 12 cells as being adjacent (C).

A

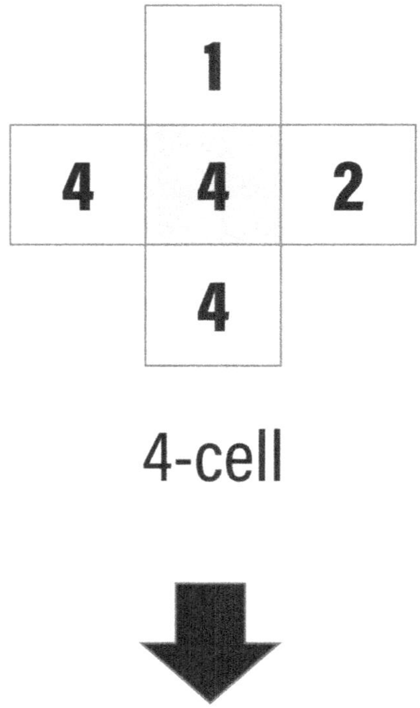

B

Land Use Class	Number Adjacent	Percent Adjacent
1 (urban)	1	25%
2 (agriculture)	1	25%
3 (grasslands)	0	0%
4 (forests)	2	50%

FIGURE 3.2. A representation of how cells are counted in a 4-cell adjacency rule (A) and are then tabulated (B).

a full-side adjacency to be a gap. Counting those cells (Figure 3.1C) that allow for a one-cell gap (designated as cells with a G) would produce a 12-cell rule for determining adjacency.

3.3 PIXEL LEVEL: ADJACENCY STATISTICS

Let us now consider the first of five ways to quantify spatial patterns in the grid environment. Adjacency tables are very simple and useful tables that count the number of like and unlike cells in categorical maps. In Figure 3.2,

	4	5	1	6	3	2
4	72634	1266	5023	4576	7741	7622
5	1266	3054	631	222	187	173
1	5023	631	4142	385	1251	1332
6	4576	222	385	2004	726	1089
3	7741	187	1251	726	3598	2723
2	7622	173	1332	1089	2723	15708

Legend
1=urban
2=ag
3=grassland
4=forest
5=open water
6=wetlands
7=barren

FIGURE 3.3. A sample adjacency table that illustrates the count of the number of cells that are next to the center pixel. Diagonals represent cells of the same value whereas those outside the diagonals represent dissimilar cell values.

note that we have a center cell or pixel (Figure 3.2A) that is coded as a forest cell. Using a 4-cell rule, we would count the number of cells in each of the four classes of land cover in the categorical map. This results in a adjacency table for the simple map in Figure 3.2A that is shown in Figure 3.2B. Note that we have 2 forest cells, 1 cell of urban, and 1 cell coded as agriculture.

If we use a geographic information system, we can create an **adjacency table** for a large raster database like the one shown in Figure 3.3. If we examine this table, we see that in the categorical raster file that has a large number of cells, the most common adjacency is forest cells that are adjacent to forest cells (e.g., 72,634). This is followed by agriculture cells that are adjacent to agriculture cells (15,708). The most common non-like adjacent cells are agriculture cells that are next to forest cells (7,741). Note that the diagonals are cells that are adjacent to their own kind. Another noteworthy aspect of this table is that the right and left side are mirror images of the tabulation so one only uses a portion of the table for a full analysis of adjacencies. For example, note that counting the number of cells coded as 5 next to a 4 is the same as counting the number of cells that are coded as 4 and have a 5 next to it (these are shown as the cells indicated by the red circles).

3.4 PIXEL LEVEL: EDGE CONTRAST

Note that the degree of contrast between adjacent cells can also be an important consideration in natural resource management at the landscape scale. Let us consider two kinds of **edges** (Figure 3.4): those that have **high contrast** and **low contrast**. Other terms that might be used are "sharp" and "smooth" or "hard" and "soft." The sharp edge could be a forest that is situated next to an agricultural field whereas a smooth edge might be an old growth forest that is next to a secondary forest that is composed of younger, and not as tall, trees.

In practical terms, landscape ecologists have quantified edge contrast by assigning magnitudes to each categorical value. Magnitudes could be assigned on the basis of either a measure of size (vegetation height) or a value associated with a process (light infiltration potential). Edge contrast is an important consideration in landscape management. High contrast edges are likely to attract different kinds of bird species (high contrast would attract predators that perch) or change the amount of solar radiation into the forest (high contrast would

FIGURE 3.4. Edge contrast comparisons here show those that might be considered as "sharp" (case study 1) versus ones that are smooth (case study 2).

have high solar radiation penetration). Edge contrast can also include cells that are natural versus those that have been altered by humans; alternatively, as inherent edges (those edges that are created from topography like forest lines in mountain areas) and induces (edges that have been created by sudden changes like wildfire). There are some features in FRAGSTATS (McGarigal & Marks, 1995) (e.g., ECON for edge contrast index) and in landscape metrics in R (Hesselbarth et al., 2019) that allow landscape ecologists to calculate edge metrics across a large raster database.

3.5 ABOVE PIXEL SCALE: PATCH STATISTICS

Patches are discrete areas that contain similar features in the environment. To determine the size, shape, and other characteristics of a patch, the 4-cell or 8-cell adjacency rules are required. To illustrate how the adjacency rule affects **patch area statistics**, let us consider a grid arrangement of a focal class that is coded as black in Figure 3.5.

Note that we get an entirely different number of patches if we apply the 8-cell rule (Figure 3.6). Counting cells that are either diagonal or sharing full sides means that the arrangement of black cells in this grid produces a pattern where all cells are connected by a diagonal or side adjacency producing a single patch (i.e., number of patches is 1).

Within the category of patch statistics, two kinds exist. These are (1) **patch area** or size or (2) **patch shape.** For patch area, a few statistics are calculated (Table 3.2). These include patch size, size of the largest and/or smallest patch, or the measure of gyration.

Consideration of what constitutes a core area is also dependent on whether one uses the 4-cell or 8-cell adjacency rule. Figure 3.7 shows how an **edge** and **core** cell are defined using a 4-cell rule. Note that the edge is defined as those cells that border the outside cells with a full edge. Those that are remaining are then considered edge. In this example with three kinds of cells, we can see that edge cells (E) share full sides with outer cells (O), but cells in black that are only diagonal do not count as an edge; thus there are only two core cells (C) using the 4- cell rule.

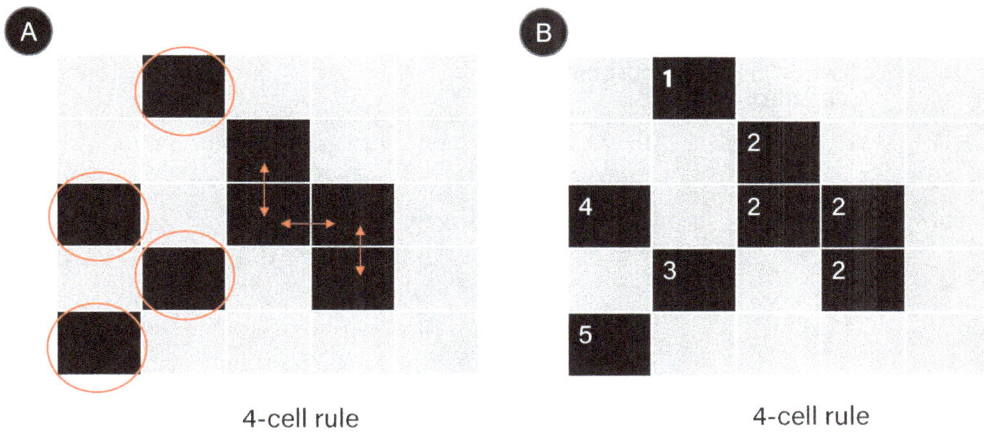

FIGURE 3.5. The application of the 4-cell adjacency rule for determining what constitutes a patch. Note that for the 4-cell rule, the upper cell and the three on the left are separate patches. The four that share common sides are grouped into a single patch. The counting of separate patches is shown in B; note that there are five patches in this 5x5 grid.

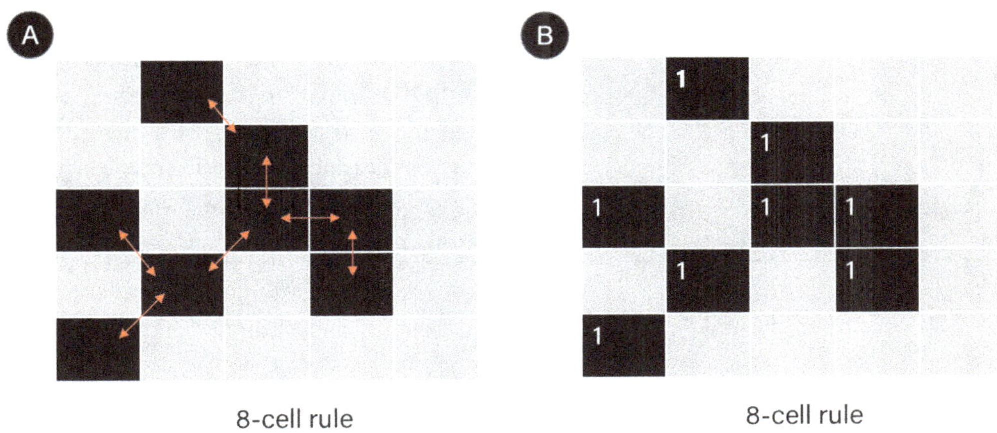

FIGURE 3.6. The application of the 8-cell rule for determining the number of patches. Note that given the same arrangement of cells as in Figure 3.5, we create a spatial arrangement that results in all cells sharing either a diagonal or full side spatial adjacency (A), giving us the number of patches of black cells of 1 (B).

If we apply the 8-cell rule to the same arrangement of black cells sitting within gray cells, we still have the eight cells composing the black area. But applying the 8-cell rule means edges are counted differently (see Figure 3.8).

Patch shape metrics are measures of the complexity of the arrangement of cells within a patch. These statistics commonly use measures of area and perimeter. For example, the simplest measure of shape is the ratio of the perimeter to its area. This unitless measure, as it is a ratio of length to area, reflects **shape complexity**.

TABLE 3.2. *A summary of the most common patch area statistics used by landscape ecologists. Adopted from McGarigal & Marks (1995).*

STATISTIC	COMMONLY USED ACRONYM	DESCRIPTION
Patch Size	AREA	The area, in a given unit such as m2, or the number of cells (area = cell size * number of cells).
Largest Patch Size	LPI	Often called the largest patch index, it is the size of the largest patch within the grid of cells reported in either areal units (e.g., m2), number of cells, or proportion of the grid that is taken up by the largest patch.
Patch Gyration	GYRATE	This is the average distance an animal would have to move, given a random starting point within the patch, to encounter an edge.
Core Area	CORE	The area or size of the core of a patch. For example, if an edge is defined as one cell deep, then all of the cells not along the edge are considered core (based on the 4-cell or 8-cell rule).

TABLE 3.3. *A summary of the most common patch shape statistics used by landscape ecologists. Adopted from McGarigal, K., & Marks, B. J. (1995).*

STATISTIC	COMMONLY USED ACRONYM	DESCRIPTION
Perimeter to Area Ratio	PARA	A unitless measure of the length of the perimeter to the area. Often calculated with the same units (e.g., meters).
Shape Index	SHAPE	The perimeter to area measure of a patch divided by a theoretical perimeter to area measure for a specific shape (square or circle).
Fractal Dimension	FRAC	This is the average distance an animal would have to move, given a random starting point within the patch, to encounter an edge.

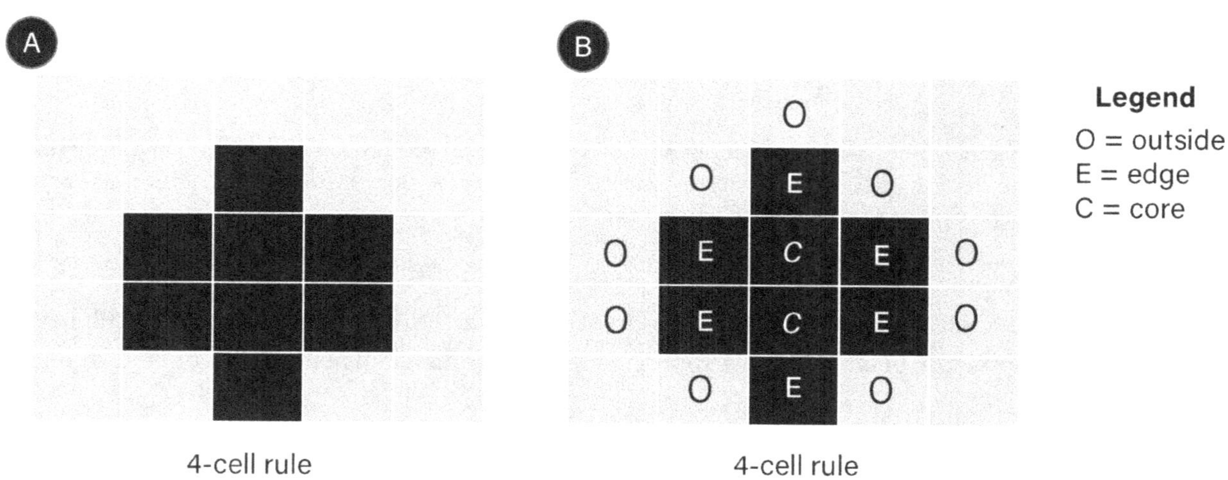

FIGURE 3.7. Defining edge and core cells in a patch where these two kinds of cells are defined using the 4-cell rule. The 8-cell patch colored in black (A) has six of its cells counted as edge and two as core (B). Key: O = outer cell, E = edge cell, C = core cell.

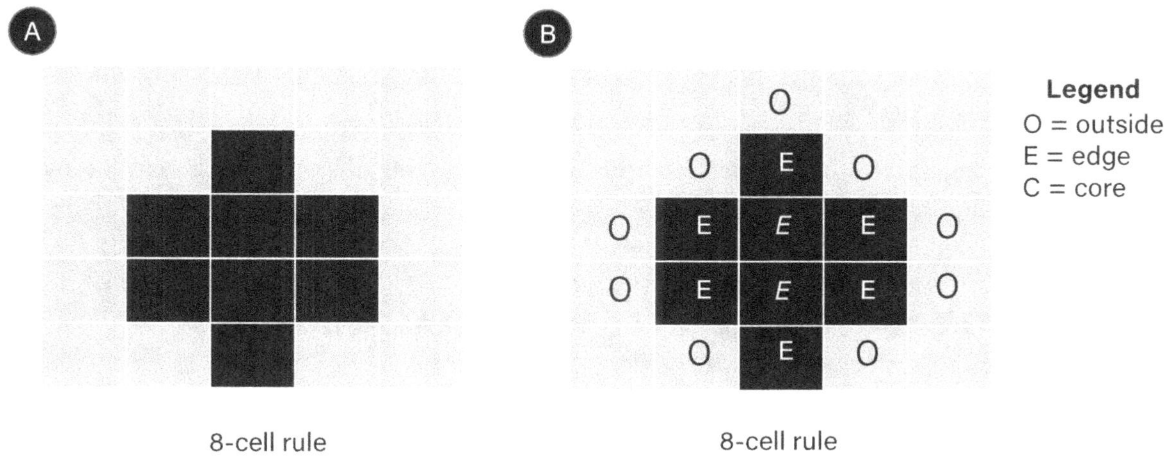

FIGURE 3.8. The number of cells that are considered outer, edge, and core when the 8-cell rule is applied. The eight cells in the example to the left (A) are the same as in Figure 3.7. But note that since the two cells in the middle share a diagonal arrangement, then these are counted as edge. Therefore, there are no core cells in the patch on the right (B). Key: O = outer cells, E = edge cells.

A patch that has the shape of a circle is one of the simplest of shapes; compared to a patch that has the same area that is shaped like a long "hot dog" means that the perimeter of that "hot dog"–shaped patch is much large than the circle and thus has a lot more edge than core habitat for animals to exist in. Landscape ecologists often use a shape index metric, which compares the shape of the specific patch to that of an ideal patch, such as a patch that is the shape of a perfect square or circle with the same area. Fractal dimension is a measure that is used a lot in mathematics and geometry. It considers measures in nature that depart from the perfect integer values of topological dimensions that are 0, 1, 2, and 3, where 0 represents the dimension of a point, 1 represents a line, 2 represents the dimension of area, and 3 represents the dimension of volume. A **fractal dimension** is where the dimension is represented as a fraction; so, for example, if a fractal measure is 2.5, it is exactly between the dimension of a line and that of a shape involving area such as a square. A classic example of a fractal dimension measurement is one of a coastline. If one were to measure a coastline using a straight line, it might be, viz., 1 km long. But if the coastline is measured using a 1-meter-long stick, following its intricate and complex shape, it is more than likely much longer than 1,000 one-meter-long sticks; let's say it is 1,234 meters. Fractal dimensions (FD) are expressed as ratios of the logarithms of values; thus, this fractal dimension is:

$$FD = \log(1234)/\log(1000)$$
$$FD = 3.09/3.00$$
$$FD = 1.03$$

which means that the length of the coastline, due to its complex shape, is greater than a simple straight line. If the coastline was very complex, let's say five times the length of an ideal straight line, then we get:

$$FD = \log(100000)/\log(1000)$$
$$FD = 5.00/3.00$$
$$FD = 1.67$$

which becomes closer to the topological dimension of area (which is 2) than the simple, one line, measured length of the coastline, which has a topological dimension of I (i.e., one).

3.6 ABOVE PIXEL LEVEL: PATCH CLASS STATISTICS

Patch class is the consideration of all patches of the same category, generally the same categorical class, often as a land use or land cover class. Patch class statistics consider summary statistics for, viz., the same category in the map such as all forest patches in a landscape. There are several groups of patch class statistics that are used by landscape ecologists: area and edge, shape, and dispersion. Summary statistics applied to each of these patch class measures include average/mean, median, standard deviation, range (largest and small value), and coefficient of variation. Table 3.4 lists some of the most used patch class statistics and their short name that is used in software such as FRAGSTATS (McGarigal & Marks, 1995).

TABLE 3.4. ***The measures, patch class metrics, short name, and description for some of the patch class statistics use by landscape ecologists. Adopted from McGarigal & Marks (1995).***

MEASURE	PATCH CLASS METRIC	SHORT NAME	DESCRIPTION
Area	Total	TA	Total area, generally in square meters or hectares, for the entire class.
Area	Total Edge	TE	The total area of the edge in all of the patches of the same type.
Area	Edge Density	ED	The total area of all edges in a particular class divided by the total area of the landscape.
Area	Mean	AREA_MN AREA_SD AREA_RA	Mean, standard deviation, and range area, typically in square meters, for all patches of the same class.
Area	Largest Patch Index	LPI	The area of the largest patch divided by the total area of the landscape. LPI is 100 if the entire landscape is composed of one class.
Shape	Summary Statistics for Perimeter-to-Area Shape Index	PARA_MN PARA_SD PARA_RA	This calculates the mean, standard deviation, and range of all perimeter-to-area ratios for each class of patches.
Shape	Summary Statistics for Fractal Dimension Index	FRAC_MN FRAC_SD FRAC_RA	This calculates the mean, standard deviation, and range of all fractal dimension indices for all patches of the same class.

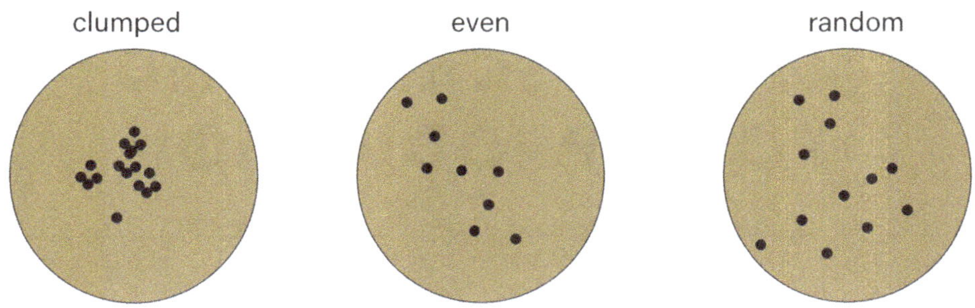

FIGURE 3.9. Three different spatial distributions of organisms: clumped, evenly spaced, and spaced with random distances.

3.7 VECTOR DATA MODELS—OVERVIEW

Biogeographers in general have several approaches to consider spatial distribution of species represented as point locations in space. These distributions can be divided into three types as shown in Figure 3.9. Each of these spatial distributions occurs in nature, but the most common is clumped. This is true because the distribution of resources that organisms rely on is also clumped in space. For example, soil moisture is often clumped in space and this distribution will govern the way that organisms respond to this environmental factor. In cases where distributions have even distances between individuals, often individuals (e.g., plants) disperse toxic chemicals that do not allow other individuals to grow. Random distributions are very rare in nature.

3.8 SUMMARY

Spatial ecologists use a variety of approaches to quantify spatial patterns for use in the management of natural resources. The review in this chapter is only a summary of approaches used in raster and vector spatial data modeling.

3.9 DISCUSSION QUESTIONS

1. Consider polygons as units that might be of value to landscape ecologists and that approaches to quantify how nearest neighbors are assessed could be useful. Given a hypothetical area composed of one polygon surrounded by five polygons, what kinds of nearest neighbor metrics could you suggest?
2. Using the example calculation above for determining fractal dimensions, compare and contrast the use of ln (natural log), $\log_2$ to that of using $\log_{10}$.

4. LANDSCAPES AND FOOD PRODUCTION

Highlights

Aims. This chapter summarizes the spatial arrangement and temporal dynamics of agricultural use in the world and in the United States. Spatial databases available for agricultural statistics are also summarized.

Learning objectives. Students should be able to describe:

- The spatial-temporal dynamics of agricultural use globally, nationally, and locally
- How water, nutrients, and agricultural policies shape the agricultural landscape and its impact on the environment and the farmer
- How spatial technologies have shaped current agricultural practices in the United States
- The historic pieces of legislation contained in the United States Farm Bill and how it has impacted cropping patterns in US landscapes

4.1 GLOBAL FOOD PRODUCTION TRENDS

According to the Food and Agriculture Organization (FAO) of the United Nations, about 37% of the total land area (4.8 billion hectares) of the world in 2021 was used for food production (Figure 4.1). By contrast, about 4.1 billion hectares are in forested land (about 32%), thus making area in agriculture the largest of all major land cover types (when excluding the land areas of Antarctica).

Within areas designated as agricultural land, about two-thirds or 67% is in pastureland, mostly supporting grazing by livestock such as cattle, sheep, and goats. The other third of the land is used to grow crops. Land to grow crops supports both human and livestock food needs. Note also that the FAO considers those crops that are planted and harvested within a year, such as corn and soybeans, to be temporary crops. Temporary crops comprise 78% of all arable cropland. Also note that permanent crops include fruit trees, nut trees, flowering shrubs, berries, vines, and other crops that span multiple years such as coffee, tea, cocoa, and rubber.

Changes in area for each of these major agricultural types have been significant in the last 20 years (2000 to 2021). For example, land in permanent meadows and pastures decreased by 5% from 2000 to 2021 while land in cropland increased by 6% during this same time period. Land in temporary crops increased by 12%, and the land area for permanent crops increased by 38%. These statistics for agricultural land change reflect the very

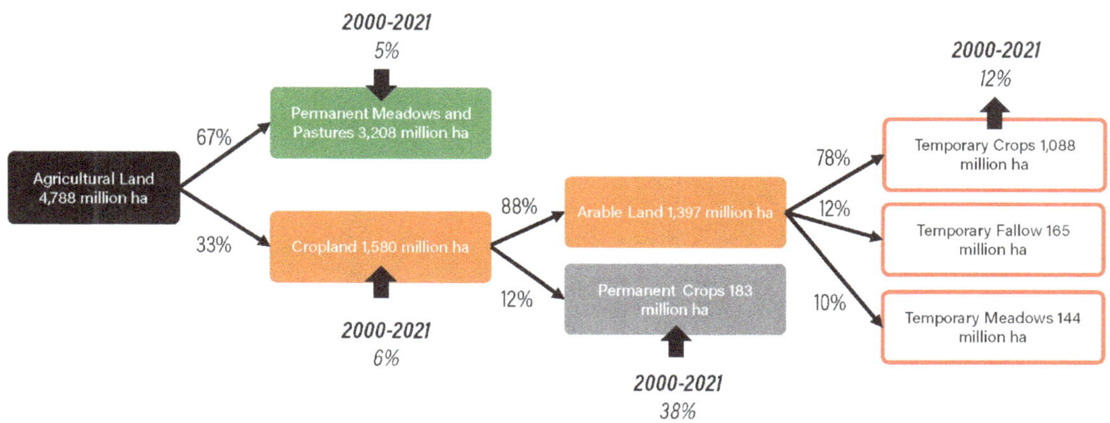

FIGURE 4.1. Amount of land, globally, in each of the major categories of agriculture. Modified from the FAO (2023). Also indicated are 21-year trends in major categories of land used for agriculture.

dynamic nature of land in each of the major categories of food production with shifts toward greater production of crops to support animals who are being supported in smaller spaces (i.e., there is a trend of increasing intensification of land used for animal production).

Another trend reported by the FAO (2023) was the large decrease in the land area in cropland per person, globally. A little less than 0.25 ha per person was allocated to cropland in 2000; this declined to 0.20 ha per person in 2021, a decrease of 18%. Other agricultural trends include: (1) an increase by 20% of the amount of land being irrigated, and (2) a threefold increase in the amount of agricultural land.

4.2 SPATIAL TRENDS IN FOOD PRODUCTION LAND USE IN THE UNITED STATES

A. AREAL AMOUNTS

The amount of land in food production (i.e., land in farms) is about 39% of the total land area of the lower 48 states. About 36% of the agricultural land is used for crop production, with 49% used as pastureland. Of the land in crop production, about 54% is in either corn (~95 million acres) or soybeans (~87 million acres). The states that comprise the corn belt (Kansas, Iowa, Illinois, Indiana, Ohio, Michigan, Missouri, Minnesota, Nebraska, and South Dakota) have around 80% of their cropland in either corn or soybeans.

Between 1997 and 2022, the total land in farms decreased by 7.8% in total area (Figure 4.2), in cropland by 1.6%, woodland by 6.2%, and pastureland by 14.2%. The decrease in pastureland is attributed to the growth of animal confinement production systems (Zulauf et al., 2024).

The most common crops being planted in the United States include corn, soybeans, hay, wheat, cotton, sorghum, barley, oats, and rice (Table 4.1). The amount of land for the first four crops is nearly 95% of the total land area in agricultural food production. Corn grown and used for silage is a form of this crop that is processed for use by beef and dairy cattle; the whole plant is used.

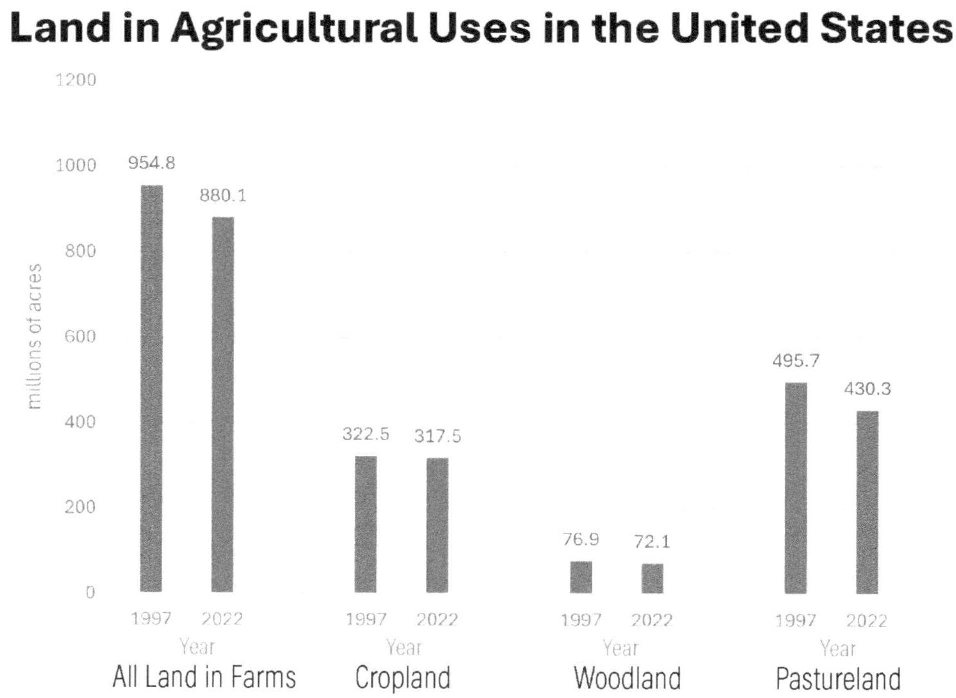

FIGURE 4.2. Land in farms, croplands, woodlands, and pasturelands in 1997 and then in 2021.

TABLE 4.1. *Amount of land in major crops for 2024 (source: USDA National Agricultural Summary Statistics). Percentages based on 320 million acres of land in cropland use.*

CROP	ACRES (APPROX.)	% OF U.S. CROPLAND
Corn (grain & silage)	~90–95 million	~28–29%
Soybeans	~83–87 million	~26–27%
Hay (alfalfa & other hay)	~50–52 million	~15–16%
Wheat (all types)	~45–48 million	~14–15%
Cotton	~11–12 million	~3–4%
Sorghum	~6–7 million	~2%
Rice	~2.5–3 million	<1%
Barley & Oats	~2–3 million	<1%
Specialty crops (fruits, nuts, vegetables, orchards, vineyards)	~10–12 million	~3–4%

B. REGIONAL CROPPING AREAS IN THE UNITED STATES

In the United States, there are about six agricultural regions that, based on annual temperature, soils, and precipitation, lend themselves to certain crops. These regions include a corn belt, wheat belt, cotton belt, dairy belt, and several specialty cropping areas (Table 4.2).

TABLE 4.2. *Major cropping regions of the United States (from the USDA National Agricultural Summary Statistics)*

REGION	CROPPING SYSTEM
Corn Belt of the Upper Midwest	corn & soybeans powerhouse
Wheat Belt of the Northern & Southern Plains	wheat, cattle, sorghum, some soy/corn
Cotton Belt & Delta	cotton, soy, rice, poultry
California & West	fruits, nuts, vegetables, dairy
Dairy Belt of Wisconsin/Midwest	milk & forages
Specialty pockets: various states	potatoes (Idaho & Michigan), citrus (Florida & California), tobacco (Kentucky & North Carolina), sugarcane (Louisiana, Florida, Hawaii)

C. CROPPING TYPES IN THE UNITED STATES

There are over 100 distinct kinds of crops grown in the United States. These crops are classified at the highest level into broad categories that include row crops (e.g., corn, soybeans, rice, wheat, and cotton), vegetables (e.g., potatoes, asparagus, green beans), fruits and nuts (e.g., apples, walnuts, grapes, citrus), and specialty crops (e.g., tobacco, sugarcane, hops, and herbs). The total amounts and spatial distribution of these crops change each year, and long-term trends are influenced by a variety of factors, such as national policies, international trade agreements, and economics. One of the most useful databases for tracking the spatial distribution of well over 100 crops in the United States is the USDA NASS Cropland Data Layer (CDL). CDL is derived from a deep learning model using remotely sensed imagery (e.g., Landsat) at 30 m resolution. The data are distributed each year, and historical data are available for downloading as TIFF images for 1997 to the current year.

4.3 AGRICULTURAL INPUTS AND OTHER CONSIDERATIONS

Spatial ecologists should consider several factors in assessing food production landscapes.

A. CROP ROTATION

There is evidence that ancient societies employed crop rotation as early as 6000 BCE in Mesopotamia (Columella, 70 BCE). Columella recognized that soil quality decreased if the same crop is grown for many years, if the soil is not "rested" in some years, or if the crop is not rotated with legumes, which fix nitrogen in the soil. In the United States, rotation of corn and soybeans has occurred since the 1800s; many Native American communities have been documented as practicing crop rotation, especially for corn and soybeans. One of the major reasons that the areas of corn and soybeans are nearly identical each year (although corn is always a bit higher) is that most land in either of these crops is rotated on a regular basis. The most common interannual crop rotations are summarized in Table 4.3.

TABLE 4.3. *Common Crop Rotations in United States agricultural systems*

MAIN CROP	ROTATION PATTERN
Corn	Corn one year → Soybeans the next
Wheat	Wheat → Fallow or pulses (peas, lentils)
Barley & Oats	Barley/Oats → Beans or Clover
Cotton	Cotton → Soybeans or Corn
Root Crops	Potatoes / Sugar Beets → Wheat / Barley

B. IRRIGATION AND WATER

About 18% of cropland in the United States is irrigated (Winters-Michaud & Hrozencik, 2024). Most of the irrigated land in the United States occurs west of the 100th meridian (-100° latitude) where rainfall is not typically adequate to grow many of the crops planted in these regions. States with the largest percentage of crops being irrigated, in order of the greatest to least, include California, Nebraska, Arkansas, Texas, Idaho, and Kansas. There are a couple forms of irrigation (Hrozencik & Aillery, 2021); the most common are (1) **surface irrigation**, which is gravity-based distribution of water through furrows, and (2) **pressurized irrigation**, which includes sprinkler distribution of water. Of the latter, center-pivot irrigation technologies are common in many areas, and these patterns (i.e., circular green cropping areas) can be observed easily in satellite imagery or aerial photographs. These irrigation systems often draw water from large groundwater aquifers that, in many areas due to long-term drought patterns, are being threatened. Most of the pivot irrigation uses 1,250- to 1,300-foot-long booms that cover an area of about 130 acres (USGS, 2017).

C. INTERCROPPING

Another long practiced agricultural management approach is intercropping. Intercropping is the practice of growing multiple crops together to increase production, reduce soil degradation, and reduce the number and amount of inputs to crops (Huss et al., 2022). Most patterns of intercropping (ATTRA, 2025) involve alternating row crops of different crops, called **row intercropping**; **strip intercropping**, which is growing wide strips of crops alternating across space; **mixed** or **polyculture intercropping**, where crops are mixed together irrespective of the rows or the crop; and **relay intercropping**, which is planting crops in different stages so that they are not growing at peak rates at the same time. Intercropping has also become one of several sustainable agricultural management practices used in vineyards.

D. MULTI-USE AGRICULTURAL LANDSCAPES

The growing demand for non-fossil fuel–based energy in the United States has resulted in a large growth of wind and solar installations occurring in rural areas, especially co-located with cropland and pastureland. Between 2012 and 2020, about 90% of all wind installations and 70% of solar installations occurred on croplands or pasture-rangeland (Maguire et al., 2024). Most solar and wind installations are in croplands in the Midwest and in pasturelands in the western United States. In 2024, it was reported that solar power generated

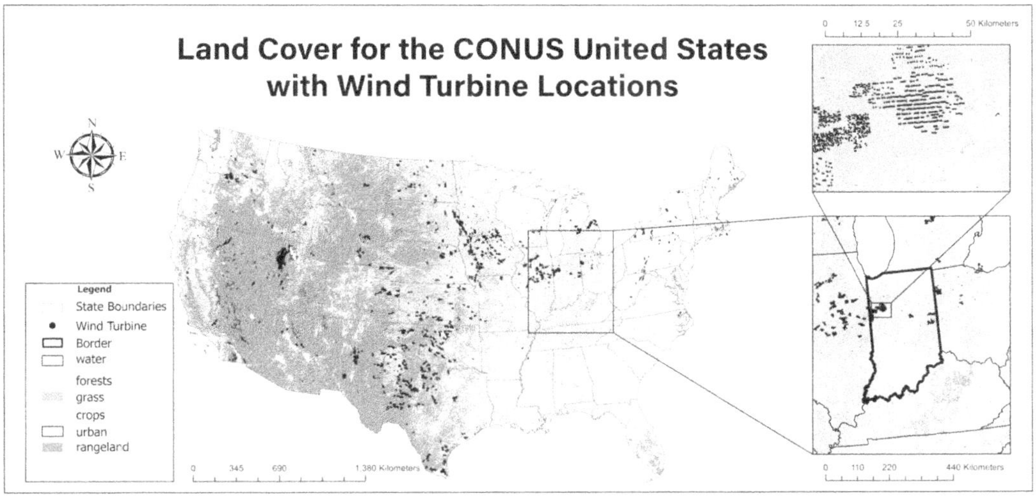

FIGURE 4.3. Distribution of wind turbines in the United States in relation to major land use/covers.

7% of the country's electricity, whereas wind power generated 10.1% (Bird et al., 2025). There are several wind farms that are very large, from 30,000 acres to well over 70,000 acres (Kaiser, 2024). Figure 4.3 shows the distribution of over 76,000 wind turbines in the United States with surrounding land use.

E. EROSION, SOIL FERTILITY, AND TILING

In cropping and pasture agricultural systems, several factors related to the soil affect productivity of the land and eventually the ability of humans to use the land sustainably. **Soil erosion** is the process of having wind, water, or ice separate soil particles from their original location to another place. Researchers have developed numerous spatially explicit soil erosion models (see Nearing, 2013 for an excellent summary) to estimate the amount of erosion based on vegetation cover, soil type, slope, and the amount of rainfall. Globally, soil erosion is one of the most important factors leading to the loss of agricultural productivity of the land; some estimates are that soil loss in cropland landscapes is occurring at 13–40 times the rate of soil renewal (Pimentel & Kounang, 1998). Also, an important consideration for sustainable agricultural systems is soil fertility. **Soil fertility** is a broad term that reflects the relative "health" (Lehmann et al., 2020) of the soil as it relates to the amount of erosion, organic carbon, nutrients (mostly nitrogen and phosphorus), its level of pH compared to its natural state, salinity, soil biological community diversity and composition (e.g., microbes), its physical structure (e.g., whether soils are more compacted) compared to its natural state, and whether the soil harbors toxic substances (see Bünemann et al., 2018 for an excellent review). Finally, an important water management practice in agricultural fields is tiling. The practice of **tiling** is to install perforated pipes underground that move water from the fields and into ditches, creeks, or wetlands in poorly drained areas. Tiling aids in reducing soil erosion, increases crop productivity (i.e., yields), and allows for greater oxygenation and nutrient retention of the soil. The practice of tiling is often done in areas that are very flat or have high water tables, and where the lack of drainage of water from fields presents problems in managing crops (e.g., planting and harvesting). Tiling is used primarily in the United States (mostly in the Midwest) and in Europe. Tiling may have negative impacts on freshwater ecosystems (see Blann et al., 2009). The USDA Agricultural Census maintains a database, by county, of the estimated acreage of tiling land; and one field in the USDA SSURGO database contains a variable, drainage class, that describes the drainage of an area across a range from excessively drained to very poorly drained (Wan et al., 2024).

F. NUTRIENTS

Many nutrients that are necessary for plant growth affect the overall productivity of crops and of grasses and forbs in pasturelands. The three most often cited are **nitrogen** (N), **phosphorous** (P), and **potassium** (K), although a few others (calcium, magnesium, and sulfur) also play critical roles in plant growth. In many cases, the use of fertilizers (containing specific ratios of these nutrients) can boost crop yields for farmers by as much as 40–60% (Erisman et al., 2008). The USDA Economic Research Service (ERS) keeps a database on fertilizer use, sales, and prices that is often used in spatial analysis of the impacts of agricultural practices on water quality and crop yields and pasture grassland primary production.

G. PEST AND DISEASE CONTROL

Other chemical inputs to cropping systems include herbicides, insecticides, and fungicides to control pests and diseases. Lime and gypsum may also be added to control soil chemistry and the physical structure of soils. Surfactants, which are chemicals that increase the adherence of chemicals to plants and the soil, are also used. An estimated pesticide use database, summarized at county levels, reported annually, is supported by the Pesticide National Synthesis Project, which is part of the National Water-Quality Assessment Program of the USGS (2025).

4.4 PRECISION AGRICULTURE AND SPATIAL ECOLOGY

One of the most impactful recent developments in the last 20 years has been the advancement of a host of technologies that support the optimization of agricultural productivity through managing lands at high resolution and precision. **Precision agriculture** involves the incorporation of technologies on farm equipment (e.g., harvesters) with that of spatial technologies like high resolution maps from space, planes, and drones. Precision agriculture generally involves six main components (Barbosa Junior et al., 2024):

1. Sensors and sensor networks. Sensors that collect, near real time, information about soil moisture, crop biomass (e.g., through NVDI measures), temperature, and soil parameters such as pH.
2. GPS and GIS. These two associated technologies are often integrated into farm equipment that can allow farmers to view their lands in high precision on maps overlaid with common shapefiles and raster file data.
3. Variable rate technology (VRT). Having sensors, combined with GPS and GIS technologies, apply inputs (e.g., fertilizers, water) in variable amounts with spatial and temporal precision based on computer application models.
4. Technology connectivity (IoT). Supporting connectivity of sensors, GPS, maps of applications in high resolution, which can be posted on a cloud data server and then analyzed by farmers using a computer connected to the internet is another part of the precision agriculture technology toolkit. This general connectivity model is called the Internet of Things (IoT), where sensors are critical parts of a highly integrated set of technologies.

5. Decision support systems (DSS). A variety of platforms are being developed that integrate data streams from sensors connected by IoT and that leverage agricultural management models that use VRT; some of these incorporate data streams such as future weather forecasts to assist in decision making at the farm field scale.
6. Artificial intelligence (AI). Many AI technologies such as artificial neural networks (ANNs) and convolutional neural networks (CNNs) are being used to support the real-time analysis of precision agriculture (PA) data, which tends to be massive and hyperdimensional.

4.5 AGRICULTURAL POLICIES: THE US FARM BILL

In the United States, the **US Farm Bill** is published every five years; it describes funding and policy aims for agriculture production and practices in six main areas: (1) crop pricing support; (2) insurance programs; (3) conservation programs (to protect the environment within and around agricultural areas); (4) biofuel and other renewable energy program support efforts; (5) nutrition assistance; and (6) research and land grant university funding programs. Some of the most impactful farm bills in US history include the following (from Monke, 2024):

1. Agricultural Adjustment Act (AAA) of 1934. This program was one of the first to establish economic and policy programs that supported farmers' ability to stay on their land despite economic hardships occurring during the Great Depression.
2. Farm Security and Rural Investment Act of 1985. This Farm Bill increased the Conservation Reserve Program (CRP), which encouraged farmers to retire or "rest" land, especially marginal land, to reduce soil degradation.
3. Federal Agriculture Improvement and Reform (FAIR) Act of 1996. Also known as the "freedom to farm" act, which gave farmers greater control over what crops that they grew. This bill also increased crop insurance programs.
4. Food, Conservation and Energy Act of 2008. This Farm Bill focused on supporting farmers in their efforts to contribute toward biofuel crops (ethanol for corn and biodiesel for soy) as well as support for the infrastructure (processing facilities). This 2008 Farm Bill was developed in response to the Energy Independence and Security Act (EISA) signed into law by President George W. Bush in December 2007.

4.6 SUMMARY

Agriculture is the largest land cover in the world. In broad terms, food production landscapes are used to support croplands and pasturelands. Most of the cropping land is used to grow crops for livestock, primarily for beef and dairy cattle. In the United States, over 100 different crops are grown, although most land is used to grow five major crops: corn, soybeans, hay, wheat, and cotton. Many areas of the United States are also irrigated (in the west) or tiled (in the east and north) to manage water resources for crop production. A new trend has emerged of multiple uses for rural agricultural land to support food production and renewable energy. Precision agriculture is also an emerging trend that uses a variety of spatial technologies to manage lands at small spatial extents. Finally, in the United States, the five-year Farm Bill has been used to direct policies and financial resources to a host of programs that support farmers, researchers, and people needing nutritional assistance.

4.7 DISCUSSION QUESTIONS

1. Describe some of the main elements of what is called precision agriculture. How important are spatial technologies? Be able to provide some specific examples.
2. Describe the trends in agricultural footprints over time and be able to compare current agricultural footprints to those of other major land covers.
3. What are the common inputs and the environmental consequences of the food production system?

5. LANDSCAPES AND WATER RESOURCES

Highlights

Aims. This chapter summarizes the main concepts of water movement and water features that exist in association with the landscape.

Learning objectives. Students should be able to:

- Describe the hydrologic cycle in terms of the major compartments and processes that exist globally and how it interfaces to the land
- Compare and contrast the water processes that are part of the water cycle
- Describe how watersheds are coded using at least two approaches
- Know how to identify stream orders in a stream network using the Stahler topological order numbering system
- Compare and contrast the Cowardin and Ramsar wetland classification systems

5.1 THE WATER CYCLE AND LAND

The **water cycle** is an impressive natural process occurring all over the world. This cycle, also called the **hydrologic cycle** (Figure 5.1), involves movement of water molecules into and out of several major compartments; involves being in one of three states—solid (as ice), liquid, or gas; occurs in the atmosphere, on land, and underground; and interfaces with the geophysical, biological, and chemical dynamics of the Earth. There are six major water compartments (i.e., reservoirs)—**ocean, surface water, ice/snow, groundwater/soil/aquifers, biospheric** (in plants, animals, and microbes), and the **atmosphere**—where water resides in any of the three forms. The ocean stores about 97% of all of the world's water, and only about 3% is in the form of freshwater (technically defined as water that has less than 1,000 mg/l of dissolved salts). There are a host of hydrologic processes in this cycle, which involve either movement or phase changes (e.g., melting of snow to water).

Important hydrologic processes (i.e., fluxes) in the water cycle, which are movements of water between the compartments, include:

1. Condensation. This part of the hydrological cycle is a process by which water vapor changes into small water droplets that are gathered into atmospheric formations of clouds or fog.

2. Precipitation. This process is the formation of water as a solid (ice, sleet) or liquid (rain); the form is based on air temperature and then falls from the atmosphere to the land or water surface.
3. Evaporation. This is a phase change that involves water changing to water vapor in gas, generally from heat energy from the sun. Evaporation from oceans, surface water bodies, and from the soil are the most common.
4. Transpiration. The release of water from plants during respiration.
5. Evapotranspiration. Ecologists recognize the combined transpiration of plants and the evaporation from the land surface to be an important process called evapotranspiration or ET.
6. Infiltration. The process in which water hits the land surface and then enters the soil and surficial portions underground.
7. Percolation. This is the movement of water down underground after it has infiltrated the surface. Infiltration and percolation are important processes that replenish groundwater compartments.
8. Advection. Through atmospheric movement, water moves through space (e.g., from over the oceans to over land) through a process called advection.
9. Sublimation. The process of a substance changing from solid directly to gas is called sublimation. In the case of water, this is the transition from ice to water vapor. The reverse process of going from water vapor to ice is called deposition (or desublimation).
10. Surface runoff. This is the movement of water over land that does not infiltrate the subsurface. Water from surface runs often moves into surface water bodies (e.g., lakes, rivers, wetlands, ocean).
11. Subsurface flows. This is the movement of water through the surficial ground that generally connects infiltration back to surface water bodies; the latter process is called recharge.
12. Groundwater outflows. This is the movement of water from underground to the ocean.
13. Snowmelt. The process of water changing from snow and ice to liquid water is often an important event in many ecosystems where winter creates large amounts of snow (i.e., snowpack) that then melts during the early spring.

The water cycle involves all of these processes working often in sequence—for example, condensation, precipitation, infiltration, and then percolation—and across multiple compartments (atmosphere, lakes, and then underground).

5.2 SURFACE WATER

Surface water can reside in several hydrologic bodies—lakes, rivers and streams, and wetlands. In ecology, those forms of surface water that are "still" are called **lentic** systems and those that move are referred to as **lotic** systems. Lotic systems are also called **fluvial systems** by hydrologists to emphasize that water is flowing or moving.

In fluvial systems, the geometry of the river is referred to, in general, as the **stream or river channel**. The sides of the channel are the **banks**, and the bottom of the channel is called the **bed**. Channels can be straight, braided (i.e., composed of a mixture of multiple channels and islands), or meandering (i.e., S-shaped). Streams can be intermittent (i.e., occurring only during certain periods of time) or permanent (i.e., they are present all the time).

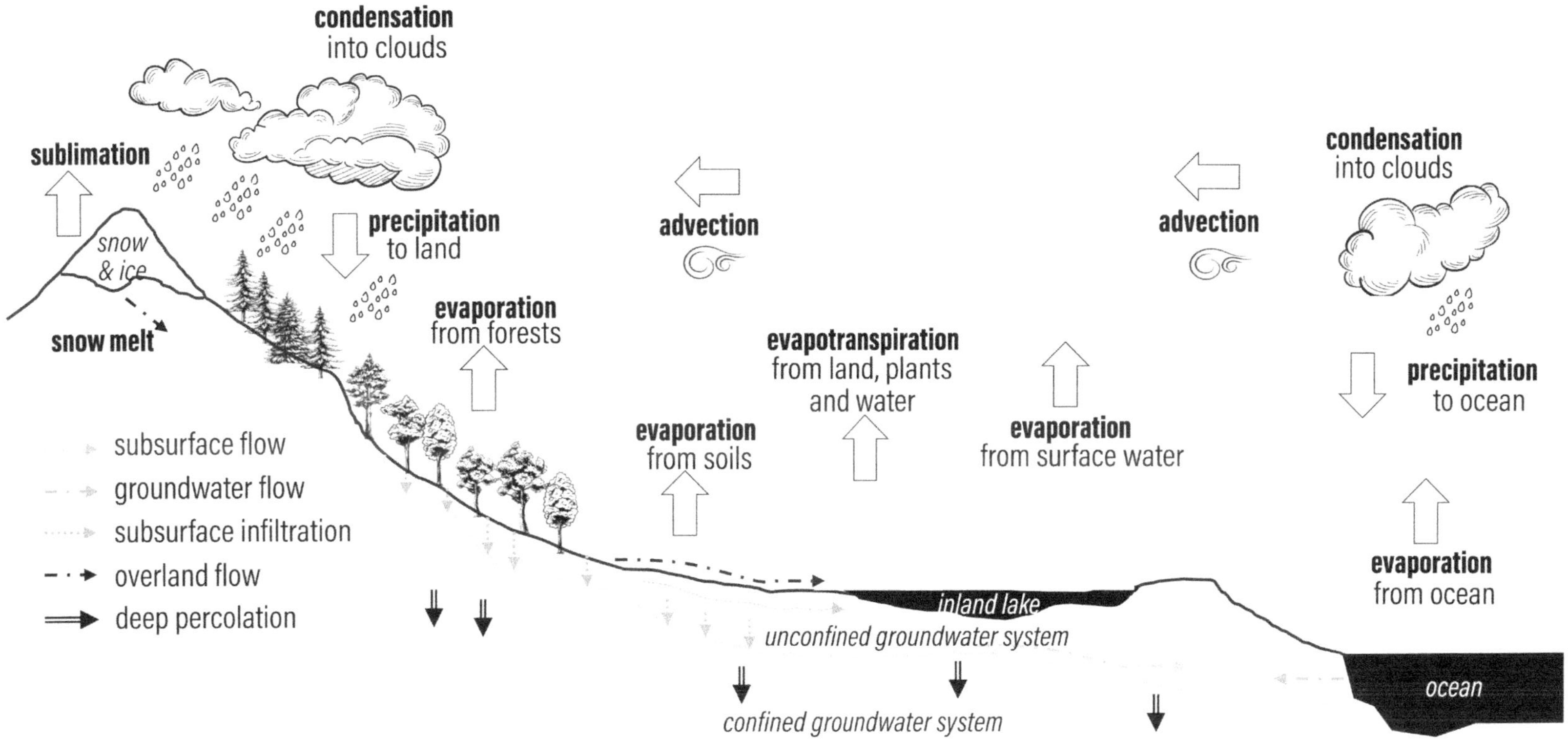

FIGURE 5.1. *The water cycle, which forms the context of how to consider water in landscapes. Modified from the USGS Natural Water Cycle by Howard Perlman and John Evans.*

5.3 WATERSHEDS

A **watershed** is the surface area of land where all water drains to a common outlet, such as a lake, ocean, or river. As such, watersheds are ecological units that unify hydrologic processes at landscapes and larger spatial extents. Watersheds can also be referred to as drainage basins or catchments. Hydrologic processes common in watersheds include overland flow, and river and stream flow that also interface with that of groundwater (e.g., water movement underground). Since water is an excellent solvent, the water that moves within a watershed will also carry with it chemicals (e.g., nutrients such as phosphorous and nitrogen, or hazardous chemicals) along with physical matter (e.g., sediment) and/or biological material (e.g., seeds, pollen, and leaves).

Watersheds are defined by boundaries, which are created by the **topography** of the surface. Topography refers to the landforms (the areal distribution of surface heights created by its geomorphology) of the Earth; it represents the 3D surface of the landscape. Locations of maximum elevation create catchment surfaces whereby the drainage of water occurs on either side of any peak. Let us consider a cross-section (i.e., side view) of land that is composed of three hills, labeled hill #1, hill #2, and hill #3 in Figure 5.2. Note that the peaks, also called ridgelines, of the hills, as indicated by locations A, B, and C, create breakpoints where water flows down the sides of the hills and then eventually to the lowest points, which for watershed #1 is defined by drainage divides, A and B, that eventually form a lake that is labeled D. Note that the neighboring watershed #2, defined by drainage divides B and C, forms a river that is labeled E. If the water stays on the surface of the land, that process is called surface runoff or **overland flow**, represented by all arrows on this figure. Drainage divides A and B create watershed #1, whereas drainage divides B and C create the neighboring watershed, watershed #2. Note that the rivers for watersheds B and C could combine at lower points along the landscape to form a single, larger river, in which case these two watersheds would, at a larger organizational scale, be part of a shared larger watershed.

5.4 WATERSHED CODING SYSTEMS

A. HYDROLOGIC UNIT CODING SYSTEM

The spatial extent of watersheds, viewed as a map, are called **hydrologic units** by the United States Geological Survey (USGS). Each hydrologic unit is assigned to a unique value so that it can be referenced (Verdin

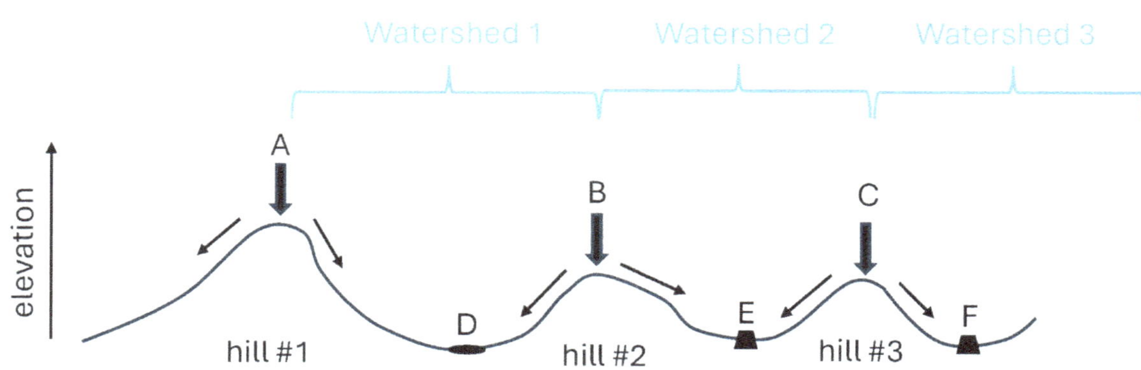

FIGURE 5.2. A side view of hypothetical terrain composed of three hills, each with ridgelines (A, B, and C) that direct water down the sides of the hill and into water bodies (D, E, and F). Watershed boundaries are defined by the peaks for the three watersheds shown.

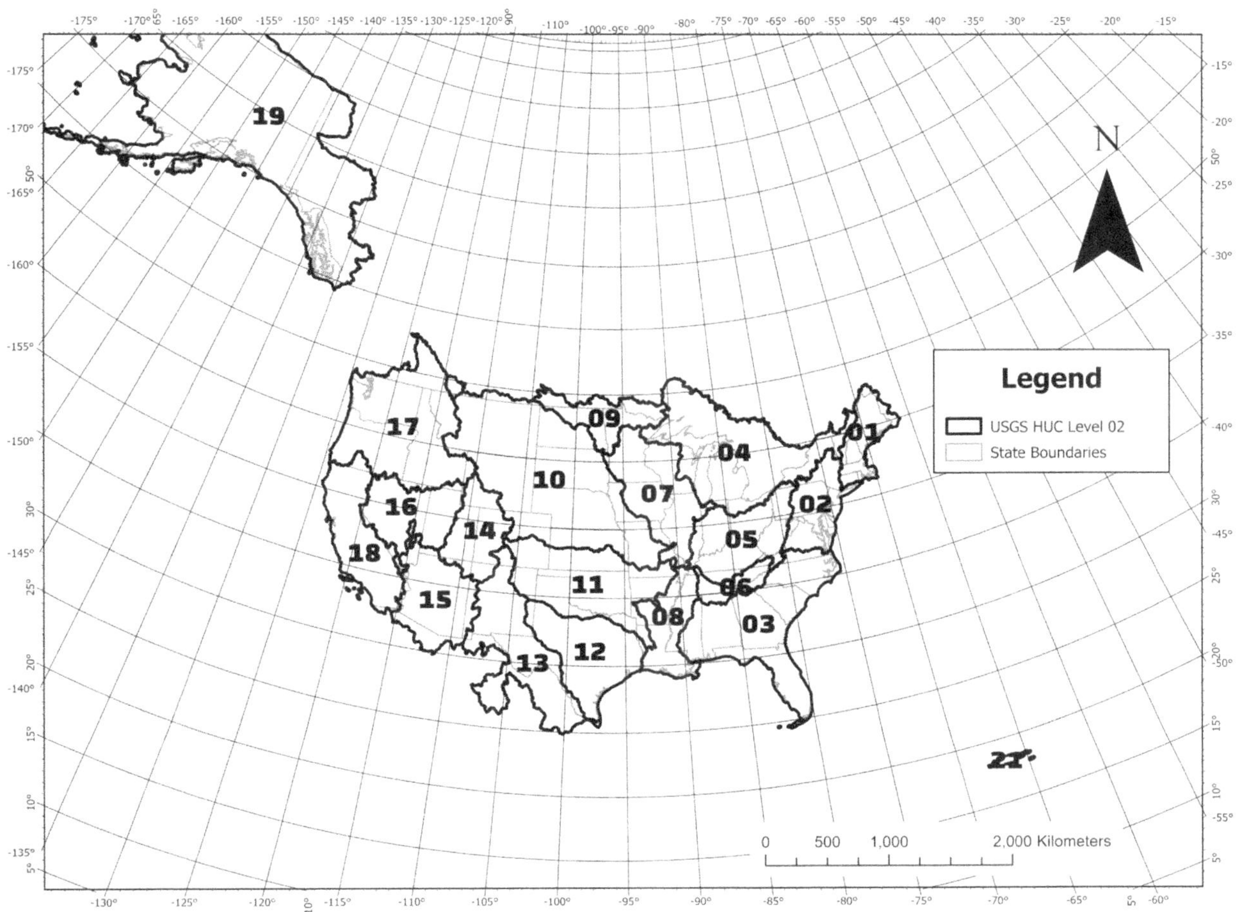

FIGURE 5.3. 2-digit hydrologic units and their codes for most of the United States including Puerto Rico.

& Verdin, 1999). This coding system recognizes that watersheds are hierarchically organized into hydrologic units that are nested within one another. Natural resource management agencies in the United States and in some other countries have designated these hydrologic units, at different levels of organization, using **hydrologic unit codes** or **HUCs** (and within GIS offices, often spoken of as "HUCs" to refer to these databases or places). This hydrologic unit system recognizes six levels of watersheds, each level designated using a set of 2-digit codes (values can range from 01 to 99).

Figure 5.3 shows the boundaries of the largest hydrologic units, which are for 2-digit levels, for most of the area of the United States. The largest hydrologic unit is coded using 2 digits and is also referred to as **basins** or **regions**. There are 21 (22 if you include Puerto Rico) 2-digit-level HUCs in the United States, some of which extend into Canada and Mexico.

A closer view of the 2-digit Ohio Region (HUC 05) with HUC 8-digit level hydrologic units is shown in Figure 5.4. Within this 2-digit HUC, there are 120 8-digit-level HUCs that span 11 states (e.g., IN, IL, KY, TN, VA, NC, OH, MD, NY, PA, WV). The complete list of 2-digit HUC codes, their official name and approximate size is provided in Table 5.1.

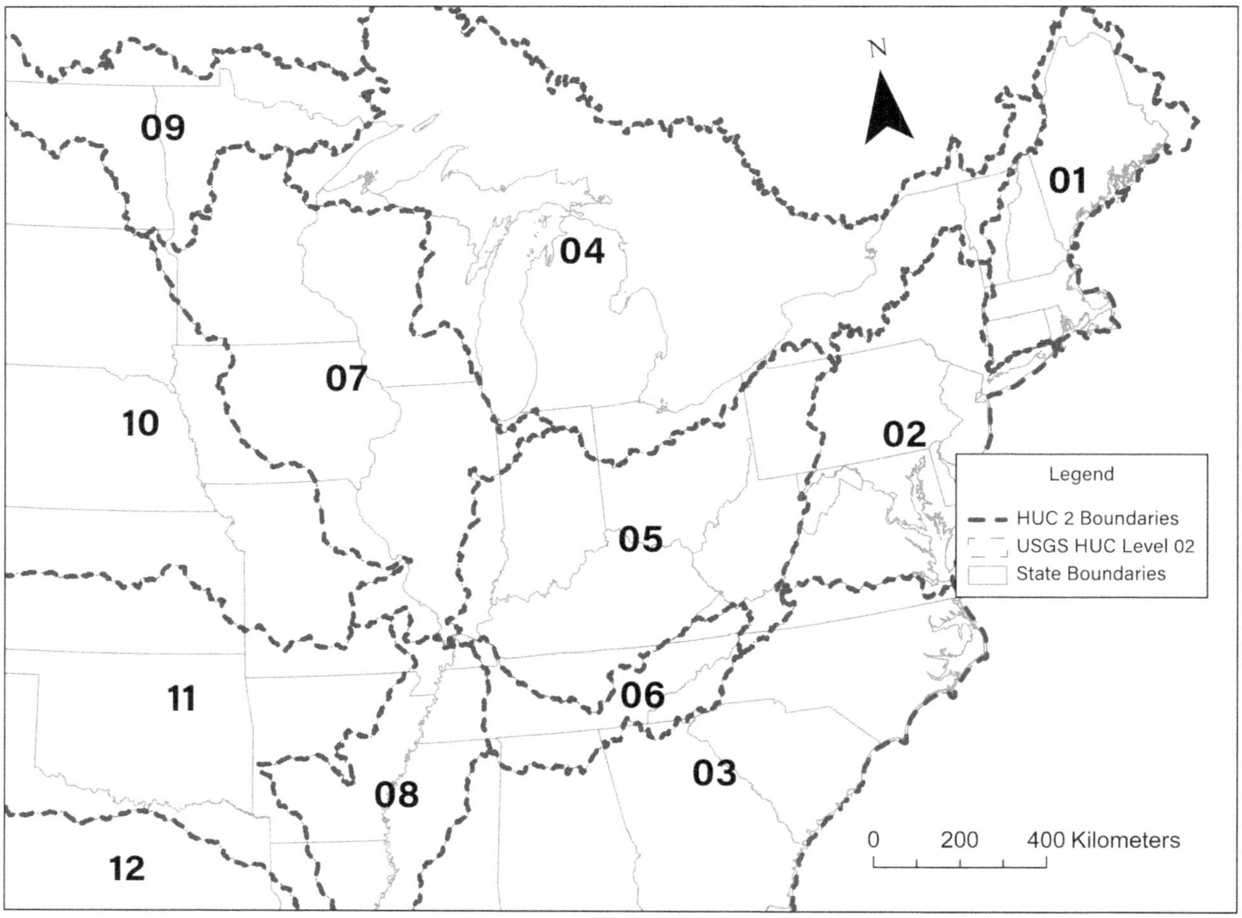

FIGURE 5.4. Hydrologic units for the eastern portion of the country, highlighting the 2-level HUCs and 8-level HUCs.

The hydrologic unit coding system, including HUC values and names, was established in the United States by the United States Geological Survey (USGS) in coordination with the Water Resources Council in the mid-1970s (Seaber et al., 1987). This system uses a hierarchical numbering system for 21 regions (not including Puerto Rico) that include 222 subregions, 352 accounting units (referred to as basins), and 2,149 cataloguing units. Table 5.2 summarizes this topological (spatial numbering scheme) system by HUC level, providing example hydrologic unit code values at each level. The names used in the HUC system for those in the United States have been vetted by a standards committee and then stored in the Geographic Information Names System or GNIS (Jones et al., 2022).

Many watersheds in the United States are managed at the 8-digit hydrologic unit level, although 10- and 12-digit hydrologic units are common managed extents. Figure 5.5 shows the arrangement of 10-digit HUCs for an area within the Great Lakes, with their names, along with the location of the 8-digit HUC for the Muskegon River Watershed, which serves as one of the book's instructional case study areas.

TABLE 5.1. *Major Hydrologic Unit Regions of the United States including their codes, names, and sizes. From the United States Geological Survey.*

2-DIGIT CODE	NAME	SIZE (IN SQUARE KM)
01	New England Region	198,829
02	Mid-Atlantic Region	276,483
03	South Atlantic–Gulf Region	739,946
04	Great Lakes Region	842,528
05	Ohio Region	421,966
06	Tennessee Region	105,948
07	Upper Mississippi Region	491,925
08	Lower Mississippi Region	276,038
09	Souris-Red-Rainy Region	259,230
10	Missouri Region	1,349,419
11	Arkansas-White-Red Region	642,213
12	Texas–Gulf Region	474,543
13	Rio Grande Region	597,884
14	Upper Colorado Region	293,569
15	Lower Colorado Region	424,356
16	Great Basin Region	367,049
17	Pacific Northwest Region	836,517
18	California Region	436,625
19	Alaska Region	2,209,448
20	Hawaii Region	28,972
21	Caribbean Region	15,567
22	South Pacific Region	8,195

TABLE 5.2. *A summary of HUC systems in the United States with an example of coding for a small (12-digit HUC) located in the Muskegon River Watershed in the lower Peninsula of Michigan*

HUC LEVEL	HUC LEVEL NAME	NUMBER OF UNITS AT THIS LEVEL	HUC NUMBER EXAMPLE	HUC NAME (FOR EXAMPLE AND ITS SIZE)
2-digit	Region	21 or 22	04	Great Lakes Region (size = 842,528 sq km)
4-digit	Subregion	222	0406	None used
6-digit	Basin	352	040601	Northeastern Lake Michigan Basin
8-digit	Cataloguing Unit or Subbasin	2149	04060102	Muskegon River Watershed (size = 7,062 sq km)
10-digit	None used	~22,000	0406010204	None used
12-digit	Subwatershed	~160,000	040601020406	Cranberry Creek–Muskegon River (size = 152.68 sq km)

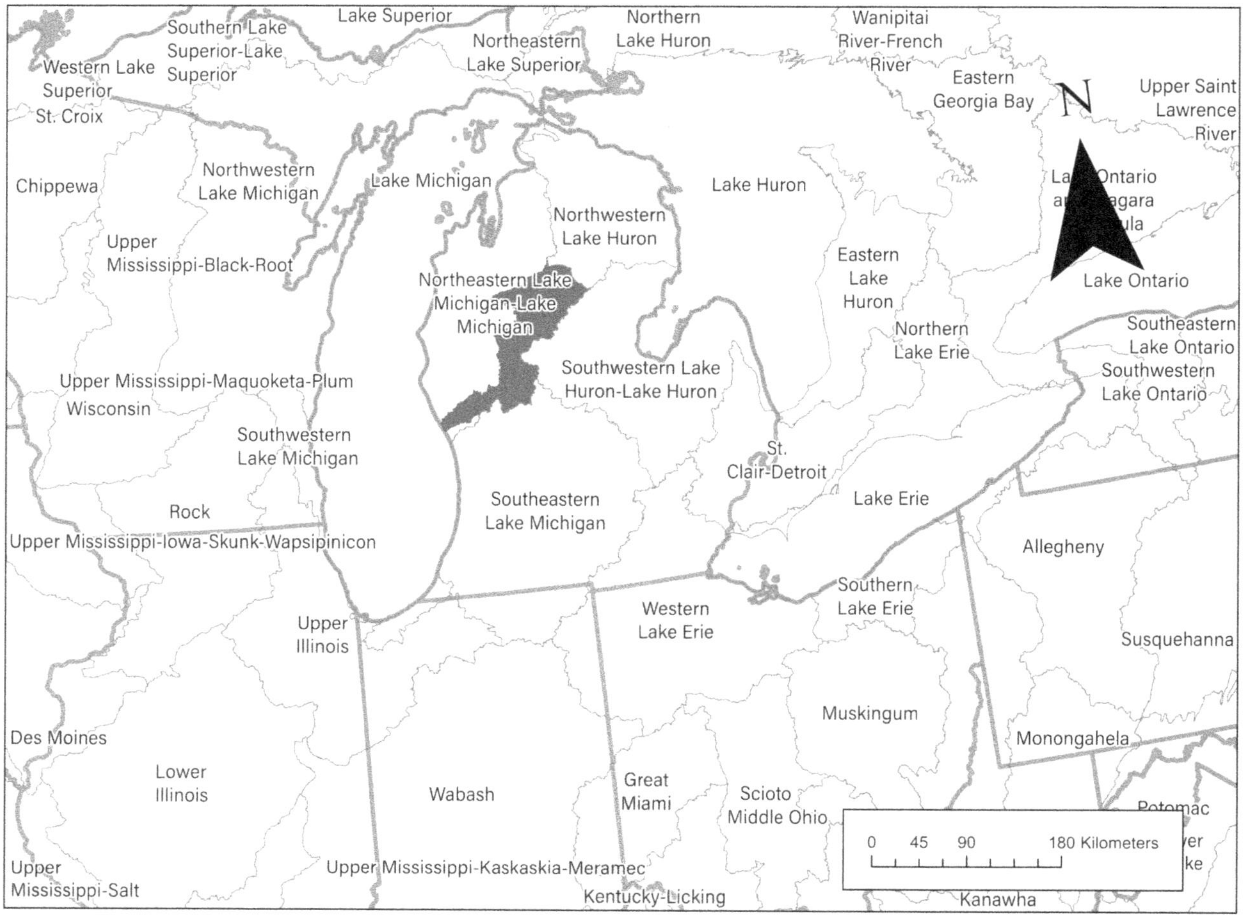

FIGURE 5.5. Distribution of HUC 4s in the Upper Great Lakes Region also shows the location of a HUC 8, the Muskegon River Watershed.

B. PFAFSTETTER CODING SYSTEM

The **Pfafstetter Coding System** for watersheds was proposed by the Brazilian hydrologist Pfafstetter in the 1950s and later documented in 1989 (Verdin & Verdin, 1999). This coding system uses a slightly different topological numbering system that assigns a value of 1 to the main drainage basin (lowest in elevation) and then values of 3, 5, 7, and 9 to drainage areas for tributaries (called **basins**) and values of 2, 4, 6, and 8 to drainage basins of stems (called **interbasins**). Figure 5.6 shows how this numbering system works, with numbers assigned based on how areas drain to tributaries or to stems. Note that, just like the HUC system used by the USGS, the numbering system is hierarchical. For example, the tributary denoted as 8 is composed of smaller basins and interbasins.

Basin spatial data for the entire world are available in this format. These data are often distributed at groupings close to that of continents (there are nine global distributions, not including Antarctica; there are obviously seven continents). A map of level 1 or major basins (coded 11–18) for Africa is shown in Figure 5.7. Also included are level 5 basin boundaries of the continent. As with the HUC system, the basins are hierarchically arranged so that there are dozens to hundreds of basins that "sit" within the larger basin. For North America,

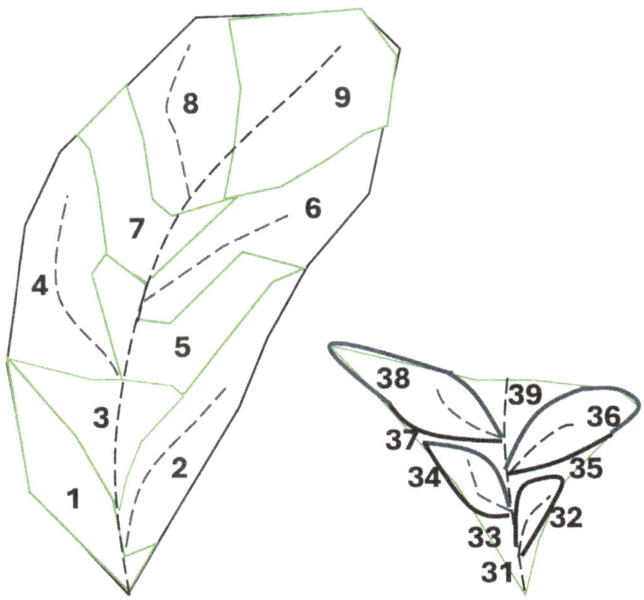

FIGURE 5.6. Illustration of the topological numbering of basins and interbasins using the Pfafstetter Coding System for hydrologic units.

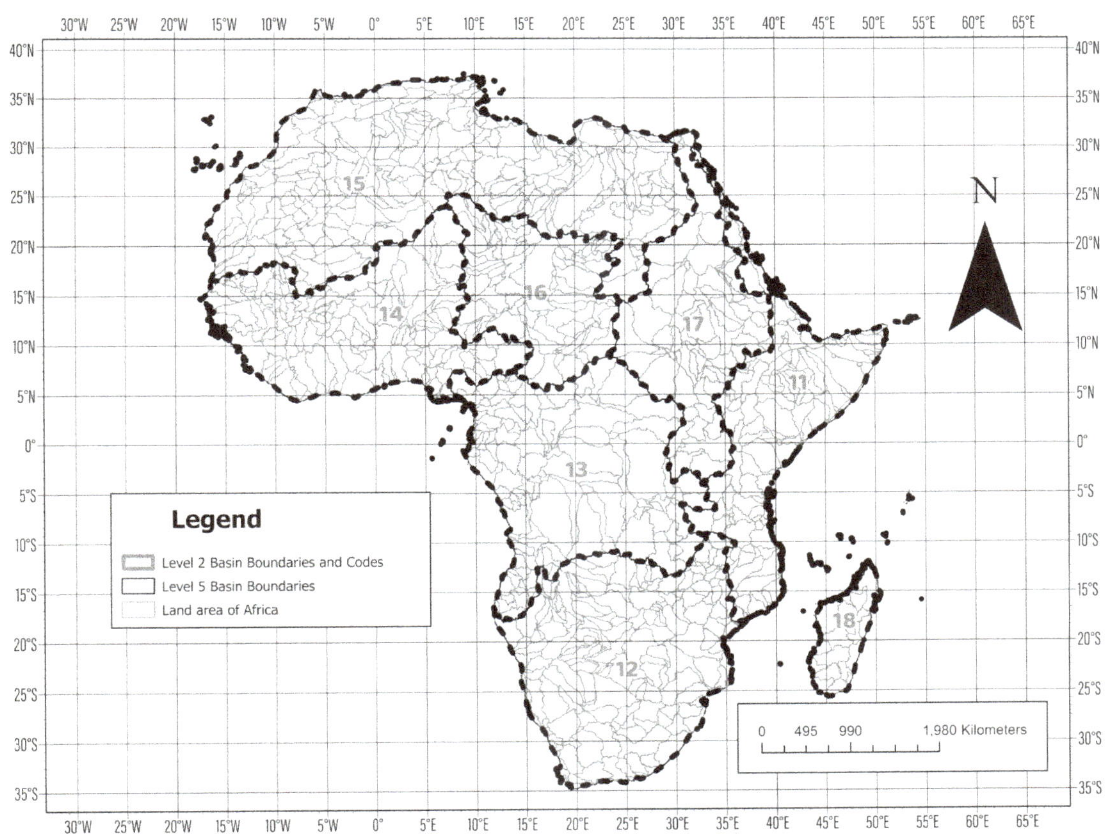

FIGURE 5.7. Map of Pfafstetter basins at level 2 and level 5 for the continent of Africa.

there are nine major basins that span Canada, the United States, and Mexico but these data do not include northern parts of Canada and most of Alaska (these are in a separate Arctic distribution).

5.5 STREAM NETWORKS

As all water in catchments essentially flows to the lowest point in the landscape, this process eventually produces a network of rivers and streams, with segments that are recognized as **stream orders**. There are several methods that are used to label stream segments based on their position within the river network. The network numbering system considers that streams start with **first-order**, or primary, streams that are then joined to one or more other primary streams to form a **second-order** stream. Second-order streams then combine to form **third-order streams**, and so on. First-order streams are often referred to as **headwater streams**.

One of the most common stream order topological numbering systems is the one proposed by Strahler (1952). This system considers segments and nodes and a set of simple rules for labeling stream orders. These include:

Rule 1: If a node has no children (it is a terminal segment), it has a Strahler number of 1;
Rule 2: If a node has two or more children of Stahler number i, then the Strahler number of the node is $i + 1$;
Rule 3: If the node has one child with Strahler number i, and all other children have Strahler numbers that are less than i, then the node is labeled as i again.

Figure 5.8 shows a stream network composed of segments with stream orders 1 through 4, with arrangements showing how rules 1 through 3 are implemented in the numbering system. The largest stream order segment of a network is also called its **mainstem** or **trunk**. The **stream network tree number** is the largest stream order in the network. The stream network tree number or trunk value is 4 in Figure 5.8.

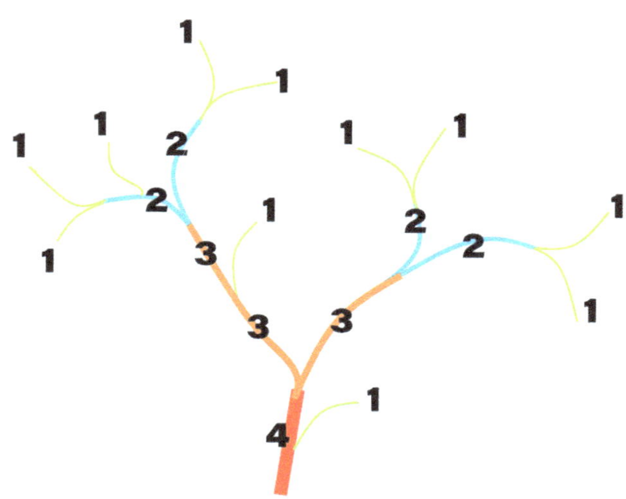

FIGURE 5.8. Strahler stream orders showing at least two ways that stream orders of 1 can be assigned (following rules 1 and 2) and two ways that stream order 3 are labeled (following rules 2 and 3).

5.6 WETLAND CLASSIFICATION SYSTEMS

There are a variety of wetland classification systems used around the world. In the United States, the most often used is the Cowardin wetlands classification system, which has been adopted by the US Department of Interior. Globally, the **Ramsar Classification System**, which arose out of the 1971 Ramsar (Iran) Convention of Wetlands, has been in use since the late 1970s and is most often used by conservation organizations. A summary of each of these classification systems follows.

A. COWARDIN WETLANDS CLASSIFICATION SYSTEM

This system developed to assess characteristics of wetlands is a hierarchical classification system (Figure 5.9) that has five levels of description: (1) system; (2) subsystem; (3) class; (4) subclass; and (5) modifiers. This wetlands classification system was first described by Cowardin et al. (1979). A more detailed description of these five levels are:

1. System: At the highest level of organization, wetlands are classified as marine, estuarine, riverine, lacustrine, or palustrine. Marine systems are those open water areas that are exposed to the full force of ocean wave action. Estuarine systems are those that border land along marine systems and that have a mixture of freshwater and marine water and are influenced by tides. Riverine systems are channel habitats associated with flowing water. Lacustrine wetlands are dominated by open water and are in the form of lakes, ponds, and reservoirs. Palustrine wetlands, on the other hand, are dominated by vegetation and are often referred to as marshes, swamps, fens, bogs, and other areas that are exposed to still or flowing water.
2. Subsystem: Each subsystem's descriptions vary by system; there are between 0 to 4 (most have 2).
3. Class: This level most commonly describes the composition of the bottom and/or the shoreline of the wetland.

	System	Sub-system
Wetlands	Marine	Subtidal
		Intertidal
	Estuarine	Subtidal
		Intertidal
	Riverine	Tidal
		Upper Perennial
		Lower Perennial
		Intermittent
	Lacustrine	Limnetic
		Littoral
	Palustrine	None

FIGURE 5.9. First two levels of the Cowardin Wetlands Classification System.

4. Subclass: At this level, classes are subdivided into plant life form (e.g., broadleaf deciduous) and/or substrate composition (e.g., rock bottom).
5. Modifiers: This level allows natural resource managers to add important information to each wetland including water chemistry, pH, physical regimes (where the wetland is located in the landscape), and soil characteristic modifiers.

B. RAMSAR WETLANDS DESIGNATION SYSTEM

This international classification system is a wetland designation system that was developed to identify and classify those wetlands that serve various roles in conservation. These are wetlands that encompass "areas of marshes, fen, peatland or water, whether natural or artificial, permanent or temporary, with water flowing or static, fresh, brackish or salty, including areas of marine water the depth of which at low tide does not exceed six meters" (Ramsar Convention Secretariat, 2016). The Ramsar designation system recognizes five basic wetland types: lacustrine, riverine, palustrine, marine, and estuarine. A global database, the Ramsar Sites Information Service (RSIS), contains information for over 2,500 internationally significant wetland sites. Nine possible criteria are included for the sites, which include those that (1) have species or communities of concern (i.e., near extinction) that rely on the site; (2) are unique in and of themself; and (3) support large and diverse populations of bird, fish, and/or mammal species. Currently, there are more than 40 Ramsar sites in the United States and the United States Fish and Wildlife Service coordinates Ramsar site designation with the international organization.

5.7 RIPARIAN ZONES

Riparian zones are areas of the landscape where a variety of biogeochemical processes occur at the interface of the land and water bodies—the latter being streams, wetlands, and lakes. Riparian zones are thus unique areas on Earth (Naiman & Rogers, 1997) that support important ecosystem services, most importantly, the maintenance of freshwater resources. The spatial delineation of riparian zones is challenging, as they vary across the landscape because hydrological processes are spatially variable. For example, in first-order streams, the riparian zone is likely to have a smaller extent and have a near 100% canopy cover, whereas in areas where there are wide channel rivers such as third-order streams, the riparian zone would encompass a large part of the upland areas buffering the stream.

A general depiction of a stream riparian zone is shown in Figure 5.10. There are three important areas in the landscape to consider (1) the **upland area**, (2) the riparian zone and the (3) **river channel**. Within a typical riparian zone are trees, sedges, grasses and herbs, with uplands composed of natural areas such as forests and grasslands. When streams are at their lowest point, this is called their **baseflow** height (dry height). Most streams will vary in their heights due to precipitation, and some will have higher levels at **flood stage**. The areas of the riparian zone and even potentially the upland area that would be impacted by river flows is referred to as a **flood map** or **floodplain**. Flood maps in the United States are created using digital elevation maps and climate and hydrological models to predict the spatial extent of the channel given the probability of precipitation events. The US Federal Emergency Management Agency (FEMA) is responsible for creating and updating flood maps. Many flood maps are produced using 1 in 100-year probability events; these are referred

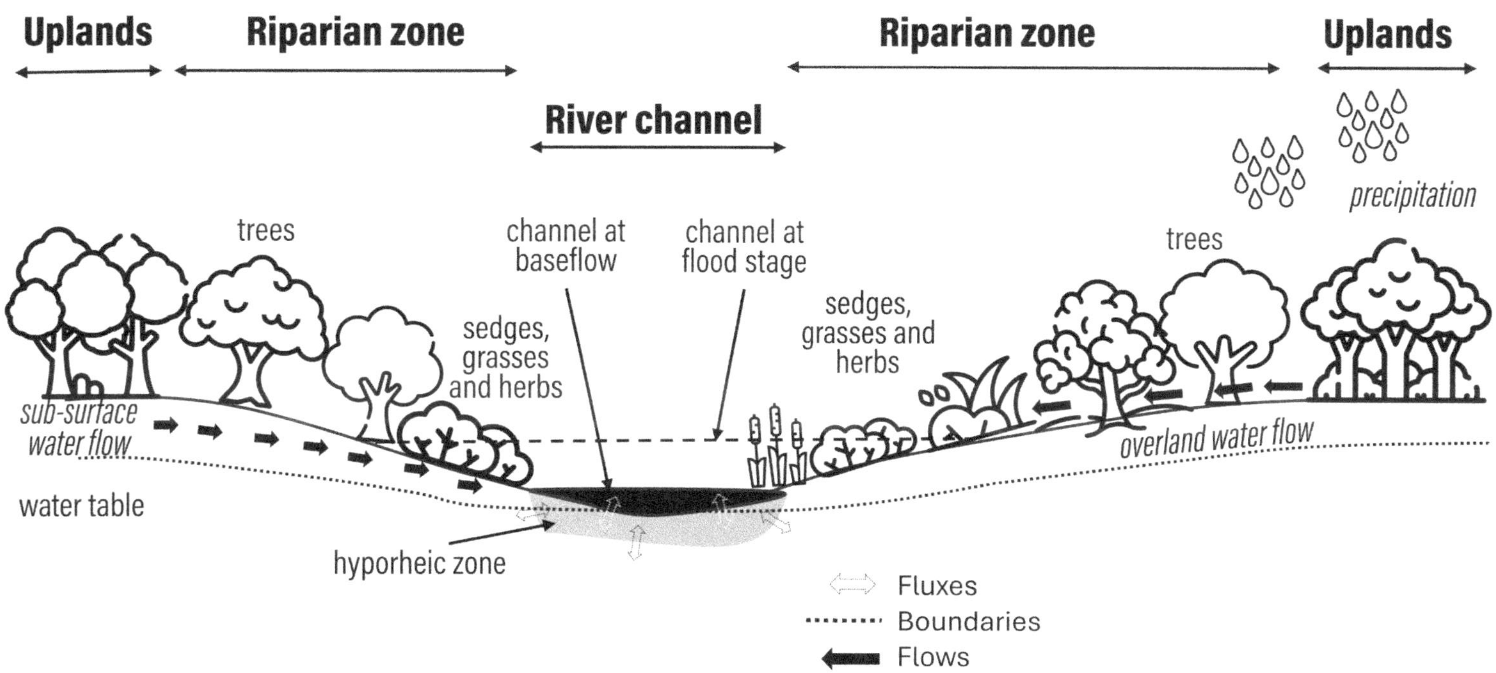

FIGURE 5.10. Cross-section of an area that forms a riparian zone, a key land-water interface between a river channel and the upland area. Note the several processes occurring within the riparian zone and river channel. See text for further details.

to as the 100-year floodplain, special flood hazard area or SFHA. Finally, an important area of these boundary systems is the **hyporheic zone.** This zone lies just beneath the stream and is a location for unique biological organisms and biogeochemical processes such as cycling nutrients, moderating stream temperatures, providing a physical filter of materials flowing through the river, and also acting as a barrier to groundwater systems.

Four forms of one-directional water flows are important and are indicated here. The first is **overland flow**, which is the hydrologic process of water moving over land and draining to a surface water body such as the stream shown here. The second water flow is **subsurface water flow**, which is principally underground. The third form is **infiltration**, which is movement of water from the surface into the top of the soil. Finally, **recharge** is the movement of water downward, eventually entering the groundwater system. The groundwater system is often composed of a variety of unconfined (surficial) and confined (located below an impermeable layer of rock) areas.

There are also a variety of areas of important hydrologic **fluxes**, locations of these systems that have two-way exchanges of water and materials. Two shown here are water moving from the subsurface system directly to the stream as subsurface/stream channel fluxes. This water movement is known to take days to months and contributes to the baseflow of the stream. The second area of flux is between the hyporheic zone and the surrounding underground materials of soil and rocks.

5.8 WATERSHED MANAGEMENT APPROACHES

In the United States, the Clean Water Act of 1972 contained provisions for allocating funds to local and state groups in developing and executing watershed management plans. This provision was described in Section 319 of the Clean Water Act. The overarching goal of Section 319 was to provide federal funds to groups to identify **nonpoint source pollution** (e.g., nutrient runoff from farm fields and urban areas). Non–point source pollution is pollution that does not originate from a point source but rather from a diffuse set of sources distributed over large areas of land. Measurement programs, local stakeholder meetings (e.g., city planners, farmers, recreation groups), and implementation projects (e.g., dam removal initiatives, stream bank erosion projects) are funded through the coordination of local and state government groups. In some instances, water quality models and land use forecast models are built and then used to identify high-impact locations across a watershed (see Tang et al., 2005). Many watershed management plans focus also on developing the spatial data resources necessary for communities to understand issues related to nonsource pollution (see Annis Water Resources Institute 2002 and 2014 watershed management plans).

5.9 SUMMARY

The interaction between land and water is complex. In terms of spatial distributions, water is distributed on the surface in lakes, rivers, and wetlands. Ways to manage natural resources are often defined by boundaries that are called watersheds. Watersheds are configured around the world as hierarchical spatial units, and a coding system called hydrologic unit codes (HUCs) are common ways that guide the organization of management units in the United States. International approaches to classifying wetlands and watersheds differ, although they share some features with those used in the United States.

5.10 DISCUSSION QUESTIONS

1. What are the major compartments and processes of the water cycle? Describe, in relative terms, the size of the water cycle compartments (i.e., which one is largest and which one smallest)?
2. What kinds of animals would you find in each of the major areas of the riparian zone?
3. If you were given the following 12-digit HUC value, 020700070200, which is for the subwatershed called the Shenandoah Basin, how would you decipher this code as it applies to federal and local management of these water and land resources?

6. LANDSCAPES AND SPECIES DISTRIBUTIONS

Highlights

Aims. This chapter summarizes how biodiversity is measured across space. It presents and summarizes alpha, beta, and gamma diversity measures and the use of Hill numbers, and describes the special phenomenon of hyperdominance. The classic species-area relationships are also presented. All of these concepts are illustrated using several small example datasets summarized as tables for species and their abundances at several sites.

Learning Objectives. Students should be able to:

- Compare and contrast alpha, beta, and gamma diversity
- Know how to calculate each of these diversity measures
- Explain how to calculate Hill number orders for diversity and interpret these values
- Describe how species richness changes with increasing area
- Use a log-log plot to calculate parameters for a species-area curve

6.1 TRENDS IN SPECIES GLOBALLY

It is estimated that life on Earth is currently the most diverse that it has ever been in its long, 4.5-billion-year-old history (Ceballos et al., 2010). So, it is not surprising to learn that Earth is currently very speciose (i.e., having a lot of species). There is a large range in the estimates for the number of species on Earth. The range is large due to several factors. First, we only have a small fraction of existing species catalogued. Second, the estimates increase in number and in uncertainty due to whether one includes noneukaryote species (these are bacteria, archaea, and viruses); if we do, the estimates increase by over 100,000-fold. Third, how we define a species is still, not surprisingly, under debate.

Two of the most recent efforts to estimate the number of **eukaryote** (organisms that have membrane-bound organelles) organisms place the number at 8.75 million (Mora et al., 2011, see also Wiens, 2023). If one includes **prokaryotic** organisms, the estimate for the number of species on Earth ranges from 8 million to as high as 1 trillion. This upper limit estimate (Locey & Lennon, 2016) was based on applying a simple model of 35,000 samples taken of species catalogues that contained both eukaryote and prokaryote species that were known and unknown and then scaling this to the entire world. The important part of most estimates (Wiens, 2023) is that

TABLE 6.1. ***The number of species, organized by a level similar to that of Kingdom (from Mora et al., 2011)***

ORGANISMAL GROUPS AND APPROXIMATE NUMBER OF SPECIES
≈ 7.77 million animal species (a majority are insect species)
≈ 298,000 plant species
≈ 611,000 fungi
≈ 36,400 protozoa (e.g., amoeba, ciliates, sporozoans)
≈ 27,500 chromists (e.g., single-celled organisms that are plant-like)
Total (eukaryotes): ≈ 8.74 million species

TABLE 6.2. ***Major IUCN Red List Status categories with a short description for how species are placed into these categories. Adopted from the IUCN (2025).***

THREATENED CATEGORIES
Critically Endangered (CR): An extremely high risk of extinction in the wild.
Endangered (EN): A very high risk of extinction in the wild.
Vulnerable (VU): A high risk of extinction in the wild.
NON-THREATENED CATEGORIES
Near Threatened (NT): Close to qualifying for or is likely to qualify for a threatened category in the near future.
Least Concern (LC): Widespread and abundant, with no immediate threats to their survival.
OTHER CATEGORIES
Extinct (EX): There is no reasonable doubt that every individual has died out.
Extinct in the Wild (EW): Exists only in captivity or as a naturalized population outside its historic range.
Data Deficient (DD): Not enough information is available to assess its risk of extinction.
Not Evaluated (NE): The species has not been assessed against the Red List criteria.

nearly all approaches must use ratios of (1) known to unknown species in a sample; (2) one group (e.g., plants) to another (e.g., fungi); or (3) body sizes of known species, especially those that are large compared to those that are small. These ratios have large ranges as well, making an estimate of total number of species challenging.

Mora et al.'s (2011) estimate of the number of species on Earth (Table 6.1) gives the number of animal species as about 8.7 million. What is also interesting is that, even though life on Earth most likely originated in the ocean, 2.2 million species are estimated to be marine, which is about 25% of all of the species estimated to be living now on Earth. Recent estimates (Anthony et al., 2023) also suggest that 59% of terrestrial species live in the soil. The number of tree species is currently estimated to be around 31,100 species (Gatti et al., 2022).

A group called the **International Union for the Conservation of Nature** (IUCN), established in 1948 as an organization of scientists, policy experts, government agencies, not-for-profit and nongovernmental organizations, has produced an IUCN Red List of Threatened Species that catalogues the status of known species. The status levels of known species include Threatened (Critically Endangered, Endangered, Vulnerable) and Non-Threatened (Near Threatened, Least Concern); these are further broken down into subcategories as summarized in Table 6.2. These categories are based on specific criteria related to population decline, shrinking geographic ranges, population size that is estimated to be critical, and species whose distributions have become fragmentated.

Currently, the IUCN has a list of 169,000 known species of organisms on Earth. Of these, 47,000 species are listed in the **threatened** categories, which represents 27.8% of all known species. This threatened category of species includes 41% of all amphibians, 38% of all trees, 26% of all mammals, 26% of all freshwater fishes, and 12% of all birds. Of these groups, birds have been the most well studied and tracked, both by professionals and bird watching enthusiasts. A recent estimate (Rosenberg et al., 2019) of bird population abundances estimated that the size of breeding bird populations has decreased by ~29% in the last 50 years, which amounts to a reduction of nearly 2.9 billion individual birds. Grassland birds have been affected the greatest, with a decline of over 50% for all grassland bird populations occurring during this period. Habitat loss, pesticide use (impacting birds directly and indirectly through reduction of insects, a major food supply), and climate change have been proposed as the major factors contributing to the reduction in population levels.

Some prominent scientists have also argued that we are now living in **Earth's sixth extinction** event (Ripple et al., 2017). The prior five extinction events, occurring over the last 600 million years, were brought on by a variety of significant environmental changes. According to these scientists, our current epoch, which some call the Anthropocene to reflect the dominance of humans on Earth, has extinction events at least 100 times the long-term background extinction rate. Many scientists (e.g., Cardinale, 2012) have also argued that the large loss of species has dire consequences as many are pollinators; regulators of pests, clean air, and water; and help regulate climate change impacts, which ultimately affect the well-being of society. Some scientists have also argued that a potential cure for some critical human diseases, such cancer, is likely found in plants in the Amazon (Ferreira et al., 2023).

6.2 MAPPING SPECIES DISTRIBUTIONS IN SPACE AND TIME

A. DIMENSIONS OF SPECIES DIVERSITY

Ecologists (e.g., Whittaker, 1972) have used several approaches to measure the diversity of organisms. These include measuring the diversity of one location, differences between the diversity of two or more locations, and regional diversity (e.g., measures of diversity of multiple locations). This section reviews the most common ways to measure these dimensions of diversity.

B. SINGLE SITE MEASURES: α-DIVERSITY

Quantifying the diversity of species at one site is referred to as α- (alpha) diversity. Several indices can be used to quantify diversity. One of the most common is to simply count the number of species; this index is called **species richness** and is designated as **S**. Species richness can be considered for a single taxonomic group (e.g., only for birds or trees) or a broad set of organisms across multiple taxonomic groups (e.g., vertebrates, thus counting mammals, birds, reptiles, and amphibians).

Another common approach is to use **Shannon's index**, which is derived from information theory. Shannon's index uses both species richness and relative abundance (also known as evenness) to quantify diversity as an index. Shannon's index, denoted as **H′**, is calculated as:

$$H' = - \sum p_i \ast \ln(p_i)$$

TABLE 6.3. *Sample data for abundances of three species at two sites and for Shannon's index calculations*

SPECIES	NUMBER OF INDIVIDUALS	PROPORTION OF TOTAL (P_i)	LN(P_i)	P_i*LN(P_i)
SITE 1				
Rhinoceros hornbill	80	80/230 = 0.348	-1.055	-0.367
Whitehead's Trogon	30	30/230 = 0.130	-2.040	-0.265
Blue-headed pitta	120	120/230 = 0.522	-0.65	-0.339
Sum	230	1.0		-0.972
Final Value				0.972
SITE 2				
SPECIES	**NUMBER OF INDIVIDUALS**	**PROPORTION OF TOTAL (PI)**	**LN(PI)**	**PI*LN(PI)**
Rhinoceros hornbill	228	228/230 = 0.99	-0.009	-0.009
Whitehead's Trogon	1	1/230 = 0.005	-2.040	-0.03
Blue-headed pitta	1	1/230 = 0.005	-0.65	-0.03
Sum	230	1.0		-0.069
Final Value				0.069

where p_i is the proportion of species i in terms of **its relative abundance**, ln is the natural log, and S is the number of species (i.e., species richness). Table 6.3 shows how relative abundances affect H' but also how one goes through the steps to calculate the values for two hypothetical sites.

Comparing these sites using H' values directly is difficult since there is not a linear relationship across values of H'. Instead, ecologists use **Hill numbers**, or effective number of species given relative abundances, to compare sites in terms of species diversity across different levels, more technically **orders**, of diversity. Hill numbers are denoted as ${}^{q}D$ where q is the order, which is assigned values of 0, 1 or 2. For Shannon's index, q is 1, and the calculation for the Hill number is:

$$ {}^{1}D = e^{H'} $$

Thus, for each site in Table 6.3, we get:

$$ {}^{1}D_1 \text{ for } H' = 0.972 \text{ is } e^{0.972} = 2.64 $$
$$ {}^{1}D_2 \text{ for } H' = 0.069 \text{ is } e^{0.069} = 1.07 $$

which means that the first site has an effective number of species of ~2.6 with the second close to being composed of only ~1 species; in other words, the first site is about 2.5 times more diverse than the second site.

An alternative to Shannon's index is **Simpson's index**, which gives greater weight to the most abundant species. Simpson's index, often denoted as D or λ (lambda), has the following calculation:

$$ D = \lambda = i\ \Sigma p_i^2 $$

TABLE 6.4. *Shannon's diversity calculation for the same two sites listed in Table 6.3*

SPECIES	NUMBER OF INDIVIDUALS (N_i)	P_i	P_i^2
SITE 1			
Rhinoceros hornbill	80	80/230 = 0.34783	0.34783^2 = 0.12100
Whitehead's Trogon	30	30/230 = 0.13043	0.13043^2 = 0.01702
Blue-headed pitta	120	120/230 = 0.52174	0.52174^2 = 0.27222
Sums	230		0.41024
Simpson's Index (λ or D)			0.41024
Simpson's Diversity (1 - D)			0.58976
SITE 2			
Rhinoceros hornbill	228	228/230 = 0.99130	0.99130^2 = 0.98268
Whitehead's Trogon	1	1/230 = 0.00435	0.00435^2 = 0.000019
Blue-headed pitta	1	1/230 = 0.00435	0.00435^2 = 0.000019
Sums	230		0.98272
Simpson's Index (λ or D)			0.98272
Simpson's Diversity (1 - D)			0.01728

TABLE 6.5. *Hill numbers for sites 1 and 2 above for species richness, Shannon's and Simpson's diversity indices*

HILL NUMBER AND ITS DESIGNATION	SITE 1 VALUE (EFFECTIVE NUMBER OF SPECIES)	SITE 2 VALUE (EFFECTIVE NUMBER OF SPECIES)
${}^0D$	3	3
${}^1D$	2.64	1.07
${}^2D$	2.44	1.02

Applying the field abundances data from Table 6.3 to the steps in calculating Simpson's diversity index would yield the values shown in Table 6.4. Also note, which tends to be confusing, that Simpson's can be reported as a diversity value (as opposed to as an index), and when it is, one uses 1-D.

To make this more confusing (and "it is what it is"), the Hill number for Simpson's is $q = 2$ and is denoted as ${}^2D$ (yes, there is another D), which is given by the inverse of λ, such as:

$${}^2D = 1/\lambda$$

which gives us ${}^2D = 1/0.408 = 2.44$ for site 1, and ${}^2D = 1/0.9826 = 1.0176$ for site 2.

Note that the Hill number for species richness is designated as an order of 0, and its value is just species richness, S, giving us Hill numbers for species richness as $q = 0$ and for both sites 1 and 2 as ${}^0D = 3$. Table 6.5

summarizes the Hill numbers for sites 1 and 2. As you can see, Hill numbers help one to interpret Shannon's and Simpson's indices, especially in relation to species richness values.

C. ß-DIVERSITY

Another way that ecologists, biogeographers, and conservation biologists measure diversity is to characterize how two or more sites differ in their diversity. In other words, we often desire to know how species composition is similar across space. Some conservation biologists might be more interested in how species *differ* in space, and so are interested in the dissimilarity (i.e., the opposite of similarity) in species between sites. The measure of how species are replaced across space is termed **species turnover** (Whittaker, 1961). ß (**beta**) diversity is the measure of species turnover or changes of species composition across space.

In its simplest form, ß-diversity is calculated as the total species richness for all sites—otherwise known as **gamma** or γ-diversity—divided by the average α-diversity for all sites. Thus, ß-diversity, as a measure of similarity, becomes:

$$\text{ß-diversity} = \gamma\text{-diversity}/[\text{average }(\alpha\text{-diversity})]$$

so that higher ß-diversity means greater dissimilarity between sites with a value of 1.0 means that all sites are the same.

We can use several forms of α (alpha) and γ (gamma) diversity to make the calculation for ß (beta) diversity. The first one, the simplest, is to use species richness values. For this example calculation, let us consider two additional sites so that we have four total sites to consider for beta and gamma diversity measures with the first two sites remaining the same (Table 6.6).

Using species richness values, ß-diversity becomes:

$$\text{ß-diversity} = 5/[\text{average}(3 + 3 + 4 + 5)]$$
$$\text{ß-diversity} = 5/3.75 = 1.33$$

which means that all sites collectively are 33% more diverse than any individual site.

TABLE 6.6. *A five-site species diversity assessment example that adds two more sites to Table 6.4 to help illustrate how beta diversity is calculated and how sites get compared using two forms of the Jaccard dissimilarity index as described in the text*

SITE	LOCATION	SPECIES A	SPECIES B	SPECIES C	SPECIES D	SPECIES E	SITE TOTAL
1	Protected Area	80	30	120	0	0	230
2	Protected Area	228	1	1	0	0	230
3	Outside	40	40	40	40	0	160
4	Outside	1	30	30	40	4	105
All Sites Total							725

TABLE 6.7. *Hill number values for the four sites and the pooled site total*

SITE	LOCATION	^{0}D	^{1}D	^{2}D
1	Protected Area	3	2.6444	2.4378
2	Protected Area	3	1.0575	1.0176
3	Outside	4	4.0000	4.0000
4	Outside	5	3.4985	3.2265
All Sites Total		5	3.4897	3.0053

ß-diversity for similarity can also be calculated using Hill numbers. Using the equations above for Hill number calculations (Table 6.7), ^{0}D, ^{1}D, and ^{2}D, we can then summarize ß-diversity calculation for these four sites in this example. Doing so for the sites in Table 6.7 would yield the following for $q = 1$:

$$^1\text{ß-diversity} = {}^1\gamma\text{-diversity/average } ({}^1\alpha\text{-diversity}) \text{ alpha}$$
$$^1\text{ß-diversity} = 3.49/\text{average } (2.64 + 1.06 + 4.00 + 3.49)$$
$$^1\text{ß-diversity} = 3.49/2.8 = 1.25$$

which is to say that given abundances, the pooled sites have 1.25 times greater than average diversity than all individual sites.

It is also useful to know whether there is a difference between sites in terms of species richness or species diversity using Hill number values. Consider now that sites 1 and 2 are within a protected area that is being managed for biodiversity, and sites 3 and 4 are just outside the protected area. You might want to know if the protected area harbors greater diversity (in terms of species number and abundances) than the areas outside the protected area. A common approach is to use the **Jaccard index of similarity**. There are several forms of the Jaccard similarity index (note that there are also versions of the Jaccard index of dissimilarity, which "flips" many of the calculations to give measures of dissimilarity). Two of the Jaccard index forms of similarity are presented here to illustrate their usefulness. Let us consider the first, which compares sites (usually two at a time) in terms of species presence and absence. This form uses set theory notation:

$$J(A, B) = \frac{|A \cap B|}{|A \cup B|}$$

Where $J(A,B)$ is the Jaccard index of similarity for two sites A and B, $|A \cap B|$ represents the number of elements shared by both sets (i.e., its intersection) and $|A \cup B|$ represents the total number of unique elements in either set (i.e., its union). If we consider sites within the protected area versus outside the protected area, we see that both sites share 3 species in common (i.e., its intersection) and that there are 5 species in total (i.e., its union). Let us write these out for protected (P) and unprotected (U) areas:

$$J(P, U) = \frac{|P \cap U|}{|P \cup U|}$$

$$J(P, U) = \frac{|3|}{|5|} = 0.6$$

TABLE 6.8. *Summary of the species abundances and the steps in arithmetic necessary to calculate the Ruzicka Jaccard index of similarity*

SPECIES	PROTECTED_POOL (P_I) ABUNDANCES	UNPROTECTED_POOL (U_I) ABUNDANCES	MIN(P_I, U_I) ABUNDANCES	MAX(P_I, U_I) ABUNDANCES
A	10	6	6	10
B	4	7	4	7
C	0	3	0	3
D	8	8	8	8
E	5	2	2	5
SUM	27	26	20	33

which comes to a Jaccard index of similarity of 0.6. This states that 3 out of 5 species are shared between the two sites or that 60% of the species are both within the protected area and outside the protected area.

Another form of this index is the **Ruzicka Jaccard index of similarity,** J_r, which considers both species presence and absence along with species abundances.

$$J_r(P, U) = \frac{\sum_{i=1}^{S} \min(P_i, U_i)}{\sum_{i=1}^{S} \max(P_i, U_i)}$$

The numerator sums the minimum number of individuals for any species found at both sites (*P* or *U*), which is then divided by the maximum number of individuals for any species found at both sites (*P* or *U*). Table 6.8 illustrates how this calculation is made, giving the pooled sites for inside and outside the protected area. This takes a little effort to understand as the way to consider minimums and maximums is not all that straightforward.

This gives a Ruzicka Jaccard index of similarity, J_r, of 20/33 or 0.61. This states that there is slightly less than a two-thirds similarity in the *abundances* of the shared species for these two grouped or pooled sites (comparing within and outside the protected area).

D. HYPERDOMINANCE

Hyperdominance is the occurrence of high abundances in a few species. This has been reported as a common pattern in tree species in tropical rainforests, such as the Amazon. ter Steege et al. (2013) found, for example, that 227 species of trees (1.4% of all species present) accounted for over 50% of all individuals in the Amazon basin. In contrast, ter Steege et al. (2013) also found that about 11,000 of the rarest tree species accounted for only 0.12% of individuals.

There are several ways to quantify hyperdominance. The most common, proposed by ter Steege et al. (2013), is to select those species that make up 50% of the total number of individuals and then report the number of individuals in this list compared to the total number of individuals present for all species. The equation for this is relatively simple:

$$Hyperdominance = N\_top\_50\% / N\_total$$

where *N_top_50%* is the number of individuals for the top species, usually using a 50% of all individuals present as a threshold, and *N_total* is the total number of individuals for all species that are present. A larger hyperdominance score reflects that most individuals are from only a few species.

Using data from Table 6.6, we can calculate hyperdominance scores for each of the two sites as:

$$\text{Hyperdominance}_{\text{site 1}} = 120/230 = 0.522 \text{ (number of species needed = 1 out of 3)}$$
$$\text{Hyperdominance}_{\text{site 2}} = 228/230 = 0.991 \text{ (number of species needed = 1 out of 3)}$$
$$\text{Hyperdominance}_{\text{site 3}} = 120/160 = 0.750 \text{ (number of species needed = 3 out of 4)}$$
$$\text{Hyperdominance}_{\text{site 4}} = 70/105 = 0.667 \text{ (number of species needed = 2 out of 5)}$$

We can interpret this as that site 2 is very hyperdominant by species A. In contrast, site 1 is moderately hyperdominant by species C; site 3 is even, as it requires 3 species to exceed 50%; and site 4 is slightly hyperdominant by species B and D.

E. RARITY

There have been many ways that spatial ecologists have quantified **rarity**. One of the most common ways has been the approach of Rabinowitz (1981), who defined rarity based on three dimensions: (1) geographic range size (narrow or wide); (2) habitat specificity (broad versus restricted); and (3) local population size (large versus small). Using these three dimensions, Rabinowitz (1981) created a matrix of these three dimensions that yielded the "seven forms of rarity." These are summarized in Table 6.9.

Another way to measure rarity is to count the number of species that are singletons (i.e., occur only once) or doubletons (i.e., species that occur only two times). Finally, organizations such as the IUCN (2022) often use areal-based measures to quantify rarity. For example, the **extent of occurrence** (EOO) is the area of the minimum convex polygon that encompasses all known occurrences of a species in the **area of occupancy** (AOO) for a global grid composed of 2 x 2 km cells; the number of occupied cells is the measure of distribution, and few numbers of cells reflect rarity. Exercise 14 focuses on the use of ArcGIS Pro and spatial data to calculate EOO and AOO.

TABLE 6.9. *Rabinowitz's Seven Forms of Rarity*

GEOGRAPHIC RANGE	HABITAT SPECIFICITY	LOCAL POPULATION SIZE	RARITY CHARACTERISTICS
Narrow	Restricted	Small	Extremely vulnerable ("double or triple rarity")
Narrow	Restricted	Large	Locally abundant but geographically precarious
Narrow	Broad	Small	Hard to conserve due to low detectability
Narrow	Broad	Large	High local density but poor range security
Wide	Restricted	Small	Habitat-restricted despite wide potential distribution
Wide	Restricted	Large	Common where conditions occur, but specialized
Wide	Broad	Small	Generalist but low density throughout its range

F. OCCUPANCY MODELING OF A SINGLE SPECIES

There are several ways that ecologists, such as landscape ecologists, conservation biologists, and biogeographers, map species distributions. One way is to map a single species or a group of species that are a focus for management. For the latter, these examples could include the distribution of an invasive insect defoliating species and its plant host or the distribution of endangered species and nesting sites based on plant species and surround habitat characteristics. Technical approaches to conducting this kind of mapping include **species occupancy modeling** (SOM). **SOM** can be carried out to map the distribution of single species or its population levels, or used to map communities (groups of species) across space and time.

6.3 INVASIVE SPECIES

An invasive species is a non-native species that has spread rapidly and dominates ecosystems. Blackburn et al. (2011) defined **invasive species** as "those that have overcome the barriers to survival, reproduction and dispersal, and that are able to spread over substantial distances." Invasive species are those that are accidentally or purposefully introduced by humans into a landscape with timelines that consists of several phases: transport, introduction, establishment, spread, and impact (Blackburn et al., 2011; With, 2002).

Conceptualization of invasive species spread was first made over a half century ago by Elton (1958) and modeled in space by Skellam (1951) using simple diffusion models. Figure 6.1 shows a simple diffusion model of spread considered by Skellam (1951), which is based on the diffusion of heat models in physics. Note that after a breeding population of organisms is introduced in space (indicated by A), it can then disperse each year at a constant linear rate of spread (indicated as r_c) that is indicated by the radius arrows. For the first year, it spreads along a front with a radius of r_1. After moving at the same linear rate of spread, the next radius for the population distribution becomes r_2. Finally, after three years of moving at the same linear rate of spread, the areal size of the population increases as the square of the radius, reaching the area enclosed by r_3. If we let r_c be 1.0 (whatever units you would like), we would have $r_1 = 1$, $r_2 = 2$, and $r_3 = 3$. Using the equation for the area of a circle, we have the following:

$$A_1 = \pi(1)^2 = \pi \times 1 = 3.1416$$
$$A_2 = \pi\,(2)^2 = \pi \times 4 = 12.5664$$
$$A_3 = \pi(3)^2 = \pi \times 9 = 28.2743$$

which shows that the area occupied by the organism, with a constant rate of linear expansion, increases quadratically (not exponentially) through time.

For lab Exercise 7, a hands-on set of exercises is described that allow you to use point data of trap counts of the invasive spongy moth in Michigan that is then interpolated for each year, and the risk to habitat (i.e., trees) is accessed over time based on population levels.

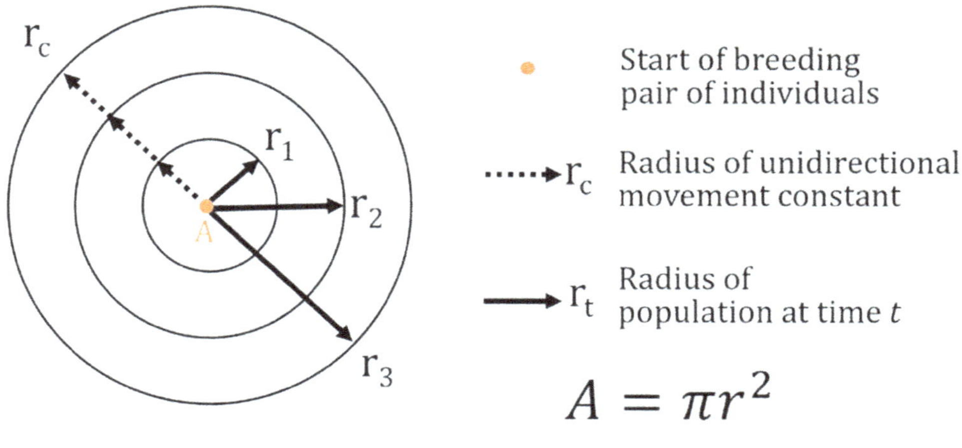

FIGURE 6.1. Simple demonstration of how a linear constant rate of spread of an organism through time affects the area occupied by the population (for three time steps). See text for further details and simple calculations.

6.4 SUMMARY

In spatial ecology, quantifying biodiversity across space can be considered in a lot of different ways. Fundamentally, the number of species at a site or group of sites are very simple ways to characterize diversity. However, spatial ecologists also know that not all species are represented at a site in equal numbers, also called abundances. Considering the number of species and abundances together is a more robust way to characterize abundance. There are also approaches to compare two or more sites in terms of species presence or absence and also including abundances. The Jaccard indices of similarity provide easy ways to describe two sites in their composition. Measures of hyperdominance or rarity are also used. Finally, species-area relationships have been long recognized as following what biologists call a "power law," which is log-log relationships that fit linear models well. Comparing the slopes of log-transformed species-area relationships is another useful approach to determine how sites differ in species composition.

6.5 DISCUSSION QUESTIONS

1. Compare and contrast the usefulness of alpha and beta diversity in making assessments of biodiversity in protected areas. What are some important considerations?
2. Compare and contrast the orders of diversity as reflected in Hill numbers. What might be some challenges of using Shannon's and Simpson's diversity measures?
3. How are hyperdominance and rarity measures helpful?

7. LANDSCAPES AND THE SENSES

Highlights

Aims. This chapter summarizes how the major human senses are considered in landscape assessments. Some of these considerations include how various technologies are configured for sensory applications of landscapes.

Learning Objectives. Students should be able to describe:

- How GIS has been used to conduct visual assessments of landscapes
- A soundscape in ecological terms
- And compare, to other senses, how olfaction is considered in landscapes
- Various dimensions of taste that are considered in landscapes

7.1 LANDSCAPE, THE SENSES AND A PURE GEOGRAPHY

Since Aristotle, scientists—mostly psychologists—have identified five key human senses: vision, hearing, smell, taste, and touch. Some modern psychologists recognize a few others, such as proprioception (body position sense), vestibular (balance and movement), and interoception (internal body states). Sensing of the environment by humans and many other organisms generally requires at least one sense, and often more than one.

How senses are related to landscapes in a scientific way was, arguably, first described by Johannes Granö, a Finnish geographer. Granö's book *Pure Geography* (1929 in German, 1930 in Finnish, and 1997 in English) is an excellent treatise that considers a detailed presentation of a landscape and how it is perceived, although his focus is mostly on humans; some of his ideas can easily be extended to consider animals as well. To Granö, what is often observed in a landscape is the **distant environment**, which is what is seen "out of into the horizon" (i.e., sight is only used as the other senses cannot perceive those distance spaces). This distant environment contains the Earth and sky in what he calls the field of vision. This is contrasted with the **proximate environment**, the space directly around an observer, where information is received by the person using their senses of sight, hearing, smell, and touch.

Figure 7.1 illustrates how Granö's framework can be applied to conceptualize *sensory landscapes*—a systematic approach to describing landscapes through multiple sensory modalities—by summarizing the visual, auditory, and olfactory characteristics of his case study site—Valosaari Island, located off the southern coast of Finland. These include the (1) proximate **colors** (and their sources), (2) proximate **auditory** phenomena (Figure 7.1A), which maps what a person would hear and its sound sources (e.g., cow bells from cows moving,

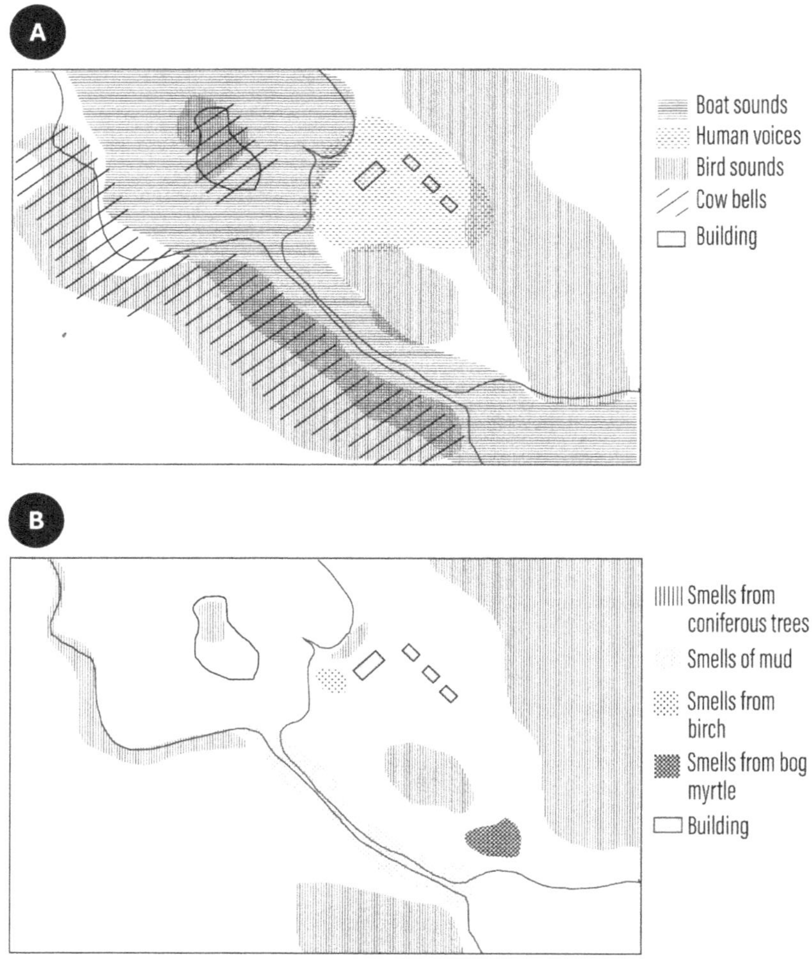

FIGURE 7.1. Close representations of the sensory maps from Granö (1929) with (A) sounds from a place and (B) smells from the same place.

the bleating of sheep, sounds from boats, noise from roads), and then the (3) proximate **olfactory** phenomena (e.g., smell of manure, odor from freshly cut hay, the smell of fish caught from the rivers) (Figure 7.1B). In short, Granö developed an extensive terminology and multisensory landscape framework that is detailed and represents one of the first attempts to characterize sight and sound in landscapes.

7.2 LANDSCAPES AND SIGHT

There are several aspects of landscapes and visual elements that should be considered in spatial ecology and that are applied in natural resource management. From a visual standpoint, what is seen is important, and so spatial analyses can allow an analyst to determine what part of the landscape can be viewed. This can be accomplished using several approaches (DeMers, 2009). In addition to what is viewed, another set of approaches considers the quality of what is viewed; the latter has been researched and formalized by the United States Forest Service for use in natural resource management.

A. VISUAL AND SCENERY MANAGEMENT SYSTEMS

In the 1970s, the United States Forest Service developed a **Visual Management System** (VMS) that focused on a set of visual characteristics related to how much harvest (i.e., clear-cut) area could be viewed by visitors from key vantage points. Perception research into how people react to changes in a landscape as well as the research in the natural dynamics of landscapes by landscape ecologists led to a shift in emphasis on how to manage landscapes for visual appeal. One of the seminal works in the area of visual perception of landscapes that drove this conclusion is that of Bell (2001, 2012), who argued that there needs to be an understanding of the couplings of landscape patterns and processes, human perception, and visualization tools using localized perception research and 3D assessments of landscapes. Bell (2001), for example, describes how landscape patterns in forests, which include harvesting patterns that mimic natural disturbance patterns, and how people respond to general visual patterns of edges, contrasts, and "flows in the optic array" lead instead to measures like the amount of area harvested. Perception research needs to be done for natural resource managers to consider how people understand and react to disturbance-based harvesting and that these two components (landscape patterns and perception) should be built into visualization tools (e.g., how one would view areas from different places, simulating this in computer visualization models).

In response to this research, the US Forest Service, following the basic framework proposed by Bell, developed procedures and protocols described broadly as the **Scenery Management System** (SMS), which is outlined in Agricultural Handbook 701 (USDA Forest Service, 1995). The SMS, as outlined by the United States Forest Service, identifies five core components as summarized in Table 7.1, that lead to desired scenic objectives for locations around the landscape.

Note that the **SIOs** (**scenic integrity objectives**) are the management targets for visual quality in different areas of the landscape. The SMS method thus builds on a landscape and ecosystem-based approach that integrates landscape character with human perception and scenic (i.e., visual) assessment scores for natural resource management.

Another important aspect of the visual appearance of landscapes is described by Nassauer (1995) in a classic paper centered around the notion of "**messy ecosystems and orderly frames**." Nassauer argues that designers and planners can use the public's perception that ecosystems are messy and seem overgrown and that managed landscapes appear neat, tidy, and orderly as part of their own landscape design approaches (applied mostly to urban landscapes). She suggest that native vegetation and areas with dead wood can be "framed" or surrounded with simple visual frames such as fences, benches, or walls. The argument here is that function and appearance have different cultural meanings when one compares a natural landscape (e.g., old growth forest with fallen trees) with that of a highly managed landscape (e.g., urban park where all trees are trimmed and leaves are taken away), and planners can use these "cues of caring" to natural areas to design landscapes that integrate natural and human-dominated areas.

TABLE 7.1. *Scenery Management System components, descriptions and purpose (based on the US Forest Service, 1995)*

SMS COMPONENT	DESCRIPTION	PURPOSE IN SCENERY MANAGEMENT
1. Landscape Character	The distinct, recognizable pattern of landform, vegetation, water, and cultural features that give an area its visual identity.	Provides the reference condition against which visible change is evaluated; anchors management in the ecological and cultural context of place.
2. Scenic Attractiveness	An assessment of the inherent visual appeal of a landscape based on variety, pattern complexity, and distinctiveness.	Identifies which landscapes are visually valuable, helping prioritize scenery management efforts.
3. Concern Levels (Viewer Sensitivity)	A classification of how much the public cares about scenery in particular locations, based on viewer type, expectations, and viewing frequency.	Ensures management reflects levels of public interest, especially along travel corridors, viewpoints, and recreation areas.
4. Distance Zones	Foreground, middle ground, and background zones surrounding key viewpoints and travel routes, each with different visibility characteristics.	Recognizes how visibility and perceptual impact change with distance, guiding design decisions accordingly.
5. Scenic Integrity and Scenic Integrity Objectives (SIOs)	Scenic integrity: the degree to which the landscape appears intact and minimally altered. Scored as very high to very low.	Provide management targets to ensure proposed activities maintain or move the landscape toward the intended level of visual intactness.

B. VIEWSHEDS

One of the most often explored topics in landscape analysis and in quantifying what can be seen from any **vantage point** is often referred to as a viewshed. A **viewshed** (DeMers, 2009) is the area that is visible to an observer, given the position of that observer and the configuration of the terrain around that person. Typical viewshed analysis is conducted using GIS tools and a digital terrain model (DTM). Figure 7.2 illustrates a few of the basic components of a viewshed. Viewshed calculations are straightforward. First a position or set of positions (the **vantage point** Figure 7.2A) is selected, including an offset (Figure 7.2B) height of the observer on top of the height of elevation data). Then, a **line of sight** (LOS, Figure 7.2C) calculation is made for every cell that is in the DTM. Upper and lower lines of sight can also be calculated, and then the proportion of the viewable area that falls on land versus the sky can be calculated. If that cell is not blocked by terrain features, then that cell is labeled as **viewable** (Figure 7.2D). If a cell is blocked by, say, a hill, then that cell is labeled as not viewable (i.e., **hidden**). Calculations are made across the entire landscape and then cells are set to visible or hidden (Figure 7.2E). **Distance classes** (Figure 7.2F) can also be assigned and then the **percent viewable** area or distance calculated (Figure 7.2G). Fisher (1994) considered these types of calculations in a viewshed analysis but extended these simple concepts to include "fuzzy" assessments, as parts of the landscape may include visual obstructions such as tall trees or buildings, and so values between 0.0 and 1.0 are integrated into the viewshed analysis to provide analysts with the ability to characterize a view given uncertainty in the characteristics of terrain features.

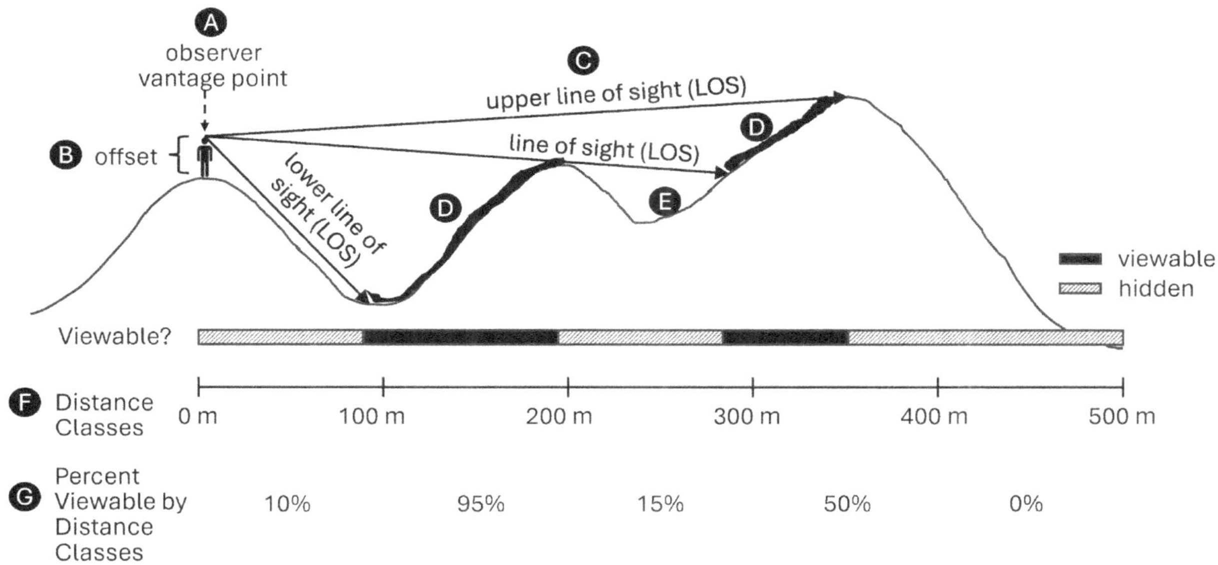

FIGURE 7.2. Components of a viewshed analysis that include observer vantage points and the locations that are viewable from that point. See text for details that describe each of the components labeled (A–G).

Quite a few GIS software packages contain viewshed analysis tools. Esri's ArcGIS Pro, for example, contains several sets, one that allows for the simple LOS calculations using viewable and hidden output of the landscape to those that are more complex, which consider the probabilistic nature of views and the potential for visual obstructions as well as the inclusion of distance zones, which would impact the quality of the view from the vantage points. Options built into these viewshed analysis tools can also quantify the amount of viewable features. For example, it might be the goal to determine how much of a lake can be viewed from a hilltop based on the configuration of the lake and the vantage point. Indeed, when Bell (2001) suggested that natural resource managers consider visualization tools in assessments of the quality of views, these kinds of GIS-based viewshed analysis tools were on his suggested list.

7.3 LANDSCAPES AND SOUND

A. SOUNDSCAPES

Sounds are a common element of landscapes (and seascapes and cityscapes). Indeed, there are always a variety of sounds that emanate from a place at any given time. Pijanowski et al. (2011a, 2011b, 2024) describe all sounds from a place at any given time as a **soundscape**. A soundscape is all the sounds—biological, geophysical, and from human-created objects like machines—that exist at a place and that are perceived by an organism. The soundscape is a sonic reflection of the patterns (e.g., network of rivers) and processes (e.g., flowing water in streams making sounds) that exist in a landscape (Pijanowski et al., 2011a). Soundscapes are what organisms sense and also perceive (i.e., process meaning) and are a source of information about how the environment is changing.

B. CHARACTERIZING THE PHYSICAL PRESENCE OF SOUND IN LANDSCAPES

Many ecologists are now using a variety of tools to characterize the soundscape (Pijanowski, 2024). These include **automated recording units** (**ARUs**), which are acoustic sensors deployed in networks and operated for long periods of time supporting what many refer to as **passive acoustic monitoring** (**PAM**). The collection of tools used to process the resulting massive amounts of data are referred to collectively as **soundscape analytics** (Pijanowski et al., 2024), which define **analytics pipelines**—a set of tools typically sequenced as a workflow to conduct an analysis. It should also be noted that acoustic recording is a form of remote sensing, and thus, by extension of the summary of light- and thermal-based remote sensing presented in Chapter 4, utilizes the same analytical approaches as these other forms of remote sensing.

Soundscape ecologists study a variety of topics that are relevant to spatial ecology. Soundscape ecologists use sound to characterize ecosystem health and/or animal **biodiversity** at a place (*in sensu* Sueur et al., 2008), determine how the soundscape of a place contributes to healthy or unhealthy individuals (humans or wildlife) (i.e., Pijanowski et al., 2025), study how the sounds of the natural landscape are used for cultural purposes (e.g., have spiritual meaning or inspire expressive sonic practices like song or have embedded traditional ecological knowledge), or understand how soundscapes contribute to human–nature connectedness (Francomano et al., 2022).

The fundamental tool for understanding a soundscape is called the **spectrogram**. The spectrogram (Figure 7.3) visualizes the three main dimensions of sound: time (x-axis), frequency (y-axis), and amplitude (z-axis). Here, there are two channels (i.e., this is a stereo recording) with the left (Figure 7.3A) and right (Figure 7.3B) channels. Time (Figure 7.3C) is always displayed in a spectrogram as the x-axis and frequency (Figure 7.3D) on the y-axis. The amplitude as a function of time is displayed (Figure 7.3E) often with the spectrogram; this

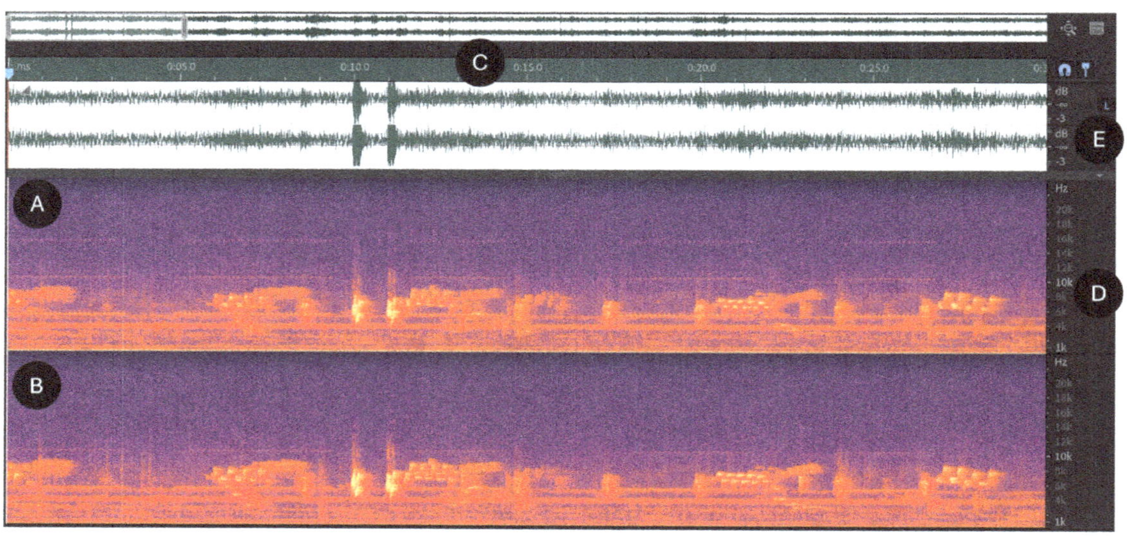

FIGURE 7.3. A spectrogram of 30 seconds of a recording from the Sundarbans mangrove forests of Bangladesh on March 2, 2020, at 7:00 a.m. local time.

plot is called the **waveform** or **oscillogram**. In this recording, two bird species and about three different insects are **stridulating** (rubbing appendages to create sound), yielding sonic patterns that are **harmonics** (repeated sounds that are at separate frequencies apart). Humans can hear sounds in the range of 60 Hz to about 20 kHz (i.e., 20,000 Hz). A **hertz** is the number of cycles per second of the sound wave as it moves through space. Most ARUs record sounds in the audible range (~60 Hz to 20 kHz) and store that information in a digital form called a WAV file. Most ARUs record sound in samples; most do this at 48,000 samples per second, which is also reported as the **sampling rate** (i.e., in this example, it is 48 kHz). As it turns out, acoustic data are the most temporally intensive data in all of ecology.

In assessments of ecosystem health and/or biodiversity assessments, two broad sets of acoustic tools are used. These are described in the following paragraphs.

C. ACOUSTIC INDICES

Dozens of **acoustic indices** have been developed that characterize the intensity, diversity, and complexity of soundscapes (Pijanowski & Rivas Fuenzalida, 2024). Simple measures of intensity, expressed as some measure of dB (decibel, the standard unit for the amount of energy in a sound clip), have been used to determine, for example, how much energy is in frequencies of sound that are characteristic of traffic sounds and other low-frequency sounds from machines versus sounds from animals, which tend to be higher frequencies. The ratio of those measures of amplitude reflect the relative energy in the soundscape of biological sound sources to those of machines. A second set of acoustic indices examines the complexity of sound by calculating how variable the soundscape is through small windows of time (i.e., 1 second) and frequency (i.e., up and down frequencies across a range of considered frequencies). Finally, another set considers how diverse the soundscape is across frequency ranges and quantifies these as entropy measures using portions of the acoustic profile.

D. SOUND-SOURCE DETECTION TOOLS

In the past few years, advances in AI have led to the configuration of tools that can identify species of animals in massive soundscape libraries. In particular, the BirdNET (Kahl et al., 2021) tool from Cornell University and collaborative organizations has revolutionized the way that soundscape recordings can now be used to create a catalogue of bird species that exist in dozens to millions of recordings. These catalogues of bird species can then be used to calculate species richness measures at a variety of sites to compare these spatially, and also across time.

Finally, it should also be noted that many soundscape ecologists do assessments of ecosystem health and animal biodiversity measures of a place using features of the surrounding landscape. Thus, the tools of GIS such as calculating buffers and then extracting the landscape features (e.g., percent forest, length of all rivers, slope) can be used to characterize **landscape-soundscape** relationships (Sangermano, 2022). Figure 7.4 illustrates the spatial analysis that is typically performed that links features of the landscape to that of the soundscape. Note how buffers (Figure 7.4A) are created that are then used to extract summary statistics (Figures 7.4B and C) from within the polygon feature. Low frequency sounds like traffic sounds from roads travel farther than higher frequency sounds like those from birds and insects. The energy of a sound wave as it moves through space decreases as the square of the distance as it travels, but low frequencies are not absorbed by the atmosphere as are mid-range and high frequencies. In addition, detection of sounds is not the

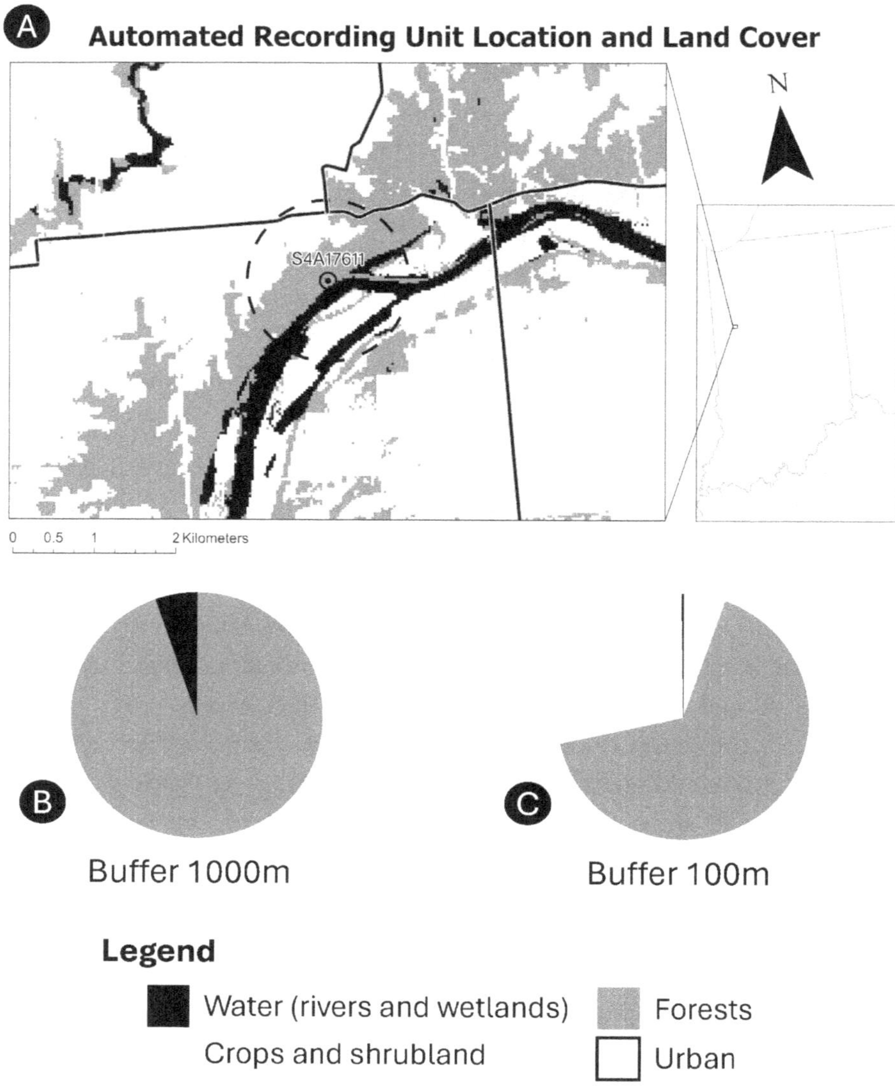

FIGURE 7.4. How a soundscape is related to a landscape. Note (A) that land covers, uses, terrain features, rivers and streams, and road networks are within the 1,000 m buffer of an acoustic sensor (the areal extent for low frequencies from road sounds) but are missing from 100 m buffers (distance that sound would travel for a bird call). A GIS is used to compute the percentage of land covers within each of the two buffers sizes, for 1,000 m (B) and for 100 m (C).

same across all frequencies; humans are most sensitive to sounds in lower frequencies (500Hz to 2kHz), and so hearing-sensitive and physical factors like atmospheric absorption are all frequency dependent. Table 7.2 summarizes some common sound sources in a soundscape recording from the Midwest in the United States. Note that, given equal sound pressure levels (i.e., sound intensities, measured in dB) for all sound sources, the distance that each travels is substantially different, thus buffer sizes of landscape features need to consider the frequencies and travel distances of these sound sources in the analysis.

TABLE 7.2. *Typical sound sources of a soundscape in the Midwest United States and their frequencies (dominant), the sound intensity at the source (using the standard reference of 1 m away), the travel distance as adjusted for human detection sensitivities, and notes on habitat and atmospheric attenuation of the sound source. Note that sound sources that are emitted as 90 dB are 100 times more (as dB scales nonlinearly) intense than a 70 dB sound wave. SNR = signal to noise ratio, SPL = sound pressure level (usually measured in dB (decibels).*

SOURCE (1 M FROM SOURCE)	DOMINANT FREQUENCY BAND*	APPROX. SPL AT 1 M (DB)	TYPICAL MAX DETECTION RANGE IN QUIET CONDITIONS	NOTES ON PROPAGATION
Highway/road traffic (mix of cars)	100–1,000 Hz main energy, with broadband up to ~5 kHz	~85–90 dB	0.5–1+ km	Strong low-frequency component propagates well; multiple vehicles create quasi-continuous "roar" that can be heard far away. Terrain shielding (cuts, barriers) matters a lot.
Coyote howl	300–800 Hz fundamentals; harmonics to ~2–3 kHz	~90 dB	1–3 km (sometimes more, with favorable wind/ temperature)	Long, tonal, low-frequency vocalization that couples well into the boundary layer; night-time temperature inversions can enhance sound and extend range.
American robin song	~2–4 kHz (strong harmonics to ~6–8 kHz)	~70–75 dB	100–200 m	Mid-high frequencies give good localization but attenuate faster than low-freq; in forests, leaves/branches scatter higher frequencies, shortening range.
Prothonotary warbler song	~4–7 kHz	~70–75 dB	50–150 m	Very high-pitched song optimized for fine-scale territory signaling in cluttered riparian forest; high atmospheric absorption and foliage scattering.
Spring peeper call	~2.5–3.5 kHz	~80–85 dB (choruses much louder)	200–500 m for a chorus; 100–200 m for one male	Calls are loud for a small animal and at a frequency that carries reasonably well; mass choruses create high SNR so you can hear wetlands from far away.
Dog-day cicada	~4–7 kHz	~85–95 dB	100–200 m	Very loud, but energy concentrated at high frequencies; atmospheric absorption and foliage reduce effective range compared to equal-level low-frequency sounds.
Common katydid (Midwest)	~8–15 kHz	~75–85 dB	20–50 m	Ultra-high frequencies; energy is quickly absorbed in air and scattered by vegetation. Often feels "up close" even if fairly loud.

E. CHARACTERIZING PERCEPTION OF SOUND IN LANDSCAPES

Perceiving the environment through sound is a multilayered process (Pijanowski, 2024). It generally involves a set of experiences that follows the order of being in an acoustic field → identifying auditory objects → making an immediate appraisal → reacting with a behavior. Many of these experiences also shape human-place relationships and thus they have a strong spatial component to influencing the perception of the soundscape.

Perhaps one of the most detailed and well-cited works in the area of human perception of the soundscape is that of Axelsson et al. (2010). These sets of studies, conducted in urban areas and thus involving the inclusion of urban sound sources (road noise, construction sounds, sounds from alarms and sirens) with those of natural sounds (e.g., birds and insects), found that, using an extensive list of qualifiers of the soundscape (>60), people responded to various soundscapes using three major "axes" of **pleasantness**, **eventfulness**, and **familiarity**—with pleasantness and eventfulness accounting for the majority of variance and forming a "circumplex model" now embedded in what is now known as the ISO 12913 soundscape framework for human perception. It should be noted that the ISO framework was developed mostly in European cities and lacks a broader assessment of how people in other cultures and places (e.g., purely natural landscapes) perceive their environment.

7.4 LANDSCAPES AND SMELL

There has been limited research and theoretical development in the area of landscapes and smell. Indeed, Granö specifically describes odor as a key sensory element of a landscape, but little work on this area occurred in the fields of spatial ecology and geography until recently. Most of the recent literature focuses on three facets of odors: (1) the chemical nature of the odor; (2) how odors are sensed; (3) specific behaviors that are related to odors specifically as opposed to other sensory fields (e.g., sound).

There are many animals that rely on scent to determine how their environmental surroundings are changing, to find a mate, to mark territories, to navigate, and to find prey (Table 7.3). Indeed, many studies have demonstrated that smell is often the most important sense that an animal relies on for any particular task, like becoming oriented to a breeding site (e.g., salmon orient to scents in water). In a classic study of the roles of senses to an animal, Wells and Lehner (1978) examined the hunting behavior of coyotes and showed that at long distance, individual coyotes used vision and sound to detect prey but relied on olfaction for final attach and inspection of the species. Hartman and Abrahams (2000) did a similar assessment of several species of fish and found that most species use smell along with vision and sound as part of sensory trade-offs for navigating through aquatic spaces. Corfield et al. (2015) examined the size of the olfactory bulb in 135 species of birds and found that there were larger olfactory bulbs in nocturnal and semi-aquatic avian species (e.g., ducks, petrels), suggesting that olfaction was an important sense to rely on if the birds were active at night.

In fact, there are several terms that have been proposed for this area of study, each emphasizing a different dimension of space and odors. Finnerty et al. (2022) describes an **olfactory landscape** as "real-time dynamic olfactory contours reflecting the patchy distribution of resources and risks" that guide animal movements and their associated behaviors. Entomologists who study insect pheromones have long examined how odor plumes create spatially structured cues for movement and behavior. These "contours" can be thought of as patches and edges of the spatial distribution of odors. Smellscapes, or sometimes odorscapes, also reflect the natural-cultural contexts of a place. Porteus (1985) describes a "**smellscape**," based on human perceptions, as the

TABLE 7.3. *List of animal groups, the environment they live in, and the main purpose of odors. From Leal (2013), Bett and Hinch (2016), Baeckens et al. (2017), Baker (2017), Cardé (2017), Bradbury and Vehrencamp (2018), and Emonet and Vergassola (2024).*

MAJOR GROUP	REPRESENTATIVE TAXA	MAIN ODOR FUNCTIONS	ENVIRONMENT
Mammals	Wolves, foxes, dogs, cats, deer, mice	Mate choice, territory, prey, social cues, navigation	Terrestrial
Bats	Fruit bats, insectivorous bats	Mother–pup ID, roost, mates, navigation	Terrestrial
Primates	Lemurs, lorises, humans	Social cues, mate choice, kin cues	Terrestrial
Bears	Black bear, brown bear, polar bear	Foraging, find mates, avoiding rivals	Terrestrial
Seabirds	Albatrosses, petrels, shearwaters	Foraging at sea, nest finding	Marine (surface)
Other birds	Vultures, kiwi, songbirds	Carrion, prey, nest sites, mate odor	Terrestrial
Snakes	Vipers, colubrids, boas	Prey trails, mates, habitat choice	Terrestrial
Lizards	Skinks, geckos, iguanas	Territory, mates, species ID	Terrestrial
Turtles	Pond turtles, tortoises	Food, mates, nesting sites	Terrestrial/Freshwater
Frogs and salamanders	Treefrogs, toads, plethodontid salamanders	Pheromones, habitat, predator cues	Terrestrial/Freshwater
Ants	Leafcutter, wood, desert ants	Trails, alarm, nestmate, caste	Terrestrial
Bees and wasps	Honeybees, bumble-bees, hornets	Floral cues, queen, alarm, nestmate	Terrestrial
Butterflies and moths	Hawk moths, silk moths, swallowtails	Sex pheromones, host plants	Terrestrial
Beetles	Bark beetles, dung beetles, carrion beetles	Aggregation, mating, host detection	Terrestrial
Flies	Mosquitoes, fruit flies, blowflies	Hosts, carrion, fermenting fruit	Terrestrial/Freshwater
Termites and roaches	Termites, cockroaches	Trails, aggregation, caste/social cues	Terrestrial
Spiders and arachnids	Web spiders, scorpions	Mates, prey, habitat	Terrestrial
Crustaceans	Crabs, lobsters, crayfish, shrimp	Food plumes, mates, shelter, rivals	Freshwater/Marine
Cephalopods	Octopus, squid, cuttlefish	Prey, conspecifics, habitat	Marine
Marine gastropods	Sea snails, sea slugs	Food trails, conspecifics	Marine
Bivalves	Mussels, oysters, clams	Larval settlement, habitat cues	Marine/Freshwater
Fish (migratory)	Salmon, eels, lampreys	Homing, migration, mates	Freshwater/Marine
Fish (others)	Catfish, reef fishes, sharks, rays	Food, predators, mates, habitat	Freshwater/Marine
Aquatic mammals	Seals, sea lions, otters, beavers	Territory, mates, pups, food	Freshwater/Marine

odiferous dimension of a place where odors originate from the natural ecosystems (e.g., trees, flowers, ocean breezes, soils) and from the built environment (e.g., bakeries, factories). Some researchers have also used the term "**smellscape**" in biomedical research. Taufer et al. (2025), for example, recently demonstrated that smells from nature can have positive benefits to a person in terms of their physiological and psychological well-being.

7.5 LANDSCAPES AND TASTE

Landscapes and taste is a broad area of sensory ecology that is often missing from any treatment within spatial ecology. This is unfortunate, as taste, and a more specialized but related area of biology called **chemesthesis** (i.e., chemical sensing that is not directly taste, but involves nose, skin, eyes, etc. and may reflect a "hotness or spiciness" if something is "tasted"), is vitally important for many species of animals. **Taste**, for humans, typically involves sensing that is described as sweet, sour, salty, bitter, and umami (sense of meatiness or savoriness). Animals use the sense of taste to (1) find rare or limiting nutrients such as Na, Ca, or trace metals), (2) avoid consuming toxic plants, (3) locate high-energy sources (e.g., sugars), and (4) associate taste with what is edible, nutritious, or dangerous in the landscape. Many animals, such as large herbivores, birds, fish, and invertebrates, have developed a variety of mechanisms to "taste" or chemically sample their environment, adjusting important behaviors to the sensing of taste. Indeed, Feeney et al. (2011) have described taste as "the factor of greatest significance in food choice" that "has been described as the body's 'nutritional gatekeeper'."

There are numerous examples of how taste and landscapes are inextricably linked together, most notably in the food and beverage industries. It is well known, for example, that soil chemistry, local plants, local microbiomes, and local climate can substantially contribute to unique "flavors" found in wine, gin, tea, coffee, and herbs. In the wine industry, this focus is called "terroir" (from the French, meaning Earth, land, or soil), which is the study and practice of managing food or beverage products given the knowledge of how local environmental conditions contribute to their flavor, aroma, and texture—this is the science of the "taste of a place." Reviews of this knowledge of local environmental factors that lend themselves to taste include those for wine (Van Leeuwen et al., 2018), coffee (Williams et al., 2022), and gin, the latter especially with regard to homegrown junipers and the local botanicals and spices that are added (Pauley & Hill, 2025).

7.6 SUMMARY

The visually based sense is dominant in humans, but the other senses, such as those involving sound, smell, and taste, are important to other animals and are often neglected in ecological studies. The visual elements of a landscape have been assessed by natural resource management agencies through GIS-based tools that are well documented and studied. The new science of soundscape ecology is adding the dimension of sound (e.g., sounds of animals, of the geophysical environment, and from machines). The science of assessing taste and odor is not well formed and is likely to have a growing presence in landscape ecology research.

7.7 DISCUSSION QUESTIONS

1. Describe how you, as an analyst with a natural resource management organization such as the National Park Service, could integrate how people perceive the visual quality of a place as well as its sonic qualities.
2. In response to question 1, what kinds of spatial analyses would be needed to conduct each of these analyses? Can you specifically describe the kinds of GIS tools that you might use is such an assessment?
3. Extend the core ideas of using sight and sound into the realm of olfaction. Consider that humans have a poor sense of smell, but that many vertebrates, especially mammals, have a keen sense of smell. How might you consider "odorscapes" in any assessment of how scents, as they are distributed across landscapes, might impact animal biodiversity?

8. BIOGEOGRAPHY AND GLOBAL ENERGY BUDGETS

Highlights

Aims. This chapter summarizes the main concepts of global energy budgets and how they affect the distribution of broad vegetation and animal communities on Earth.

Learning Objectives. Students should be able to:

- Know the four major geophysical processes that contribute toward an uneven distribution of energy on Earth
- Explain how each of these processes work to drive air temperature and precipitation patterns for any location on Earth
- Explain how major biological community distributions occur, linking these to the ways air temperature and annual precipitation occurs around the world
- Understand how biological realms, biomes, and ecoregions differ but also how they are related to one another

8.1 GLOBAL ENERGY BUDGETS

The composition of biological communities, which are called **biomes** (Chapter 9 describes these in detail), is shaped principally by global energy budgets. Biomes are distributed according to patterns in the Earth's energy budget, which is driven by how much solar energy, called **insolation**, is distributed across the planet. The key controls of this energy distribution are latitude, altitude, and the axial tilt of the Earth as it orbits the sun. These global energy budgets also control precipitation patterns, wind speeds, and wind directions at the surface. Let's consider each of the main factors that contribute to the variation of energy as it is distributed across the Earth.

A. SUNLIGHT ORIENTATION

To understand one of the major features of the energy budget and how Earth, as a sphere, impacts solar radiation insolation, consider Figure 8.1, which illustrates how sunlight, traveling from the sun to the surface of the Earth, hits the surface (points A) at a direct angle (i.e., no angle at all). As we move toward the poles, at around 30 degrees north and 30 degrees south, the sun's rays begin to strike the surface at an oblique angle (points B). Even closer to the poles, at greater than 60 degrees north and 60 degrees south, the rays are at a

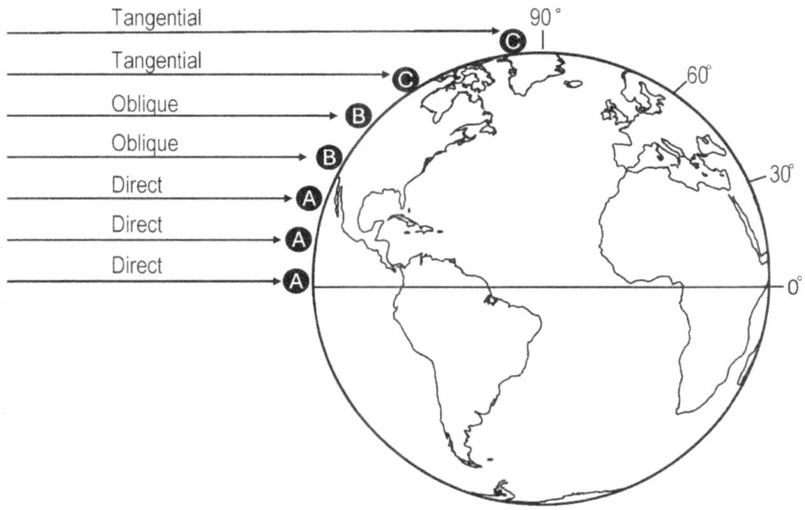

FIGURE 8.1. Angles of solar radiation as it hits the surface of a sphere. Note that as one moves from the equator to the poles, the spacing along the surface of the sphere increases despite the density (represented by the spacing of the solar rays) of radiation being the same. Three general solar radiation angles are presented, direct, oblique, and tangential, to illustrate how these angles impact solar radiation flux.

point where they "graze" the surface. What is most important is understanding that the number of sunlight rays hitting the surface is a function of the angle of solar incidence. The higher the latitudes, the fewer number of rays hit the surface. If we consider this in area (e.g., square kilometers), the density of solar rays striking the surface decreases considerably as we move from the equator to the poles. This density of sunlight rays is referred to as **solar radiative flux.** The amount of solar radiative flux at 45° angles to the incoming radiation will be about 70% of that occurring at the equator (note: the amount of flux is expressed as the angle of incidence, which is equal to the cosine of 45°, which is 0.7071; which means that about ~30% less sunlight hits the surface on a per square area).

B. GLOBAL ATMOSPHERIC CIRCULATION PATTERNS

The uneven heating across the Earth's surface creates well-known large-scale global circulation patterns based on rising and sinking air masses. Three bands of circulation in the Northern and also Southern Hemispheres form coupled circulation cells (Figure 8.2). Those that create a circular pattern from the equator to about 30° north and 30° south are called **Hadley cells,** those between 30° and 60° latitude are called **Ferrel cells,** and those between 60° and 90° latitude are called **polar cells.**

Hadley cells are created by the intense warming of the surface at the equator. This warming creates a situation where air rises. As it rises, it cools and then condenses, creating clouds and rain (which is why we refer to the forests of these regions as rainforests or cloud forests). As the cooler air rises to about 10–15 km in altitude, it also moves toward the poles. At about 30° latitude north and south, the air begins to fall back down, but this

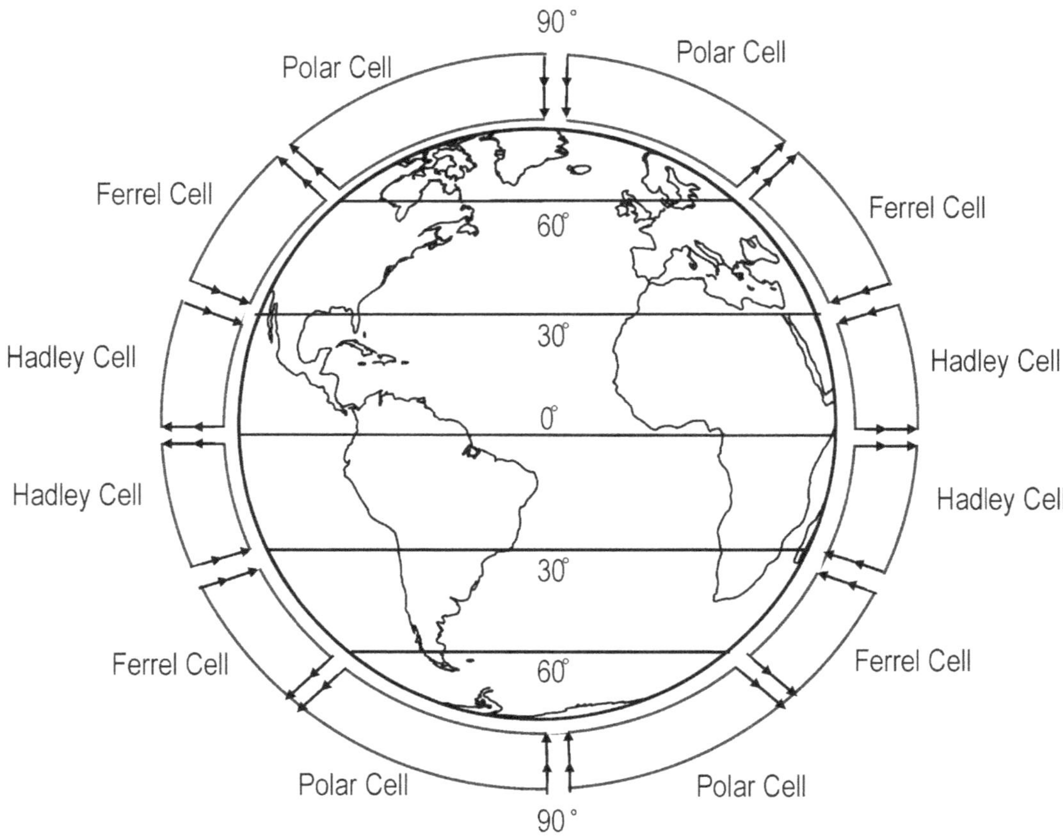

FIGURE 8.2. A very simplified view of circulation cells and the direction of air in the atmosphere. Note that closest to the equator are Hadley cells, followed by Ferrel cells and then polar cells.

air is very dry and becomes warm. This 30° band of dry, hot air creates the planet's deserts. When the air gets close to the surface, it creates strong winds that move back toward the equator. These strong winds, called the **trade winds**, are locations that historically have supported sailing ships and thus intercontinental trade. Trade winds tend to blow from east to west in this area that is just north and south of the equator.

Between 30° and 60° are the Ferrel cells. Here, air movement flows at the surface from 30° toward the poles, and, due to the spinning of the planet, the Coriolis effect pushes the air to create winds that move from west to east (called westerlies). In latitudes between 30° and 35°, winds are very weak, creating a band of wind speeds referred to as the **horse latitudes** (note: these are areas of the ocean where there are no winds to power sailing ships, and so if the ship carried heavy cargo such as horses, they would be thrown overboard to conserve water for passengers). At 60° latitude (both north and south), air begins to rise where it meets with the colder air coming from the poles.

Polar cell circulation is created when the cold, dense air at the pole sinks, creating a flow of air at the surface that moves toward the equator. Thus, all three of these bands act as overturning circulations, coupled loops, or chained circulations, which move heat energy and moisture from the equator to the poles. They also operate, along with complex forces of circular motion via the Coriolis effect, to create areas of high and low winds, and wind directions that are either easterlies or westerlies depending on latitude.

C. AXIAL TILT AND EARTH'S DAILY ROTATION

A third major feature of Earth's variable energy budget across space is caused by Earth's tilt off its axis. Earth's 23.5° **axial tilt** (also called obliquity) creates another uneven distribution of energy that gives rise to various seasons as well as differences in the amount of light per day depending on the calendar day. Figure 8.3 shows how the Earth receives more sun in the Northern Hemisphere during the calendar months of May through August. These are the same months when the Southern Hemisphere receives the least amount of sun per day. In the **mid-latitudes** (i.e., around 45° north and south), four seasons per year occur. In addition, four days of the year represent special day-night **solar turning points**. Two of them, called **equinoxes** (the vernal, occurring around March 20, and the autumnal, occurring around September 22) represent the day when all locations on Earth receive 12 hours of daylight and 12 hours of night. The other two, called **solstices**, represent the day when the Northern and Southern hemispheres experience the greatest difference in daylight and nighttime lengths in the day. The June solstice, occurring around June 20 of every year, represents the greatest amount of time the Northern Hemisphere experiences daylight per day.

Also, due to the tilt of the Earth and the trajectory of the planet around the Sun, biogeographers recognize that there are two forms of the equator. One is represented as the geographic equator, which is at 0° latitude; another is a **thermal equator**, which is the *latitude* at which the greatest mean temperature occurs on any given day. This means that the thermal equator changes seasonally, and it is often either north or south of the geographic equator of 0° latitude.

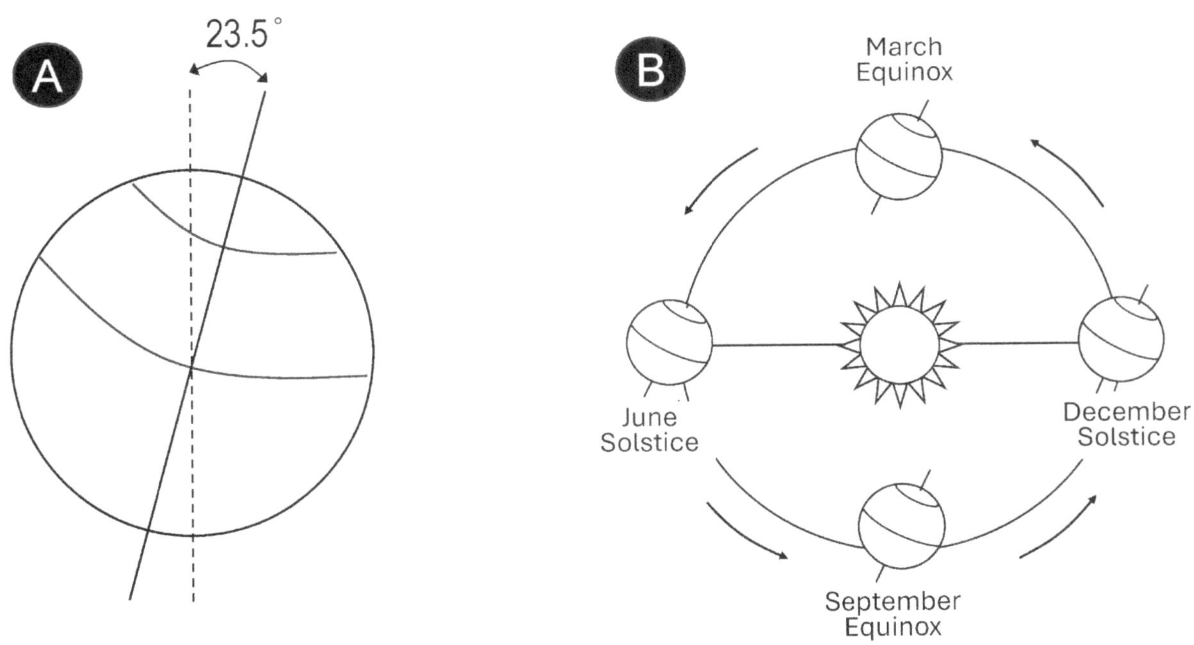

FIGURE 8.3. (A) The axial tilt of the planet and (B) the orbit of the Earth around the sun and how the axial tilt affects the seasons and the four solar turning points.

D. INTERTROPICAL CONVERGENCE ZONE

An important geophysical zone on Earth is the **Intertropical Convergence Zone**, often just referred to as the ITCZ. The ITCZ is a very broad, shifting belt created by the rising air near the thermal equator where trade winds converge from the northeast and the southeast. These winds shift the air and moisture north and south each year. As the winds bring air with moisture, they create distinct seasonal rainfall patterns. Close to the thermal equator, the seasons are typically two rainy and two drier seasons (termed a bimodal rainfall pattern), with areas more toward the poles having one rainy and one drier season (and hence called unimodal rainfall patterns). The exact position of the ITZC is greatly influenced by the complex configuration of land masses with oceans as well as the timing of the turning points. Thus, the ITZC is not a perfect circular belt, but it is a key driver of plant and animal distributions and the patterns of their life histories (e.g., breeding, flowering).

8.2 ALTITUDE AND ITS OROGRAPHIC AND ADIABATIC EFFECTS

The fourth important geophysical effect that controls temperature and precipitation is based on altitude. The key processes are important to understand. The first is a physical process called the **adiabatic effect**. The adiabatic effect occurs when the air at the surface of Earth is heated, as this is where sunlight, which contains energy, is dissipated into heat energy after it hits the surface (note: if you have ever walked on a sandy beach in the summer, you have noticed that the sand is very hot, much hotter than the air above it!). As the air is heated, it begins to rise. But as it rises, it is cooled because air higher in the atmosphere is less dense (note: air is less dense higher in the atmosphere because there are fewer molecules pressing down—this is the same effect that occurs in the oceans where the deeper you go, the greater the pressure on objects). As the air expands, this requires heat energy and thus its temperature decreases. Temperature reductions as a function of altitude is called the adiabatic lapse rate; for dry air it is 10°C per 1,000 meters and for humid air it is about 6°C per 1,000 meters. Thus, the adiabatic effect is exerted and the air is cooled by its expansion (i.e., by reducing air density) as it rises. The adiabatic effect results in having glaciers form at the tops of mountains (e.g., Mt. Kenya) near the equator (which is where the sunlight density is the greatest).

Orographic and adiabatic effects also work together to create regional climate patterns that are also important. Oro- is the Greek term for mountain and -graphic is the suffix for representations, thus, **orographic effects** translates to mountain representation effects. To understand orographic effects (and how the adiabatic effect is also combined), let us consider a mountain range as shown in Figure 8.4. As winds generated from the global circulation patterns in the mid-latitudes are westerlies (Figure 8.4A), the air that moves toward a mountain is warm and contains moisture. But as it begins to approach the mountain, the terrain pushes the air upward and it begins to cool due to the adiabatic effect. As it cools, the air is not able to hold water vapor and the water condenses into small droplets that form clouds, and then eventually further condenses into rain. Thus, the air becomes cool and dry as it moves up the mountain. Ultimately, most of the water is converted from its gaseous form into liquid (rain) but at the highest elevations, there is little rain and the air temperature is low. Vegetation types along the mountain respond to both air temperature and rainfall, such that toward the base of the mountain are located broadleaf, deciduous trees common to areas in the mid-latitudes, but near the top of the mountain are located needleleaf, coniferous tree forms typically found in higher latitudes. Eventually, the low air temperature and low annual precipitation will form a tree line where the highest altitude for tree

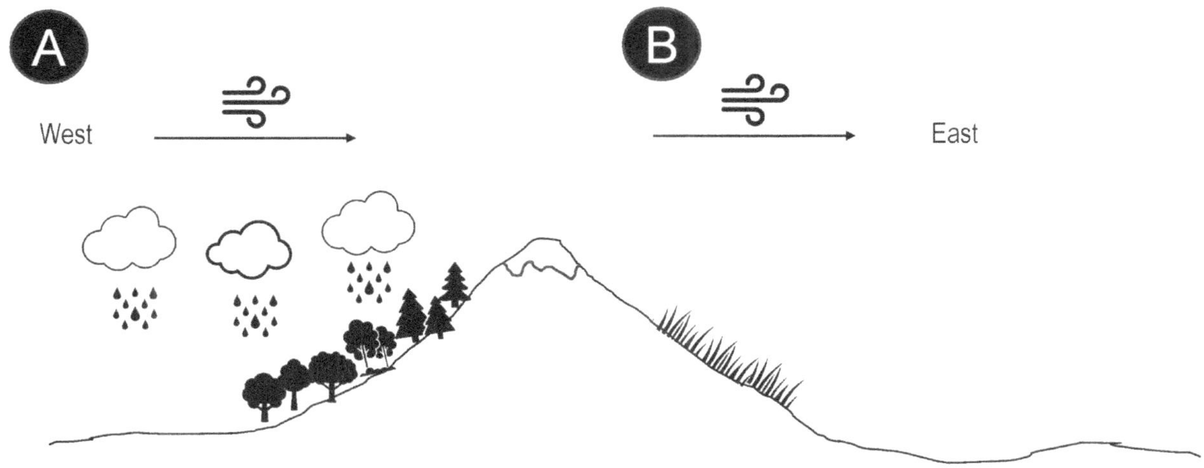

FIGURE 8.4. Orographic effects that create regional climate patterns along the westward and eastward slopes of a mountain range, also given that wind directions are from west to east (i.e., westerlies). See text for how changes in air temperature, rainfall, and vegetation vary across this spatial gradient.

growth will end, resulting an area toward the top of the mountain where only short, slow-growing herbaceous vegetation can be sustained. At the very top of the mountain there could be conditions that only support the formation of ice, which means that glaciers are formed.

The eastern slopes of the mountain (Figure 8.4B), on the other hand, experience air that contains no water vapor since that was released on the western side of the mountain. As a result, the eastern slopes are dominated by vegetation adapted to arid climates; these include shrubs and grasses, and in some cases, arid succulents (e.g., cacti). The Rocky Mountain range, a range of mountains that extends from its southern reach (starting about north-central New Mexico, United States) to its northern extent (into northern British Columbia, Canada), is an excellent example of how the orographic effect impacts air temperatures, precipitation patterns, and dominant vegetation types as one moves from the west (e.g., western Colorado) to the east (western Kansas).

8.3 SUMMARY

Global energy budgets vary around the world based on several important geophysical processes. This includes how solar radiation is dispersed at the surface based on latitude, the position of the Earth relative to the Sun during its annual orbit, the circulation patterns created by rising and falling air masses, which then alter winds at the surface, all of which redistribute heat across the planet. Some important outcomes of these patterns are the creation of the Intertropical Convergence Zone and bands of dry air that contribute to the formation of deserts at 30° N and 30° S. Finally, mountains contribute to large shifts in temperature along elevational gradients, and windward sides of mountains force the atmosphere to release water in the form of precipitation on the leeward side.

8.4 DISCUSSION QUESTIONS

1. How might the area in Chile referred to as the Atacama Desert, considered the driest place on Earth, have become so dry?
2. Note that the areas of the desert that contain mountains are often referred to as sky islands. Why do you think that they are called this?
3. Compare and contrast how rainy and dry areas differ in mountain ranges located in the Rocky Mountains versus those in the central part of South America, say in northern Chile.

9. BIOGEOGRAPHY AND REGIONAL TO GLOBAL COMMUNITY DISTRIBUTIONS

Highlights

Aims. This chapter summarizes the main concepts of global energy budgets and how they affect the distribution of broad vegetation and animal communities on Earth.

Learning Objectives. Students should be able to:

- Explain how major biological communities are distributed, linking these to the ways annual air temperature and precipitation occur around the world
- Understand how biological realms, biomes, and ecoregions differ but also how they are related to one another

9.1 GLOBAL TERRESTRIAL COMMUNITIES

A. BRIEF INTELLECTUAL ROOTS OF GLOBAL BIOLOGICAL COMMUNITY CONCEPTS

Alexander von Humboldt (Humboldt & Bonpland, 1807) is considered one of the originators of the idea that plant communities around the world are shaped by climate patterns. Humboldt is well known for his detailed treatise on nature, traveling extensively in remote areas for long periods of time across the Americas (1799–1804), including Peru, Ecuador, Colombia, Cuba, Mexico, and the United States (note: the latter was only a short visit when he met with President Thomas Jefferson). He also traveled and made extensive observations across Europe and northern Asia (e.g., Siberia). These observations led him to conclude that broad plant community forms were greatly influenced by climate and topography (e.g., mountains). Earlier, other 18th-century naturalists like Buffon, Forster, and Linneas also noted that spatial distributions of plants and animals were greatly influenced by large-scale geographic, mostly latitudinal, patterns on Earth (Lomolino et al., 2017).

Alfred Wallace (1876) is often credited with describing how animal distributions vary across broad geographic areas. He described in detail how animal distributions were influenced by large-scale climate patterns (e.g., annual temperature), land boundaries (e.g., islands) and barriers (e.g., mountain ranges, oceans), and evolution. Indeed, Wallace's core ideas eventually shaped what later ecologists would term realms and biomes.

B. BIOGEOGRAPHIC REALMS

Realms are large biogeographic regions of the world that are based mostly on continental land masses. This level of classifying living communities on Earth is at the apex (i.e., top) of all classification systems. It is an important spatial extent to consider as these realms reflect similar long-term histories as well as geographic barriers to dispersal. The World Wildlife Fund (WWF) (RESOLVE, 2017), for example, identifies eight realms: Afrotropic, Australasia, Indo-Malayan, Nearctic, Neotropic, Palearctic, Oceania, and Antarctic (Figure 9.1). Another biogeographic realm classification system, by Holt et al. (2013), contains 11 different realms: Nearctic, Palearctic, Sahara-Arabian, Afrotropical, Madagascan, Panamanian, Neotropical, Sino-Japanese, Oriental, Australian, and Oceanian. Holt et al. (2013) used a data-driven approach using measures of phylogenetic ß-diversity measures for over 20,000 vertebrate species to delineate some continental regions, such as the continent of Africa and areas in the western Pacific, into separate biogeographic realms. The WWF approach, on the other hand, was more expert driven where the biogeographic realms were delineated on the basis of major plant and animal community assemblages and relative degrees of geographic isolation.

C. BIOMES

Geographic realms are considered different from another global classification of regions that are termed **biomes**. One of the first attempts to describe a biome was by Clements and Shelford (1939), who defined a biome as regional or global biotic communities characterized by vegetation types based on life forms and structure, and its prevailing climate (i.e., annual temperature and annual precipitation). They argued that these vegetation-climate assemblages influence the animal and micro-organismal communities that adapt to these regions. Shelford (1963) later proposed that there be five major biomes: tundra, coniferous forest, deciduous forest, grassland, and deserts. Subsequently, several other biographers proposed different biome classifications. One well-known system is that of Whittaker (1975), who described nine different biomes based on vegetation form and climate; these included tropical rain forest; tropical seasonal forest/savanna; subtropical desert; temperate rain forest; temperate seasonal forest; woodland/shrubland (Mediterranean/chaparral); temperate grassland–desert; boreal forest (taiga); and tundra.

Most of the more recent biogeographic classification systems have been developed using a lot of spatial data, which the biogeographic areas distributed as GIS layers. The WWF (RESOLVE, 2017) classification system, distributed as shape files, contains a different classification of biomes; here, 14 terrestrial biomes (Figure 9.2) are nested within the 8 biogeographic realms. Note that in the case of biome distributions in these maps, biomes can be distributed across more than one biogeographic realm. For example, the tropical and subtropical moist broadleaf forests are contained in the Palearctic and Oceanian biogeographic realms.

There are several large-scale spatial patterns of WWF biomes, which are also reflected in other biome maps, that should be noted. First, deserts typically are found as horizontal bands around the world at 30° N and 30° S. The Sahara in Africa, Sonoran (e.g., Arizona), Chihuahuan (e.g., Mexico), Mojave (e.g., California), Arabian (Saudi Arabia), and the Thar (e.g., India) deserts are all located along the 30° N latitude. In the south, the Kalahari (e.g., Botswana), Namib (e.g., Namibia), Atacama (e.g., Chile), and several of the Australian deserts (e.g., Great Victoria, Great Sandy, and the Gibson Deserts) all fall along the 30° S latitude. Second, tropical rainforests occur along the equator, extending to about 23.5° N and 23.5° south; both of these latitudes represent the Tropic of Cancer (north) and the Tropic of Capricorn (to the south). Third, there are several

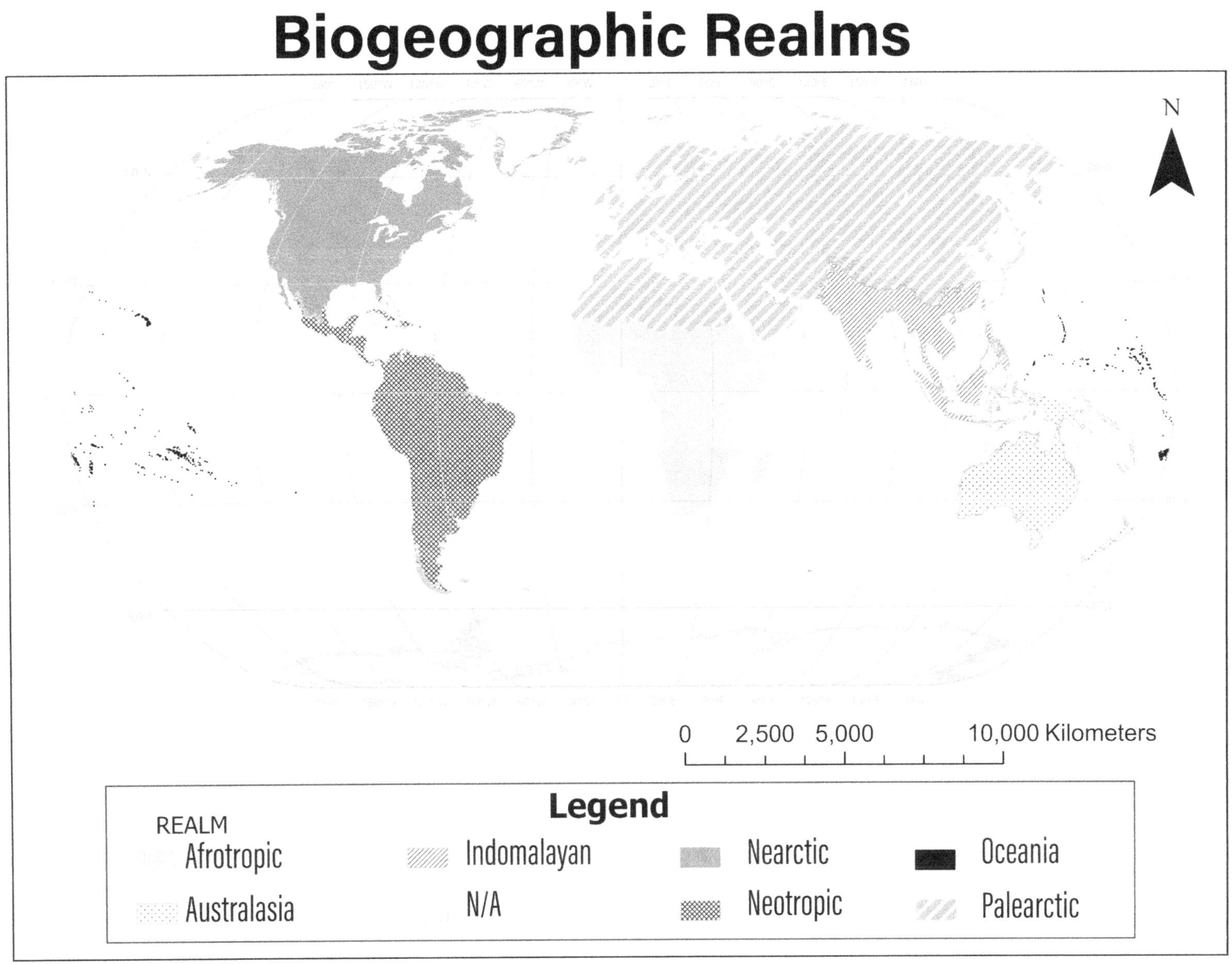

FIGURE 9.1. Biogeographic realms from the WWF (2017) that are, for the most part, organized by continental regions.

north-south-aligned belts of biomes that exist due to large-scale mountain ranges. These mountain ranges are referred to as **cordillera** or **orogenic** belts. The Rocky Mountain ranges, which extend from New Mexico to Alberta, Canada, and the Andes Mount Range, which extends from its northern extent of Venezuela down to the southern tip of Argentina, form the American Cordillera. Oher north-south mountain ranges that eventually created biome distribution patterns that aligned with these large-scale topographic features are on the Alaska-Yukon range, East African Rift Highlands in Kenya and Ethiopia, Scandinavian (Norway and Sweden), and the Southern Alps of New Zealand. The Himalayan mountain range, which runs northwest to southeast, also functions to create nonlatitudinal bands of biome distribution. Fourth, the mid-latitudes (from 30° to 60° N and S) are typically composed of plant communities that have short dormancy periods, with a strong seasonal pattern (e.g., most commonly four seasons), westerly winds, and seasonal patterns to daylight (e.g., winters have short daylight periods and summers have long daylight periods per day). Fifth, the zones that lie north of 60° N and 60° S differ in their characteristics since in the north there are large land masses in the Eastern and Western Hemispheres and very little land in the south. To the north, the biomes described as Subarctic boreal, Arctic tundra, and High-Arctic polar desert are places with extreme day/night patterns during the equinoxes with very cold temperatures during the periods when daylight length is very short. Nearly all of these high latitude regions have a condition referred to as permafrost, which is where the soil remains below freezing (0° C or 32° F) for at least two consecutive years; soil at the surface may thaw during the summer but always freezes during the winter (van Everdingen, 2005).

D. CLIMATE ZONES

Although not strictly considered biomes, there are similar biogeographic classification regions that are considered "parallel" to biomes, and these are often referred to as climate zones. The most commonly used climate zone classification system is the Köppen-Geiger climate classes. There are five major groups (described using letters A–E, and sometimes with an additional F class), and then these five or six are subdivided into 30 subtypes. Exercise 11 describes these databases more fully with some exercises designed to explore them further.

E. ECOREGIONS

Ecoregions are considered the finest and most detailed of the biogeographic zone classification system. **Ecoregions** are large areas of land, and sometimes aquatic areas, that have geographically distinct assemblages of plant and animal communities that also have similar environmental conditions such as annual precipitation and mean annual temperatures (Olson et al., 2001). These ecoregions are often classified on the basis of their natural species assemblages; in other words, it does not consider land transformations that have occurred as a result of human activities. There are several widely used ecoregion classification systems that are distributed as spatial data. These include those from the WWF (RESOLVE, 2017), Omernik (1987), Bailey (1995), and Olson (2001), all of which have been described in peer-reviewed papers. How the last three differ is described in Exercise 12.

An important ecoregion database that is well organized with biogeographic realms and biomes is the WWF Ecoregions database (RESOLVE, 2017). Here, WWF created a list of around 846 ecoregions, which are contained as spatial data (Figure 9.3A).

The WWF Ecoregions database contains a unique, potentially very useful classification of the ecoregions, which they refer to as Nature Needs Half (NNH). The NNH field (Figure 9.3B) was created using a GIS analysis

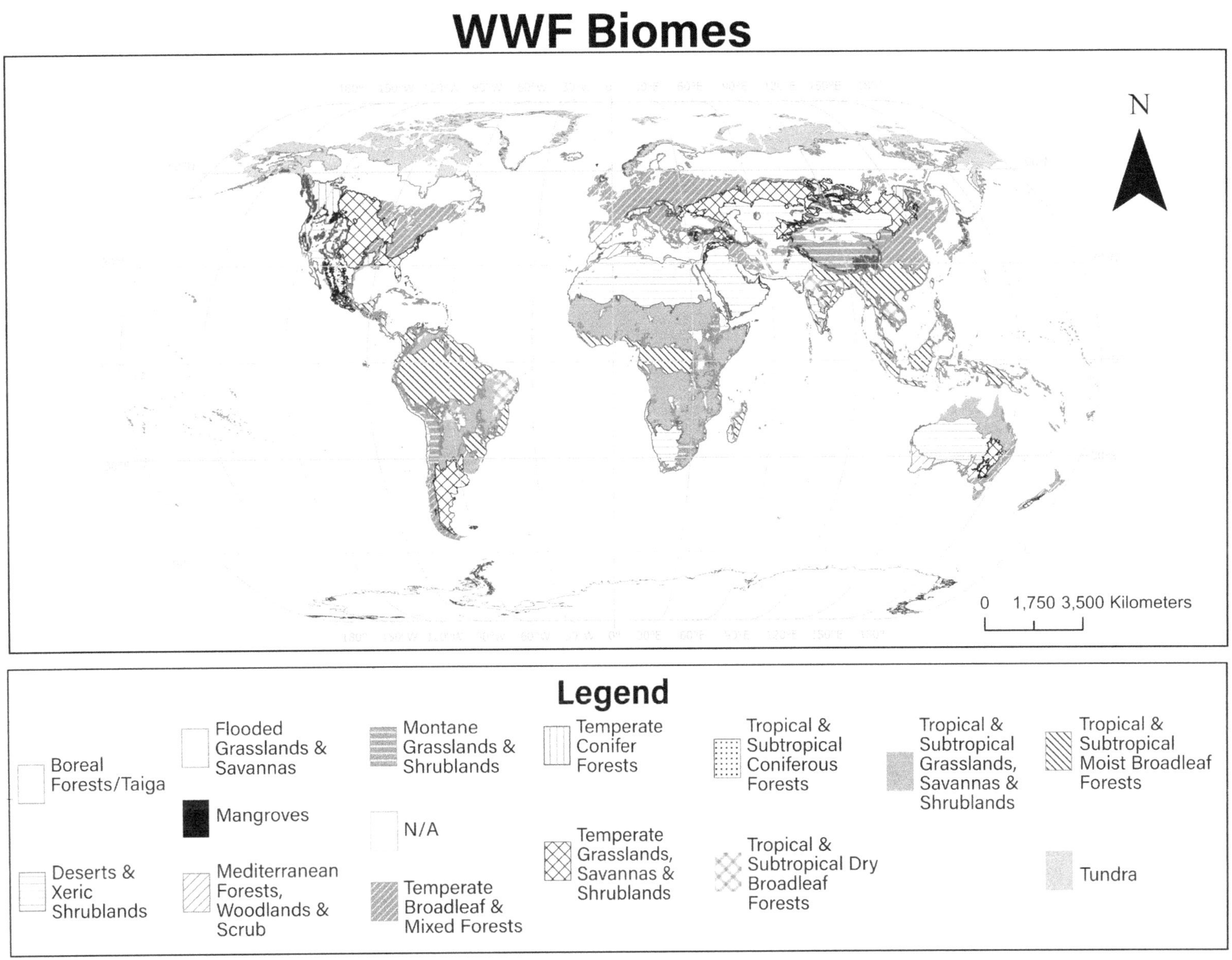

FIGURE 9.2. WWF biomes. There are 14 biomes that are distributed, and not fully nested within, the biogeographic realms. See text for more information.

WWF Ecoregions

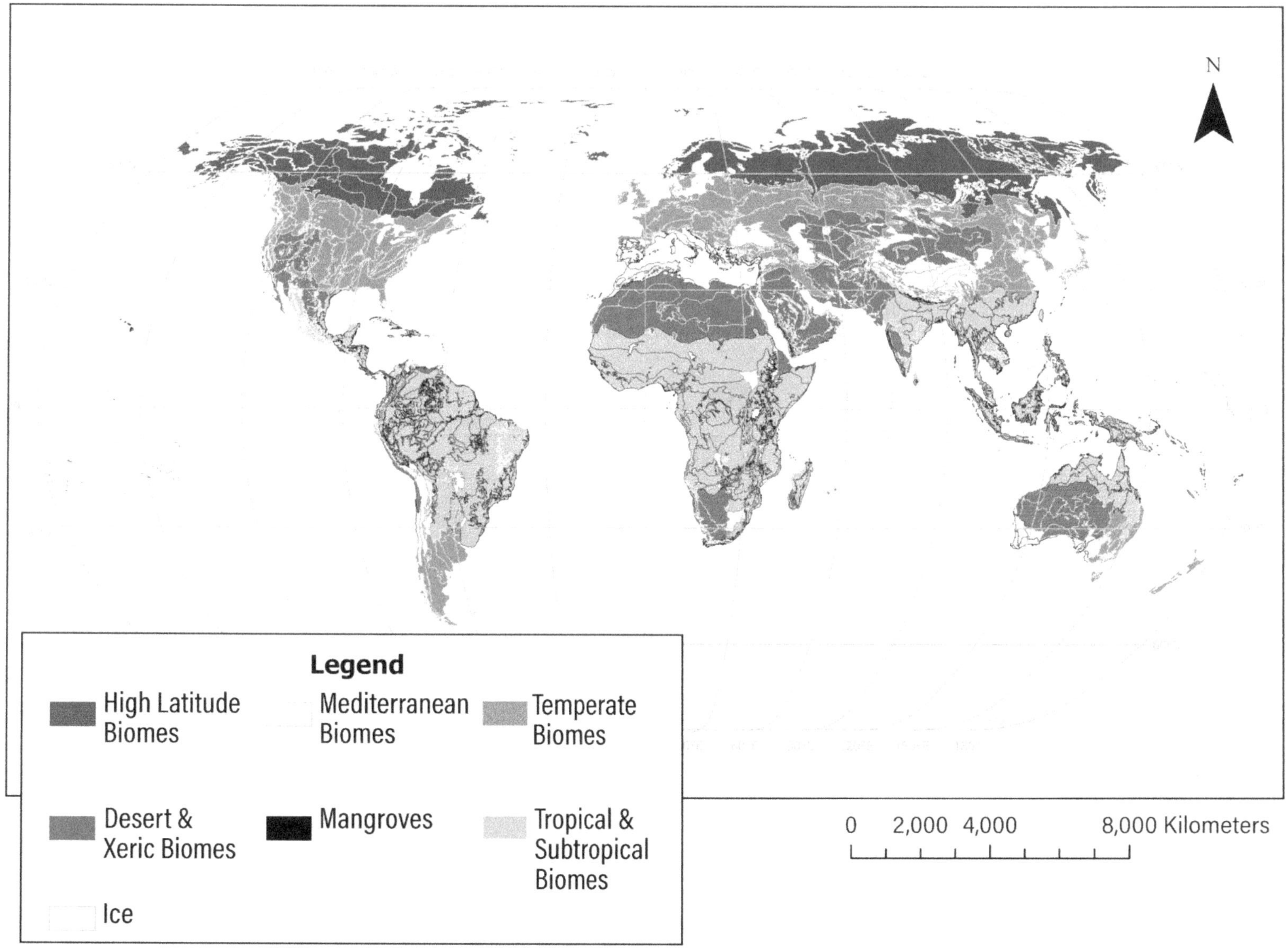

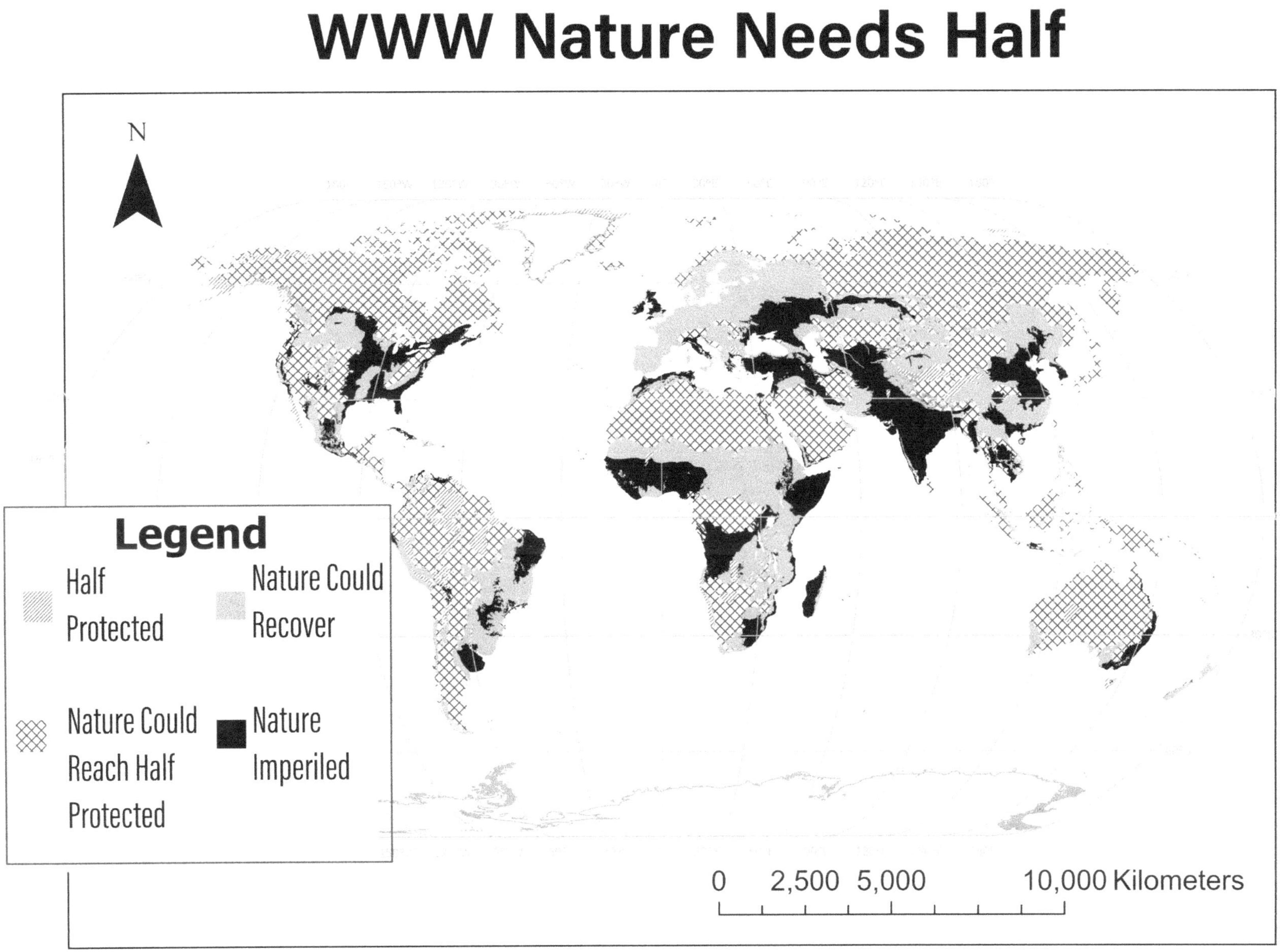

FIGURES 9.3A AND B. WWF ecoregions of the world (>840) are distributed on the terrestrial surface (A); status of the ecoregions based on several criteria (B). See text for details.

that took the boundaries of the ecoregions and intersected it with a map of protected areas for the world. This was reported as a percentage of the ecoregion that is formally protected. They also intersected the ecoregions polygons with that of the global forest change (Hansen et al., 2013) and the nonforested regions map (Anthromes v2 of Ellis et al., 2010) to determine if there was potential space for extending current protected areas. The outcome was an NNH set of classes that included: half protected (already); nature could reach half protected (i.e., there is nonhuman use space available that is not protected but potentially could be protected); nature could recover (some areas would have to be restored); and nature imperiled (protected + remaining area <= 20%).

F. LIFE ZONES

Life zones are yet another way to classify biogeographic regions. Life zones have been defined mostly based on climate conditions at very broad scales, and these tend to be more theoretically driven than ecoregions and biome classifications. Life zones also differ from climate zones in that life zones are developed to predict broad classes of vegetation and associated biota. Climate zones often do not directly consider biota.

There are two commonly used life zone classification systems, the classic one developed by Holdridge (1967) and a more recent one developed by Walter and Lieth (Walter, 1979). The Holdridge life zone classification system has become a "classic" for its creativity of using broad climate variables of annual biotemperatures (days that have averages between 0° C and 30° C) plotted against a ratio of potential evapotranspiration to precipitation measures that shows predicted life zones. Leemans (1990) developed global life zone maps that represented the current distribution of life zones and then used climate forecast models with future biotemperatures and PET/precipitation to "predict" future life zone distribution. Exercise 11 contains a set of exercises to conduct an analysis of these model forecasts.

One of the most well-cited studies on Holdridge life zones is that of Lugo et al. (1999). This study, conducted for the lower 48 states of the United States, used long-term meteorological data constructing life zones based on biotemperature, PET/precipitation, and some additional spatial data (e.g., locations of frost lines). They compared these life zone maps to those developed by Bailey's ecoregions and land cover maps developed using AVHRR (see Exercise 2 for an overview of this remote sensing data) and found that the Holdridge life zones matched reasonably well for forests but did not have a very good agreement with grassland, shrublands, and deserts, suggesting that the general approach applied by Holdridge and others in life zone mapping may be limited for large-scale conservation efforts.

One common natural management usage of Holdridge zones has been its application to understanding how elevation gradients give rise to plant community assemblages (Figure 9.4). Sometimes described as "**altitudinal belts**," these more local models of how climate affects vegetation and other biotic distributions have been applied to climate change impact studies. Often, life zones near the top of mountains are mapped and future climate projections are used to determine if these biogeographic regions would exist given these climate changes. Enquist (2002), for example, examined how life zones that are located along elevational gradients in Costa Rican mountain ranges would be impacted by changes in climate. They determined that at higher elevations, life zones would be most impacted by changes in biotemperatures whereas at lower elevations, life zone distributions would be affected by changes in PET/precipitation ratios. They also predicted that regional diversity of plant communities would be reduced for a warmer, wetter future climate, and that lower elevation life zones would most likely be impacted by future climate change.

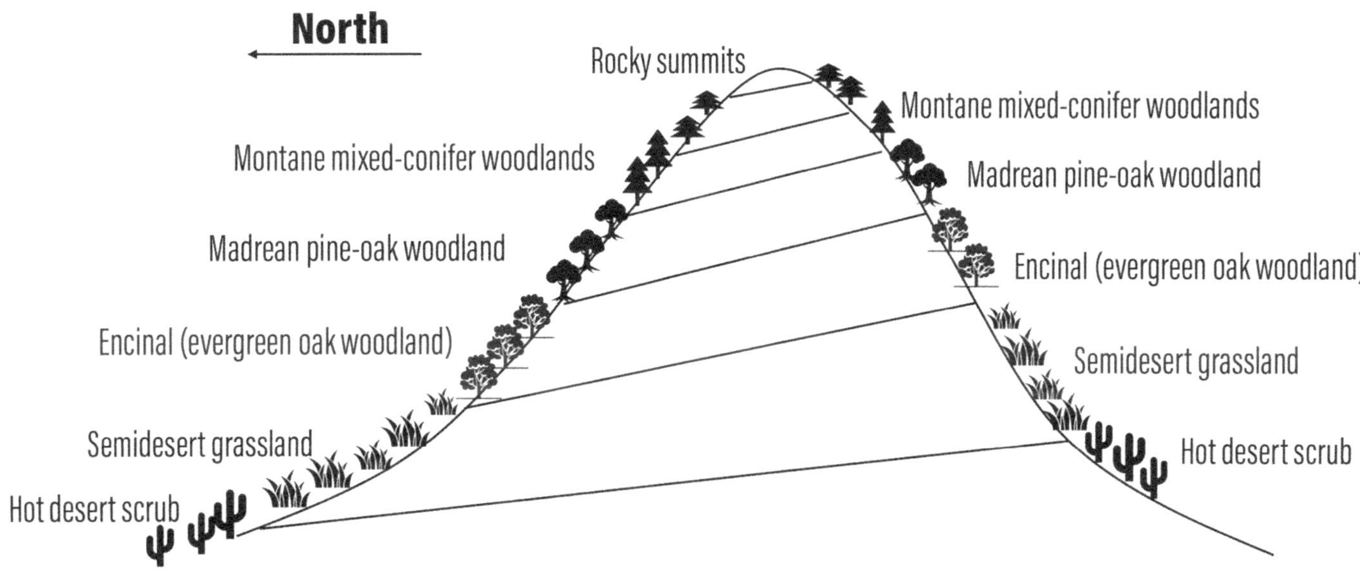

FIGURE 9.4. Altitudinal belts, often referred to as life zones, as they appear in one of the mountains of the Sonoran Desert. Note that there are about six life zones here that start at the lower elevations of the hot desert, composed of cactus and other arid-adapted species, through to grasslands, several forms of woodlands, and ending at the top with rocky summits. Also note that the southern-facing slope has different elevations for the belts compared to the northern-facing slope, as the southern-facing slope receives more sunlight (due to its latitude of around 32° N) in southeastern Arizona. Modified from Marshall (1957) as reproduced in Warshall (1995).

9.2 MARINE GLOBAL BIOLOGICAL COMMUNITIES

One of the most widely used marine ecoregions classification systems is the one developed by Spalding et al. (2007), which is referred to often as MEOW (Marine Ecosystems of the World). This classification also contains biogeographic realms (12), provinces (similar to biomes, which total 62), and ecoregions (which total 232). The major marine biogeographic realms are aligned with the major oceans and large bays and coastal areas. Also, parts of the realm classification are designated warm and cold areas within the same oceans.

The distribution of ecoregions is starkly different from that of the marine biogeographic realms. Spalding et al. (2007) defined and delineated the marine ecoregions using a variety of data on geophysical (i.e., temperature regimes, ice regimes, currents, bathymetric configurations, nature of freshwater inputs) and geographic (i.e., isolation and barriers, coastal morphology) measures, ultimately producing over 230 ecoregions all located along the coasts of continents and islands. Note that marine ecosystems border these land masses in an irregular way, mostly due to the depth and configuration of the ocean floors.

The MEOW maps of Spalding et al. (2007) have been used by a variety of international natural resource management organizations including the United Nations Environment Programme (UNEP), the Nature Conservancy (TNC), and the World Resources Institute (WRI).

9.3 SUMMARY

Identification and classification of biological communities at global and regional scales have had a long history, with many classification systems developed for use in comparing areas around the world, and most recently for global and international natural resource management. Many of these biological community classifications explicitly consider temperature and precipitation, two key variables that influence plant community types around the world. In the last 20 years, many of the boundaries of these biological communities have been developed using massive spatial meteorological data as well as land and water boundaries and features (e.g., elevation). The distribution of these boundaries as GIS shape files has facilitated global and regional research as well as applications to natural resource management.

9.4 DISCUSSION QUESTIONS

1. Compare and contrast how the concepts of biogeographic realms, biomes, and ecoregions can help address management of natural areas and the preservation of biodiversity in each of the spatial extents covered by these classification systems. Which level of biogeographic classification might be most useful at the global level of developing policies?
2. What are some of the key (in your opinion) ways that well-known scientists such as Humboldt, Wallace, and Holdridge were able to make their contributions to our understanding of biogeographic patterns of life on Earth?
3. In your opinion, what has been the role of GIS in facilitating research, education, and policy development in the area of conservation at the global and regional (e.g., national) scales?

10. BIOGEOGRAPHY AND INSULAR COMMUNITIES

Highlights

Aims. This chapter presents a summary of the integrated theory of biogeography (ITB) that is based on an equilibrium graphical model. It is considered to be one of the most significant milestones in all of ecology.

Learning objectives. Students should be able to:

- Describe some of the backgrounds of the two ecologists that co-developed this theory
- List the building blocks of the ITB
- Use the four components of ITB to build the graph that shows how species richness and species turnover equilibria values are determined
- Describe some of the shortcomings of the ITB

10.1 INTEGRATED THEORY OF BIOGEOGRAPHY: INTERESTING BACKSTORIES

A. INTRODUCTORY COMMENTS

The **integrated theory of biogeography** (ITB), also known as the **equilibrium theory of integrated biogeography** (ETIB), has two very interesting backstories. The first is how the two scholars—Robert MacArthur and E. O. Wilson—who collaborated to produce a book that outlined this new theory were educated, and then how they combined their talents to create this landmark concept. The second is the novel use of historical data that created a simple equilibrium model driven by two geographic features of islands—their relative isolation from the mainland and island size. The ITB is often cited as one of the most "elegant" presentations of any concept in ecology; it is considered to have broken new ground in the ways that it was synthesized and carefully presented. The ITB was first presented in 1963 as a paper and in 1967, in a more detailed format, as a book. Indeed, noted scholars had this to say about the ITB:

> "MacArthur and Wilson's equilibrium theory revolutionized the field of island biogeography and, to a large degree, ecology as well" (Brown & Lomolino, 2000).
>
> "The equilibrium theory of island biogeography has been one of the more influential concepts in modern biogeography, ecology, and evolutionary biology" (Wu & Vankat, 1995).

MacArthur and Wilson's *The Theory of Island Biogeography* "is one of the most influential books on ecology and evolution to appear in the past half century" (Losos & Ricklefs, 2009).

"It's absolutely clear that a pivotal point in [ecology] was MacArthur and Wilson and 'Island Biogeography'" (Professor Lord Robert May, in Powell, 2007).

B. PERSONAL BACKSTORIES

Robert Helmer MacArthur was born in Toronto, Canada, in 1930, the youngest son of a University of Toronto professor of genetics. Robert was described as very interested in the outdoors, fascinated with animals, and very quantitative—eventually in later years he would be described as a mathematician-naturalist. MacArthur would go on to receive his undergraduate degree from Marlboro College in Vermont (where his father had moved to teach), then a master's degree in mathematics from Brown University, and finally a PhD in biology from Yale University in 1957. At Yale, he studied under the renowned evolutionary biologist **G. Evelyn Hutchinson**, famous for advancing many theories in ecology and evolution including defining the concept of a niche as it is understood and used today. While a student with Hutchinson, MacArthur worked in the very specific area of theoretical population biology, attempting to advance this area through the development of foraging niches models. His research focused on outlining simple, graphically based, spatial (3D) models of foraging niches for five warbler species that coexisted during the breeding season in the forests of Vermont and Maine. His dissertation work was then distilled into a 1958 paper that is now considered a classic paper in ecology (MacArthur, 1958). After his PhD, he went on to study with **David Lack**—a world-renowned ornithologist—at Oxford University. While at Oxford, MacArthur sharpened his field ornithology skills. After his brief two-year stint at Oxford, he accepted a position as assistant professor at the University of Pennsylvania and then a few years later moved on to accept a prestigious position at Princeton University where he stayed for the rest of his career. MacArthur tragically died of cancer in 1972 at the early age of 42. The Ecological Society of America named one of its most prestigious awards the **Robert H. MacArthur Award**, which is given to mid-career ecologists for "meritorious contributions to ecology, in the expectation of continued outstanding ecological research" (ESA, 2026).

Edward Osborne "E. O." Wilson was born in Birmingham, Alabama, in 1929. He received BA and MS degrees from the University of Alabama. He then moved to study with **Frank Carpenter**, an ant biologist and paleontologist, at Harvard University. He received his PhD from Harvard in 1955 with his dissertation focusing on revising the taxonomy of a large group of ants. He went on to accept a faculty position at the Museum of Comparative Zoology and the Department of Organismic and Evolutionary Biology at Harvard University where he stayed for his entire career.

In researching and documenting the history of Wilson's contributions to ecology and evolution, Powell (2007) reports that Wilson ended up studying ants because he had vision problems. A childhood fishing accident injured his right eye, and eventually he was not able to focus at a distance. Studying at a close distance was easy (Wilson, 1994), so he spent a large portion of his career conducting behavioral, ecological, and taxonomic research on a subject he could study up close—ants. He was also an adventurist and so in his early years (e.g., late 1950s to early 1960s), he spent a considerable amount of time in the South Pacific on the island of Borneo and some of the tropical island archipelagos of Papua New Guinea and Melanesia (Wilson, 1994). Wilson's career would eventually evolve through three phases (Simberloff, 2022), from studying ants and other insects (e.g., Wilson, 1971), then into sociobiology (e.g., Wilson, 1975), and finally into a focus that had high impact in conservation science, especially the concepts of half Earth (e.g., Wilson, 2016) and biophilia (with also a strong emphasis on mentoring the next generation of scientists (e.g., Wilson, 2013), where he stressed that younger scientists should exude passion, put forth "big and bold" ideas, and not shy away from doing work with a sense of outdoor adventure.

At an **American Association for the Advancement of Science** conference in New York City in 1960, Wilson recalled (in Powell, 2007) that he and MacArthur explicitly discussed how their "old-fashioned subject" of ecology could gain the intellectual rigor of other biological fields, and they agreed that population biology was the way to do it. To accomplish this, they agreed to work together in the area of theoretical population biology using some of Wilson's data and MacArthur's ability to be more mathematically oriented, to create models that could be tested and then advanced to support a hypothesis-driven field of ecology.

10.2 BUILDING BLOCKS OF EQUILIBRIUM THEORY OF ISLAND BIOGEOGRAPHY (ITB)

There are several "building blocks" of the ETIB: (1) species-area relationships, (2) island isolation and immigration rate, and (3) species turnover rate. Let us consider each of these and then how they are integrated into a graphical, equilibrium model.

A. SPECIES-AREA RELATIONSHIPS

If we consider how species richness (i.e., species number) increases with area, we find that it typically follows a nonlinear relationship that is well known in spatial ecology (Figure 10.1). Referred to as the **species-area relationship (SAR)** curve, this association is evident when one plots the number of species and considers the area within which that count of species is made. Doing so generates a classic species-area curve that follows this equation:

$$S = cA^z$$

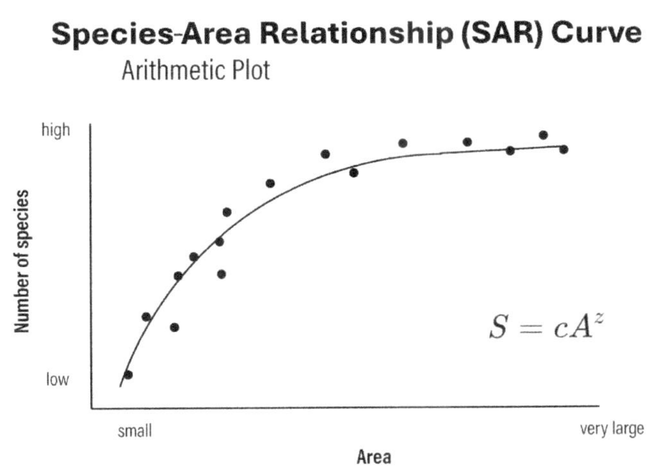

FIGURE 10.1. Species-area relationship (SAR) curve is typical for quantifying the relationship between area occupied and the number of species.

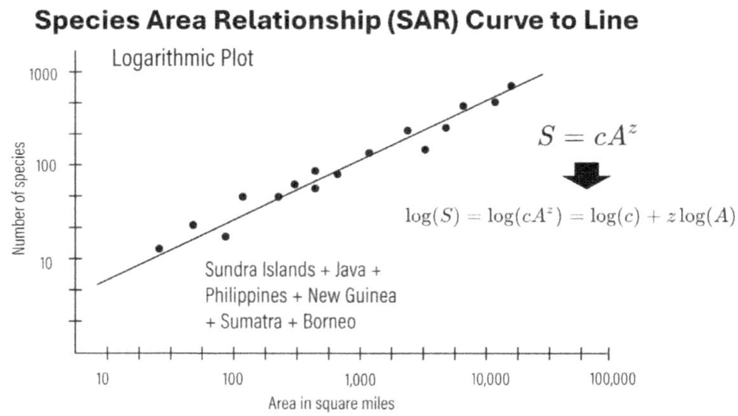

FIGURE 10.2. Close representation of species richness data plotted by island size and used in MacArthur and Wilson's integrated theory of island biogeography.

where S is species richness, A is area, c is a constant (also as the intercept or the species richness at unit area of 0), and z is the rate at which species richness increases with area. These kinds of curves are also examples of log-log relationships. Converting values of A and S into log values gives us a straight line:

$$\log S = \log c + z\log A$$

which is now interpreted as z as slope of the line with c as the y-axis intercept. The log-log plot is area (x-axis) and species richness (y-axis). In MacArthur and Wilson's (1963) classic ETIB paper, they gathered data from about a half dozen studies on the number of bird species and island size and plotted that relationship, producing a plot very similar to what is shown in Figure 10.2 for a set of islands in the southwest Pacific.

The SAR curve, they argue, influences an important pattern, which is extinction rates. They surmised, on the basis of SAR, that for small islands, extinction rates are high and for large islands, extinction rates are low.

B. ISLAND ISOLATION EFFECT

MacArthur and Wilson (1963) also summarized bird survey data from many scholars, but mostly from Mayr (1933, 1940, 1943, 1944), also a Harvard University professor, which showed that the proportion of mainland bird species on an island decreases with increasing distance from the mainland (Figure 10.3). This relationship, which they labeled as distance of island from mainland versus **species saturation** line, produced a non-log to log relationship that was logical; that when an island is very close to the mainland, the saturation of mainland species should be close to 100% (i.e., islands very near to the mainland should reflect nearly 100% of the species that live on the mainland; dispersal to the island should be "easy" and "frequent") and this saturation should decrease with increasing distance from the mainland.

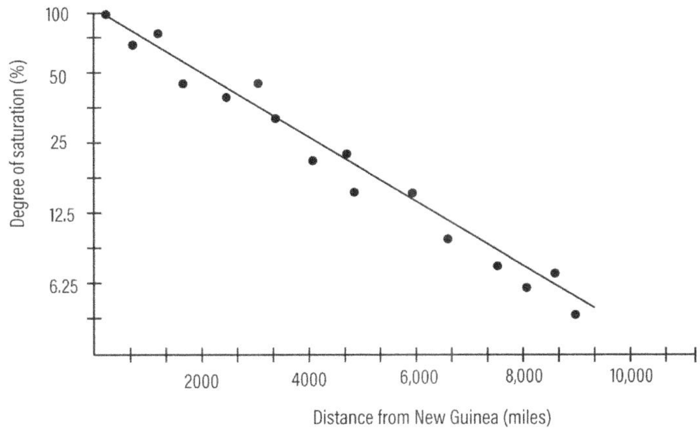

FIGURE 10.3. Close representation of data used to plot island isolation (x-axis) and the percentage of species that are found on the island that are from the mainland (y-axis, also referred to as degree of saturation).

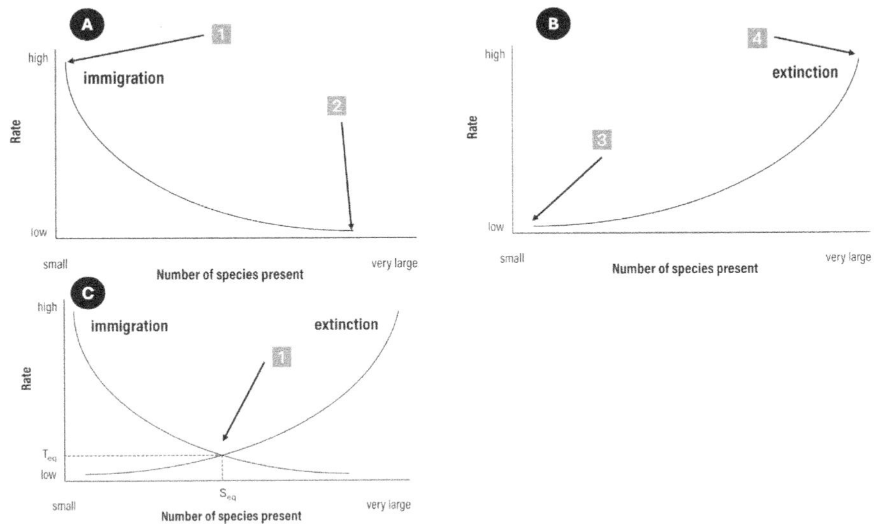

FIGURE 10.4A, B, AND C. Plots illustrate (A) how immigration rate and number of species present relate; and (B) how extinction rate and number of species present relate, both independent of island size and isolation. The location where the rates "balance out" is the equilibrium point (C).

C. EQUILIBRIUM MODEL

The third component of their ETIB is to consider how **immigration** and **extinction** rates combine to create the stable number of species present on an island. The logic goes something like this. Let us consider an island that has no species on it. When a species eventually colonizes the island for the first time, the immigration rate is 100% (Figure 10.4A, label 1). When the island has no species on it, the immigration rate is 0% (Figure 10.4A, label 2).

The next part of the logic is to consider that, when there are no species present, extinction rates are zero (Figure 10.4B, label 3). As the number of species increases, the available niches (multidimensional resource spaces) decrease and interspecies competition increases, causing some species to go extinct. Thus, as the number of species increases on an island due to immigration, the extinction rate will increase to a very high level (Figure 10.4B, label 4). Eventually, the immigration and extinction rates will come to an equilibrium to generate a stable number of species. Graphically, the opposing rates of immigration and extinction plotted as a function of the number of species, as shown in Figure 10.4C, label 1, will generate a **species number equilibrium**, denoted as $\mathbf{S}_{eq}$. Likewise, the equilibrium turnover rate, designated as $\mathbf{T}_{eq}$, can be determined as the value along the horizontal line for the equilibrium point that meets the y-axis; this is called the equilibrium turnover line or the line of constant turnover.

D. INTEGRATED THEORY WITH EQUILIBRIA

Let us now introduce the two geographic components of the ETIB: isolation (far or near) and island size (large or small) and how they impact the two forces shaping species richness, immigration and extinction (Figure 10.5). For **immigration rates**, MacArthur and Wilson (1963, 1967) surmised that for islands that are near the mainland, immigration rates should be higher than for islands that are farther away. Thus, the immigration rate curve for near islands is located above that for far islands. They also reasoned that **extinction rates** should be greater on small islands than large islands as small islands have fewer resources and fewer niches. Thus, for the originally drawn immigration and extinction curves across species numbers, the equilibrium number becomes S_{NS} since this is for near and small islands.

The final version of the equilibrium graph is to identify all of the species number equilibria for each of the four cases, near and far and large and small islands (Figure 10.5). Note that species numbers for these four cases becomes

$$S_{LN} > S_{LF} > S_{SN} > S_{SF}$$

and that if we plot out the turnover rate for these four cases, these are:

$$T_{LF} < T_{LN} \approx T_{SF} < T_{SN}$$

which one can interpret from the graph.

E. ASSUMPTIONS OF THE ETIB

As MacArthur and Wilson describe their model, in both their 1963 paper and 1967 book, some assumptions are made. These include: (1) there is a fixed mainland species pool over time, (2) immigration declines as the number of species on the island increases, (3) risk of extinction is equal for all species, (4) speciation does not impact the number of species, and (5) immigration rates and extinction rates are independent factors. In addition, they assume that islands of the same size that might have different habitat configurations will support the same number of species and that all species as they colonize have the same relative impact on the quantity and quality of resources on the island.

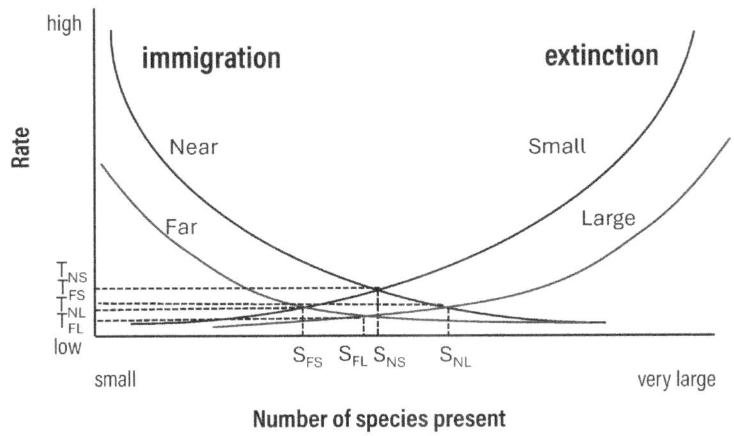

FIGURE 10.5. Plots of the four major patterns undergirding the equilibrium component of the island theory of biogeography. For (10.4C), note that this is for one island. When we begin to compare species richness and turnover for many islands, equilibria points are compared along the x-axis (species richness) and y-axis (species turnover). See text for further explanations.

F. NOTE ON HANDS-ON EXERCISE

Exercise 9 will provide instructions on the use of GIS and butterfly species lists by county to create species-area curves. From these data, users will calculate c and z for the species-area curve. Species-area curves can be used to compare two or more sites (e.g., different kinds of habitats or within and then outside protected areas). In theory you would expect greater values of z in areas where the habitat is of higher quality, thus supporting more species of, viz., animals.

10.3 SUMMARY

The integrated theory of biogeography represented a landmark work in ecology due to its rigor and appeal to a broad community of scholars in ecology, evolutionary biology, and geography. The ETIB drew from dozens of published papers reporting on island species richness statistics that were then combined with island size and isolation (i.e., distance from mainland) to predict species richness given immigration and extinction rates.

10.4 DISCUSSION QUESTIONS

1. Can you think of nonisland systems that could apply to the ITB?
2. Based on what you know of landscape ecology principles, what could be introduced to the ITB to make it relevant to what we know about patterns and processes in landscapes?
3. Is the ITB relevant today as a theory and does it have practical application in conservation science, given all the great advances in ecology and evolutionary biology today? Some say "not."

11. GISCIENCE AND ITS BRIEF HISTORY

Highlights

Aims. GIScience is a rather unique field in that it combines information science and principles and analysis in geography. This chapter summarizes its brief history, associated core ideas, and relevant technologies.

Learning objectives. Students should be able to:

- Describe the history of GIScience, clearly describing it within the context of GIS (geographic information systems)
- Describe how its history evolved with major milestones of information science and computers
- Describe the key debates in the field of GIScience, placing these within the context of spatial ecology and other technologies such as remote sensing that played key roles in the advancement of both GIScience and spatial ecology

11.1 INTRODUCTION TO INFORMATION SYSTEMS

GIS has had an interesting and storied history, evolving in parallel with several other forms of information systems over the last several decades. For the purposes of this book, GIS stands for **geographic information systems**. As originally conceived, it is part of a broader set of technologies that are simply referred to as **information systems.** Langefors (1966) was one of the first to define it as "information that is contained in a system that is organized in [a] way for everyone to legitimately use it." This definition was advanced later by Davis (1974) to include the use of machines or computers for organizing information. Current definitions of information systems, such as the often cited one by Laudon and Laudon (2018), emphasize that they are an integrated set of components that collect, store, organize, process, distribute, visualize, and allow for the analysis of information generally for the purposes of supporting decision making and coordinated sharing of information across an organization. There are many domain-specific forms of information systems—for example, these include business management information systems to track personnel; health information systems for managing appointments, prescriptions, procedures, patients, doctors, and other staff; library information systems that manage books, other holdings, and patron records; and, finally, transportation information systems that track flights, schedules, baggage, and other travel logistics.

11.2 GEOGRAPHIC INFORMATION SYSTEMS

Geographic information systems are configured for georeferenced data but also focus on decision making within a spatial context. The traditional definition of geographic information systems is often the one proposed by Burrough (1986; see also Maguire, 1991), who stated that it is "a powerful set of tools for collecting, storing, retrieving at will, transforming and displaying spatial data from the real world." GIS is also closely related to other kinds of spatial information systems, in particular CAD/CAM systems used for engineering, manufacturing, and design, and land information systems, which have been used to manage and analyze land ownership, carousel maps, and parcel boundaries.

11.3 HISTORY OF GIS WITH IMPORTANT MILESTONES

The development of geographic information systems has been accomplished through investments in software development, research in spatial sciences, and important initiatives by federal agencies, prominent universities, state agencies, and not-for-profit organizations. There has also been the articulation of conceptual paradigms and long-term initiatives that helped advance GIS as a major framework and set of tools for spatial analysis (Coppock & Rhind, 1991; Goodchild, 1992; National Center for Geographic Information and Analysis, 1989). Highlights of these developments, along with landmark events, are described here organized by decades.

A. 1960s—GENESIS AND EXPERIMENTALISM IN THE MAINFRAME ERA

Goodchild (2018) describes the origins of GIS as being multifaceted and focused on representing space in 2D with work entirely done on mainframe computers. Many scholars point to the historic work of Tomlinson (1968)—sometimes referred to as the "grandfather" of GIS—who led the development, with IBM, of the first "GIS," called CGIS (Canada Geographic Information System). **CGIS** was a vector-based system that stored land system data at a national scale; it was mostly experimental, a first attempt to work through the technical details of encoding spatial information. At this same time, the Harvard Laboratory for Computer Graphics and Spatial Analysis, which was founded in 1963, developed **SYMAP** (SYnagraphic MAPping), which stored digital maps in a raster file format. The lab also produced **GRID**, a computer mapping system for raster file formats. SYMAP eventually evolved into **ODYSSEY**, a vector-based GIS, which lived well into the 1980s. Jack Dangermond, the founder of Esri, worked at the Harvard Laboratory of Computer Graphics and Spatial Analysis and was inspired by the work done at this prominent lab. One of the first funded projects for mapping was using these tools for the computerized mapping of diseases (COM) of the northeastern United States. Another parallel development occurring during this time was in the United States Census Bureau's Division of Research. This group developed a digital mapping framework referred to as **DIME** (dual independent map encoding) files, which were a set of digital base maps for the entire country that linked geographic files with tabular data such as population, income, and other demographic statistics. At Northwestern University, geographer Duane Marble (in Coppock & Rhind, 1991) created a GIS for local transportation systems. Finally in the United Kingdom, David Brickmore, a cartographer originally with Oxford University Press who created some of the early digital forms of maps for books, started a spatial mapping initiative at the **Experimental**

Cartography Unit (**ECU**) of the National Environmental Resources Council (NREC) that led to national scale maps of natural resources for the United Kingdom (Rhind, 2008).

This early work was very challenging. Computers in the 1960s were a lot different than they are today. They stored information on tapes and data were entered into **mainframe computers**—building-sized machines—that were expensive; only large organizations could purchase and operate them. Users would have to go to rooms where the mainframe computer existed. Paper-based punch cards would contain the code (in one stack) and data (as a separate stack); the cards were entered at table-sized card machines. Visualization of the data was using crude representations of objects or spaces printed on dot matrix printers with standard characters that are found on a typical keyboard. In short, this was a decade of experimenting with developing and working with spatial data, focused in a few national labs and leading-edge universities around the world, and tended to be for the purposes of mapping natural resources.

B. 1970s—FIRST COMMERCIAL GIS PACKAGES AS CENTER ERA WITH USER-ORIENTED DATA

The 1970s witnessed a major transition from experimental, mostly federally supported work in geographic information systems, to user-oriented, laboratory-based use of spatial information. This was also a decade when digital data were being developed using large digitizing tables, frequently from paper maps, and space-based imagery was being captured at almost global scales for the first time. The decade also saw the start of some of the prominent GIS software being initiated, including Esri's **Arc/INFO**, Clark University's **IDRISI** (now **TerrSet**), U.S. Army Corps of Engineers' Geographic Resources Analysis Support Systems (**GRASS**), United States Fish and Wildlife Service's Map Overlay and Statistical System (**MOSS**), and Erdas's **IMAGINE**.

What is common to nearly all of these emerging GIS packages is that they focus on storing, managing, and analyzing digital natural resource data. Maps of land use/cover, natural features such as rivers and lakes, political boundaries, and transportation networks were commonly stored digitally by state and federal agencies. Despite all of these advances, the internet still did not exist, and computers were very expensive, so only a few groups could afford to purchase and maintain them. Indeed, the building-size mainframe computers gave rise to refrigerator-size mini-computers. Access to mini-computers was limited, as users had to visit rooms that had special terminals connected to the mini-computer. This was also the decade when investments in experimentation in the 1970s began to pay off, as much of the data and results from tools and approaches to spatial analyses could help support decision making for managing natural resources. Several states established statewide digital mapping initiatives to manage natural resources. These included, most notably, the state of Minnesota's Land Management Information Center's EPPL/EPPL7 and Michigan's Department of Natural Resources, Michigan Resources Information System (MiRIS). Support to develop and maintain MiRIS came from funds acquired as part of the state's hunting and fishing fees.

One of the most significant events of the 1970s was the launch of the first satellite devoted to monitoring the Earth. Originally called the Earth Resources Technology Satellite -1 (ERTS-1), it was later renamed **Landsat** 1, and this has been considered the "big bang" start of the Earth Observing System constellation of satellites. Launched in July 1972, it was designed to last one year and create maps of land cover at 80 m x 80 m pixels using a multispectral scanner (MSS) with four spectral bands. It was eventually able to map major land covers and then used to calculate greenness using NDVI measures; its mission ended up lasting for about five and a half years. Since 1972, there have been eight other Landsat satellites launched, with three total in the 1970s (Landsats 1–3).

What is also very interesting is that during the 1950s through to the early 1980s, with some prominent examples occurring in the 1970s, many **geography departments** began to close (Goodchild, 2007; Huntley & Rosenblum, 2020). Most notably were the scheduled closures in the late 1970s of the geography departments at the University of Michigan, University of Chicago, Northwestern University, and Columbia University. At the same time, there was the emergence of the use of GIS and space-based remote sensing in areas outside of geography departments, creating a bit of an "identity crisis" across the field of geography. These outside-geography entities at universities, at times, thrived. Most notably was Purdue University's **Laboratory for Applications in Remote Sensing** (LARS), established to conduct advanced research, training, and outreach on agricultural applications of remote sensing and GIS. Funded by NASA and the USDA in the late 1960s, it had around 120–170 staff members and was the largest research team on campus. It is renowned for being one of the first research units to create maps of the Landsat 1 data for use in agricultural monitoring. Early projects from LARS included mapping corn blight and developing open-source software like **MultiSpec** (Biehl & Landgrebe, 2002) and training conferences. In short, this decade witnessed the decline of geography departments across the United States, the emergence of high-impact applications of GIS and early data streams from space-based remote sensing, and some fundamental concepts of geography being proposed.

C. 1980s—COMMERCIALIZATION AND MIGRATION TO WORKSTATIONS

Personal computers emerged in 1981, with IBM's Model 5150 becoming the first true PC; a host of UNIX workstations were also produced, which were designed for engineering applications such as GIS, all of which contained their own monitors, separate CPU, one or two floppy disk drives, a keyboard, and a mouse. The 1980s also saw the invention of the **internet** (demonstrated famously on January 1, 1983), which allowed computers to "talk" to one another using TCP/IP protocols (internet addresses and send/receive rules), and in 1989, the World Wide Web was created that allows users to create and share web pages using hypertext markup language (html).

With the transition of computational infrastructure to personal computers on desktops, most GIS packages also transitioned to this environment. Esri released a version of their GIS software package as PC Arc/INFO. Burroughs's 1986 book *Principles of Geographical Information Systems for Land Resources Assessment* was the first to describe how the storing, managing, analysis, and visualization of spatial data could be conducted while also creating a clear vocabulary for geographic information systems that were built from principles of geography. Also during this time, debates about spatial data quality and **uncertainty** became front and center as well as the development of statistical methods that specifically focused on means and variances across space (e.g., semivariance, Moran's I; see Chapter 13).

This decade saw the development of the **Arc Macro Language** (aml) that would allow users to create scripts so that they could automate the often common, multistep processing of spatial data. The aml code also contained commands that would allow users to create graphical user interfaces (GUIs), thus facilitating the development of customized **decision support systems** (DSS).

At the national scale in the United States, original DIME files evolved into more detailed and greater spatial extent files, such as **TIGER** (topologically integrated geographic encoding and referencing) files by the United States Bureau of the Census. TIGER files were the first nationwide coverage of key spatial features, which included roads, rail, political boundaries, hydrology (rivers, lakes, ponds), and other features (e.g., places), that was structured to interface to census tabular data. The USGS also developed national **digital elevation**

models (DEM), which were derived from paper maps enhanced with spot filed survey points. These DEMs were still coarse but their value for mapping of areas like floodplains emphasized the importance of this form of data as well as the broad value of GIS and spatial analysis.

Another set of events that transpired in the 1980s was the creation of a new datum (see Exercise 9) in 1983, **NAD83**, that used very precise and updated information about the shape and size of the Earth. This datum also facilitated a greater adoption of national scale coordinate systems such as UTM and Albers Equal Area, leading many local and state GIS offices to transition from state plane systems developed in the 1930s to these national coordinate systems, as much of the spatial data in use was being originated from federal agencies such as the USGS and the U.S. Bureau of the Census.

This decade also saw some of the more unique applications for the use of GIS. In Michigan in 1984, the Michigan Department of Agriculture began a statewide monitoring program to track the spread of the **spongy moth** (then called the gypsy moth). This forest insect pest was discovered in the city of Lansing, Michigan, in the early 1950s. A trapping density, one trap per 3 square miles, was established statewide and continued for about a dozen years, eventually integrated into a national program called Slow the Spread (see Exercise 7).

Finally, one of the watershed events in this decade was the establishment of the **National Center for Geographic Information and Analysis** or **NCGIA**. Founded in 1988 with funds from the U.S. National Science Foundation, NCGIA was a multi-university consortium research, education, and outreach entity involving the University of California Santa Barbara, University of Maine, and University of Buffalo-SUNY. It focused on a variety of research initiatives and developed a suggested curriculum that would be used in geography and other departments to teach GIS and its applications.

D. 1990s—DESKTOP GIS, NATIONAL DATA INFRASTRUCTURES, AND CUSTOMIZATION

This "**desktop GIS**" decade saw the migration of GIS from a command line, aml-driven use to a menu-driven interface to most commands through GUIs that functioned like other software packages, such as those found in Microsoft applications. Arc/INFO transitioned to a package called ArcView and, although aml was still supported, Visual Basic was added as an additional language to facilitate programming and interface design for decision support systems. Many GI packages, in previous decades, required specialized engineering-class or massive computer infrastructure; this decade saw its migration to the same computers on which users had word processing and other computer applications.

This was also the decade that NCGIA started having broad impact in the geography community. The remaining geography departments in many cases embraced the teaching and use of the tools in GIS for research and outreach, thereby extending the reach of geography education to include the technical training required to use GIS and associated technologies like remote sensing. Founded in 1995, the **University Consortium for GIS (UCGIS)** started to promote a **research agenda** for GISci.

Goodchild and some of his colleagues at NCGIA also began an engaging debate on the differences and similarities of GIS as a tool and whether the major paradigms and research topics in geography should be considered a "science," the nexus of these two areas forming what they called **GISci** (see section 11.4 below).

This decade also saw the initiation of several standards and interoperability specifications. After being established in 1989 by the Office of Management and Budget of the White House, the **Federal Geographic Data Committee (FDGC)**, formed with spatial sciences experts across the federal government along with key academics and state agency representatives, began to establish guidelines and standards that facilitated the development and use of a **National Spatial Data Infrastructure** (**NSDI**) that would allow for better integration and sharing (called data democratization) of spatial data from all the agencies, such as USGS, USDA, U.S. Bureau of the Census, U.S. Army Corps of Engineers, and so on.

This decade also saw the introduction of specialized "extensions" to the basic GIS software packages, making it an **extensible package**. The most noteworthy were—for the Esri set of GIS packages—Spatial Analyst, 3D Analyst, Network Analyst, and Business Analyst, along with a massive, national transportation data package that was called StreetMaps.

E. 2000s—WEB-BASED DATA PORTALS AND OPEN-SOURCE SOFTWARE

As GIS and, for the most part, GISci are focused on conducting spatial analysis that can inform decision making, and as internet speeds and web-server protocols had advanced in the late 1990s, it is not surprising that the next decade witnessed the advent of the delivery of mapping and simple spatial analyses to users, especially decision makers, over the web. In 2000, ArcIMS, with IMS standing for Internet Map Server, was released. It provided **internet mapping services**. Four years later, a more advanced product, ArcGIS Server, was released that incorporated many of the spatial analysis services found in the ArcGIS Desktop package.

Another noteworthy advance in this decade was the release of a set of spatial tools collectively referred to as **Arc Hydro**, although this was not incorporated into the Esri products as an extension; the utilities, however, did allow users to process hydrologic data and digital elevation data to create stream networks and watershed boundaries (see Exercise 6). Developed by David Maidment at the University of Texas, these tools not only served as the foundation for how to process hydrologic data, but this paradigm for processing hydrologic data that also needed to include extensive temporal data evolved into the spatial data model known as **geodatabases**. This work was described in detail in Maidment's *Arc Hydro: GIS for Water Resources* (Maidment, 2002).

This decade was also the period when several **open-source GIS packages**, such as QGIS and rgdal, were released. **QGIS** (for Quantum GIS) was first released as a simple spatial data display tool that evolved into a comprehensive GIS package. **rgdal** (for R in Geospatial Data Abstraction Library) allowed users to have R packages to read, transform, and write spatial data for analyses. There were also initiatives that solicited **citizen scientists** to contribute toward building spatial databases (e.g., Lee et al., 2006), and the use of **mobile devices** like cell phones and tablets to enter data in the field and to connect to map services was initiated (e.g., Zhang & Mei, 2008).

F. 2010s TO NOW—CLOUD-BASED GIS

Following the trends for most other enterprise software and data-driven applications, cloud-based services emerged in the early 2010s and continue today as the major paradigm for hosting software tools and large databases. **Google Earth Engine** was released on December 2, 2010, at the UN Climate Change Conference in Cancun, Mexico. It is now a platform for researchers and analysts to extract global databases, especially remote sensing imagery and end products produced by NASA and ESA, at global extents interfaced to a host of spatial analysis tools (e.g., buffers of points). Researchers at universities and not-for-profit organizations can use the data and services for free with an account. Esri's **Living Atlas**, a cloud-based, ArcGIS Pro, and Arc Online "live" access to spatial data, was launched in 2014. This Living Atlas allows users to connect to these data through the application interface and/or download the data for use on their local computers. Finally, a host of other **cloud-service data warehouses** were launched and are maintained by national agencies, international agencies (e.g., the UN), and by not-for-profit organizations (e.g., WWF).

In recent years (2020–now), novel capabilities that are emerging in GIS applications such as Esri's ArcGIS Pro are (1) an **AI-based deep learning** set of tools that can be used for classification of raster maps and for point cloud data analysis; (2) visualization tools that allow map makers to determine if their map has **accessibility** (color vision deficiency simulation) issues; and (3) space-time **3D/4D cube data** storage and analysis that allows users to analyze rich time series of data, as there currently exists rich temporal data from space-based remote sensing as well as water quality/flow data.

Esri's ArcGIS Pro continues to expand its capabilities and also its ability to create packages for special domain areas (e.g., airport spatial data management). In recent years (2015–now), it has provided extensions for advanced analysis (e.g., 3D Analyst, Spatial Analyst, Business Analyst), data and workflows (e.g., Publisher), and for specific industries (e.g., Aviation Charting, Defense Mapping).

11.4 SUMMARY

GIS and associated applications have grown considerably in the last six decades, have evolved with all computational advances, and have expanded their reach into many areas of use and into highly specialized domains. Spatial data has gone from being entered into computers using punch cards, to having petabytes of data available through analysis engines that are free to researchers and analysts with not-for-profit organizations. Interfaces to AI are just starting and are likely to grow in the coming years. Finally, as we will see in the next chapter, the technology and scholarship that is identified as GIScience have grown together, often the big ideas and the technology moving forward in logical advances.

11.5 DISCUSSION QUESTIONS

1. Discuss what you think are the benefits of storing all data for a project in one geodatabase as opposed to having shape files as separate files.
2. Describe the importance of metadata to data management. Using an internet search, what information should go into a metadata file for spatial data?

12. GISCIENCE AND SPATIAL DATA MODELS WITH DATA PORTALS

Highlights

Aims. As spatial data come in a variety of formats, we summarize those here as they are represented as spatial data models. Spatial data models have several fundamental features and terms that are used to describe them. This high-level review of spatial data models should allow all students the ability to do a more "deep dive" on these as this area of GIScience is very technical and rich.

Learning objectives. Students should be able to:

- Describe what sources of remote sensing can be used to develop land use/cover maps and the scale at which these are collected
- Recognize the major map elements and describe their purpose
- Describe the differences between shape files and geodatabases
- Describe where some of the major data portals are and how one can acquire data
- Understand how to compare and contrast the different data sharing agreement forms

12.1 INTRODUCTION TO THE DATA MODEL CONCEPT

Data models are conceptualizations of how digital data represent real-world entities. Data model descriptions are guided by a specific design that also aligns to a computer language that is designed to retrieve, store, copy, and/or manipulate the data. Data, data models, computer languages, and general rules for storing and manipulating data are often referred to collectively as **information systems** in the field of data science.

12.2 INTRODUCTION TO SPATIAL DATA MODELS

Spatial data models are intended to structure digital representations of entities that are arranged in space (Peuquet, 2011). In general, there are two major forms of spatial data models. The first is called **vector data models**; these are digital representations of objects in space that fall into the spatial structures of points, lines, and polygons (Figure 12.1). Locations of camera traps, city centers, and cell phone towers would constitute

point (Figure 12.1A) spatial databases. **Lines** (Figure 12.1B) are spatial data that are composed of two end points that are called **vertices** (plural) or a **vertex** (singular); vertices have coordinate information (x,y) and a record ID (called an FID). An electrical line that is connected by two electrical poles would constitute a line with two vertices. A **polygon** (Figure 12.1C) is a set of lines that are connected, creating a closed shape, such as a rectangle. The lines of a polygon are also connected by locations called **vertices**. As many forms of spatial entities are complex, spatial data models also include **polylines** and **polypoints**. A collection of points in a database, such as the location of all city centers in a state, would constitute a polypoint (Figure 12.1D) database and these would be grouped as one record in the database. Polylines (Figure 12.1E) are composed of a set of connected lines with vertices that never (or at least very rarely) form closed polygons. A classic polyline is a river network composed of a series of small tributaries that combine into a major branch, which we would call a river. All vector data are composed of two main components, often as separate files. The first is the spatial data or the geometry data; the second file contains the attribute data. Records for both are linked (technically called related, see below).

The second kind of spatial data model is referred to as the **raster data model** (Figure 12.2). Raster data models code spatial information in a grid composed of cells. Raster data models are always rectangular and in nearly all cases with cells that are square. A raster file has dimensions of height and width, reported in either the number of cells (high and wide) or length (e.g., in meters high and wide). There are some spatial entities that are almost always stored as a raster. One form of a raster data model contains a header with the number of column and rows (Figure 12.2A), the coordinates of the upper left corner, cell size, and the value used to code no data (Figure 12.2B), with the rest of the file containing the values (Figure 12.2C) for each cell in the rectangular grid. Viewing an Esri raster grid table (Figure 12.2D) provides summary information that includes the value in the grid and its count. These include digital elevation (how high a location in place is) and data from remote sensing imagery, most commonly from satellites, planes, or drones.

Both vector and raster data models are composed of two kinds of representations: **geographic** and **attribute** values. Geographic values are specific coordinates in space; this is typically the locations as specified as latitude and longitude. Attribute values are either numerical and/or character values associated with the place. A numerical value for a raster data model could be the height in meters above sea level. A character value for a spatial data model could be the name of the city in the point database of city centers. For vector data models, the geographic representation (Figure 12.1F), its associated table (Figure 12.1G), and the attribute representation (or value) for the spatial entity are lined by an object ID (Figure 12.1H).

Grid-based raster file formats are also part of a general spatial data model called **tessellations** (Figure 12.3). Square (Figure 12.3A), triangular (Figure 12.3B), and hexagon (Figure 12.3C) tessellations are the most common form of grid-based raster file formats. Tessellations can also be created using tiling of more than one shape, such as forming a hexagon from six triangles (Figure 12.3D), which are called Archimedean tessellations. Another set of tessellations are created from a network of points that define the shape and configuration of irregular polygons called **Voronoi** or **Thiessen** tessellations (Figure 12.3E). Note that the network of points here is used to create areas that separate each point defined by boundaries that are exactly half the distance between nearest neighbor points. It is often argued that data stored as tessellations can scale quickly using computers if databases for the spatial entity representations (e.g., terrain features) are exceedingly large (i.e., elevation data for the Earth). In many GIS packages, tessellations are stored as spatial data models called **fishnets**, which are often used to summarize data from another spatial database (e.g., number of sample points, percentage of land cover).

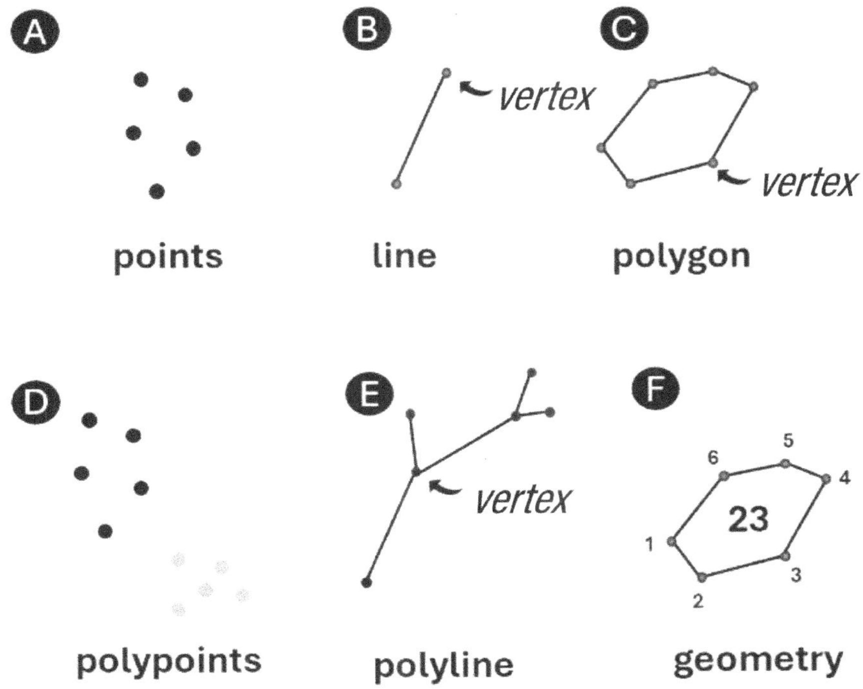

G

FID	vertex	x	y
23	1	65.3421	-96.210
23	2	65.2121	-96.333
23	3	65.2366	-96.412
23	4	65.2590	-96.456
23	5	65.4519	-96.213
23	6	65.2122	-96.489
24	1	65.3421	-96.210

H

FID	County FIPS	Population	Median Income
22	.....	......	......
23	18157	186,251	57261
24	.....	......	....

FIGURE 12.1. Spatial data models for vector data representations.

As was indicated above, vector and raster spatial data models are the most common, but they are not the only ones used. **Triangular irregular networks** (TINs) are spatial data models based on a "mesh" of irregularly shaped triangles; these are a specialized vector format. The most common TIN use is to display elevation across an area or landscape. The structure of TINS is that the data are stored as mass points for the locations of the vertices within the network (Figure 12.3F). Typically, digital elevation models are created by TINS when elevations of points are known and these locations, technically as vertices, are used to create a TIN representation of the terrain.

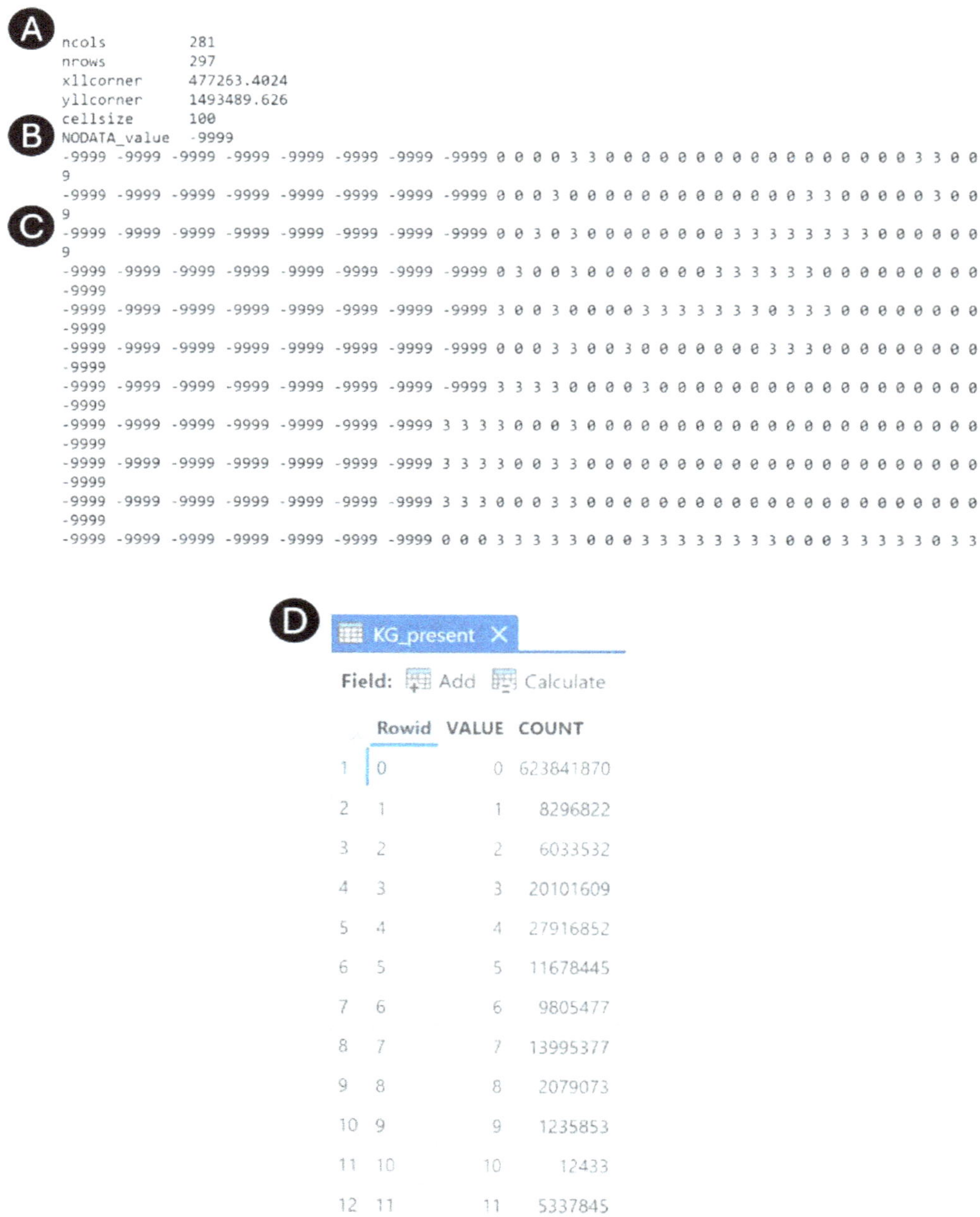

FIGURE 12.2. Spatial data models for raster representation.

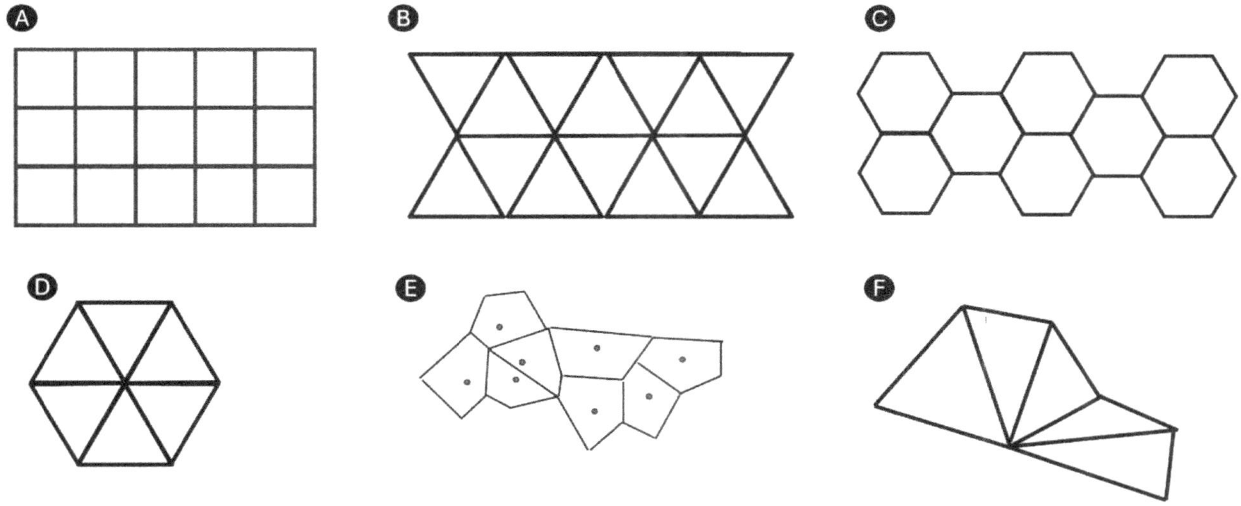

FIGURE 12.3. Tessellation shapes for raster representation of spatial data. These include (A) squares, (B) triangles, (C) hexagons, (D) hexagons formed from triangles, (E) Voronoi tessellations formed from locations of points, and (F) triangular irregular network (TIN).

12.3 BRIEF OVERVIEW OF SHAPEFILE FORMATS

The most common format for storing spatial data using the above spatial data models is called the shapefile format. Developed by Esri in the 1990s, it is a mostly open-source format to store and manipulate spatial data. The shapefile format is composed of at least three files:

*.shp (the file containing the geographic information)
*.dbf (the file containing the attribute information)
*.shx (the file containing index information)

where * is the name of the spatial entity for all three files. Each of these files is composed of a header and associated information relevant to the file's purpose. It should also be noted that the *.dbf file is stored in a format referred to as dBASE. dBASE file formats were one of the most common formats in the 1990s and early 2000s and were associated with the company dBASE (in some years, owned by the Borland Company). dBASE file formats are easily imported into Microsoft Excel or Microsoft Access.

Some shapefiles also have *.prj or *.shn files as well. The file extension, *.prj, contains specific projection information. This is sometimes not needed if all spatial data being analyzed together share the same projection system (projections and coordinate systems are covered in Exercise 9).

A commonly used term with shape files is **feature**. A feature is a single geographic item; in other words, it is a record that has a geometry (point, polygon, line) and a set of attributes. A **feature class** is a database with geometry and all of its associated attribute data. It should also be noted that all shape files have a field called the ObjectID, OID, or FID, which is a unique value, a kind of key, that is used to sort, index, and uniquely identify each record. It typically starts at 0.

12.4 BRIEF OVERVIEW OF RASTER FILE FORMATS

One of the common raster file formats is the **grid** format developed by Esri. The grid file format is quite a bit different from shapefiles in that the data are stored in two folders (not specifically as files as in shapefiles). One folder contains the name folder and the second contains the info folder. Name folders will have several files stored there, most ending in *.adf. The info folder will contain files that have *.dat and *.not extensions and start with an arc- prefix. Grids can be in integer format (most common and very efficient) or floating-point format (not common and very slow to process). An integer grid can also be used to create a value attribute table (or VAT) that lists cell values and the number of cells that are contained in the grid.

An open-source file format for raster files is the *.asc format, which is a text-only file format that contains a header followed by integer values separated by a space and organized as a grid. This format is easy to read by any program and can be input into any geographic information system.

Although there are several dozen raster file formats available for use in geographic information systems, two others have become very popular due to their ease of creation and use. These include the GeoTIFF format, which is a tagged image file format (TIFF) developed to also contain geographic information in the header. GeoTIFFs are used to distribute satellite-, plane-, and drone-based imagery. GeoTIFFs have the extension *.TIFF and may also be distributed by four other files with the following extensions:

*.tfw (the world file, which contains the geographic information)
*.aux (auxiliary information that contains important information such as projection parameters)
*.xml (usually containing metadata such as data source, data of acquisition, custodian name, etc.)
*.ovr (information that allows the geographic information system to use pyramids to process large files efficiently)

The second very common format for raster files is the ***.img** file format that is also used to distribute imagery. Originally developed for multilayer image distribution and analysis by the ERDAS IMAGINE remote sensing software company in the late 1980s, its original design allowed users to analyze multiple layers of data that are typical of imagery (e.g., photos from sensor platforms, such as NASA's Landsat satellite platforms), which can have anywhere from six to nine bands of data separated by spectral band widths that relate to physical properties (e.g., chlorophyll absorption).

12.5 GEODATABASES

Another very common spatial database format used by geographic information systems is the **geodatabase**. Developed in the early 2000s by Esri (and the basic concept adopted by other organizations), geodatabases are essentially collections of shapefiles, attribute tables, and raster files that are stored, accessed, and transferred between computers. There are two forms of geodatabases: file geodatabases and enterprise geodatabases. The former are used in small organizations where data are stored locally on a computer. Enterprise geodatabases are often used in large organizations where data are served to multiple computers at the same time and where the local computers access centralized data, avoiding the need to store multiple copies of the geodatabase.

12.6 KML FORMAT

Another common format, which is in the format of plain text, is the **keyhole markup language** or **kml** format. The kml file format was created for use in Google Earth and is principally used to exchange spatial data between GIS and other spatial analysis platforms. It stores data in vector file formats (e.g., points, linestring, and polygons). kml uses a simple xml (i.e., extensible markup language) format for data and is generally good for visualizations of simple spatial features (e.g., locations of sensors, bounding boxes for study areas). A kml file will have a header followed by information for geometry. All geometry files use WGS 84 geographic coordinates only. An example kml file for the location of the Purdue University Memorial Union Building is:

```
<?xml version="1.0" encoding="UTF-8"?>
<kml xmlns="www.opengis.net">
<Document>
<name>Purdue University</name>
<description>The main campus of Purdue University in West
Lafayette, Indiana.</description>
<Placemark>
<name>Purdue Memorial Union</name>
<description>A central hub for students and visitors.</description>
<Point>
<coordinates>-86.9192,40.4265,0</coordinates>
</Point>
</Placemark>
</Document>
</kml>
```

Note that the coordinates have latitude, longitude, and altitude (as an offset from the natural meters above sea level).

12.7 JSON FORMAT

A JSON (java-script object notation) is a simple, plain-text format for data that nests objects with headers and values. The spatial form of JSON is called a GeoJSON. A very simple representation of a one data point JSON record is:

```
{
"type": "Feature",
"geometry": {
"type": "Point",
"coordinates": [-86.9081, 40.4237]
},
```

```
"properties": {
"name": "Purdue University",
"city": "West Lafayette",
"students": 50,884
}
}
```

A JSON or GeoJSON file can be imported into ArcGIS Pro directly and will be converted into a shape file automatically. A standard JSON file will have the extension *.json or *.geojason.

12.8 JOINS AND RELATES

There are times when the data that you want is only in the format of a table. It may have a field in the database that refers to a spatial unit. For example, a table of land ownership might have owner's name, land value, property tax values, square footage of the home, and parcel ID. A shape file with polygon features may have a field in it for the is parcel ID. Thus, the two databases can be linked. One does this using a **Join** or **Relate** item set of tools in ArcGIS Pro (or any other typical GIS). The Join and Relate commands in ArcGIS Pro create a temporary link between the two databases. Joins and Relates can also be one of several forms:

One-to-one (1:1): There is one record in each of the tables and the resulting join is a table with the same number of records but with the attribute table temporarily added. GIS analysts should use `Join`.

Many-to-one (M:1): A condition where there are many features that are within the same lookup row. An example are counties that are located within a biogeographic realm. GIS analysts should use `Join`.

One-to-many (1:M): A database has one row that can match multiple rows in the second database. One site (survey location of a person standing on a corner) has many surveys. GIS analysts should use `Relate`.

Exercise 13 will introduce students to the use of Join to link an Excel spreadsheet for butterfly species richness for counties to a county shape file database.

12.9 NETCDF

Finally, a specialized format for a spatial data model that is designed for spatial-temporal data is called netCDF (network common data form). **netCDF** files were developed by the University Consortium for Atmospheric Research (UCAR) to store, and allow for the processing of, very large, temporally rich (e.g., daily weather data globally) spatial data (Rew & Davis, 1990). It has evolved over the years, but the most recent version, netCDF-4, allows for the storage of spatial-temporal information as a space-time data cube (i.e., as a set of data arrays) that can be nested hierarchically. Data can also be stored as numbers, characters, or strings, and the data format can also be compressed, making file storage and transfer easier. Esri has a special implementation of the format they call a **data cube**. These data have a proprietary header attached to the file with files containing the extension *.nc.

12.10 LEGACY SPATIAL DATA MODEL FORMATS

There are a few legacy formats for spatial data formats that are important to know, especially if GIS archived data go back to the 1980s. Some of the most common are the ***.e00** formats from Esri. This interchange file format was used to combine vector geographic data (from ARC components) and attribute data (from INFO components) into one file that can then be moved between computers.

12.11 DATA PORTALS

A. INTRODUCTION

In the 1960s through to about the late 1980s, most of the data used by a spatial ecologist was developed "in shop"; by that, it is meant that nearly all of the data used by a GIS office was created by the staff of that office. Most recently, a majority of the data used by a spatial analyst comes from large data portals, and these data are often integrated with small quantities of data produced by local GIS offices. Some recent trends with data acquisition from drones is changing the balance of the quantity of external to internally derived data, but it is important for spatial ecologists to know that extremely important data exist at a variety of major "data portals." A **data portal** is a web-based platform where data, metadata, and a variety of APIs (also known as application program interfaces) exist along with a set of user interfaces (UIs) in an effort to distribute these data widely (Máchová & Lněnička, 2017; Thorsby, 2017; UN DESA, 2026). APIs are important components of information systems as they specify the rules for interfacing humans to data, typically for data manipulation, searching, and visualization. A similar concept, **data repositories**, are locations where data are archived, often with DOIs (digital object identifiers) attached to them, which are used to associate contact information and use restrictions. Many scientific journals support data repositories for data that are associated with a published article.

B. MAJOR DATA PORTALS

Spatial ecologists should be aware that there are dozens of very useful and unique data portals that can be used for research and natural resource management. Some of the major ones, organized around the type of supporting organization, include the following:

1. <u>National and International Government Centers.</u> Since the 1990s, **NASA** has supported the distribution of its remote sensing data through platforms called DAAC (**Distributed Active Archive Center**). These DAAC support its Earth Observing System Data and Information System (EOSDIS) long-term program. As of 2026, there are 12 NASA DAACs, each focused on a specific data domain area, generally partitioned into land, water, or air data resources. The NASA Land Processes (LP) DAAC, for example, is located at the USGS's Earth Resources and Observation (EROS) Center in Sioux Falls, South Dakota; this DAAC focuses on land use/cover mapping including process (e.g., wildfire) and is co-managed with the United States Geological Survey (USGS). Other DAACs include Oak Ridge National Laboratory (ORNL) DAAC for biochemistry and carbon data and the National Snow and Ice Data Center DAAC for data related to high-latitude monitoring of these ecosystems.

   In the United States, there is also a National Science Foundation–funded set of sites that collectively are known as the **National Ecological Observatory Network** or **NEON**. NEON is a continental-scale observatory for ecological data across nearly two dozen eco-climatic domains. At each site, hundreds of variables are collected including land cover attributes, soil, animal community data, and a host of geophysical data (e.g., air temperature, carbon fluxes, etc.). Data processing includes rigorous data quality checks and attaching metadata to the near-real time datasets.

In Europe, the **European Space Agency**'s (ESA) Earth Online and the Copernicus Data Space Ecosystem support ESA data distributions, especially Sentinel data. Other national government data portals include Canada's NRCans Earth Observation Data Management System and Japan's JAXA G-Portal site.

2. NGO Data Portals. There are several large, mostly global-coverage data portals that could be useful for spatial ecological research as well as natural resource management. These include the Global Biodiversity Information Facility (GBIF), which is technically hosted by an intergovernmental body; Red List species occurrence data distributed by the IUCN; WWF's GLOBIL data portal for conservation layers; BirdLife International's Important Bird/Biodiversity Areas (IBAs) and Key Biodiversity Areas (KBA); and NatureServ's Explore and Mapping services. A summary of these data portal sites is contained in Table 12.1.

TABLE 12.1. *Major biodiversity-focused organizations, their web sites, core data types, and key variables that are distributed*

ORGANIZATION	PORTAL NAME	URL	CORE DATA TYPES	TYPICAL SPATIAL ECOLOGY USES
GBIF—Global Biodiversity Information Facility	GBIF main portal	https://www.gbif.org	Global species occurrence records (museum collections, citizen science, surveys), mostly point data with taxonomic and temporal metadata	Species distribution modeling (SDM), richness and endemism mapping, invasion ecology, climate-change impacts on species ranges
IUCN	IUCN Red List Spatial Data	https://www.iucnredlist.org/resources/spatial-data-download	Range polygons for Red List species, richness and rarity-weighted richness rasters, plus GIS tools and mapping standards	Global extinction risk assessments, gap analyses, prioritization, Red List of ecosystems work, conservation planning
WWF	WWF GLOBIL—Global Observation and Biodiversity Information Portal	https://globil.panda.org	WWF-curated geospatial layers: ecoregions, priority places, pressures, protected areas, StoryMaps, dashboards	Global and regional conservation planning, pressure mapping, landscape prioritization, communication products
NatureServe	NatureServe Explorer & Open Data	Explorer: https://explorer.natureserve.org Open data hub: https://geohub-natureserve.opendata.arcgis.com	Range maps, ecological systems, at-risk species distributions (NA-focused), threat layers, and modeled "Map of Biodiversity Importance"	Fine-scale conservation prioritization in North America, habitat modeling, assessment of imperiled species and ecosystems
BirdLife International	BirdLife DataZone	https://datazone.birdlife.org	IBAs/KBAs, species factsheets, range data, country and flyway summaries, state-of-the-world-bird indicators	Site-based conservation (IBAs/KBAs), flyway analysis, avian diversity and threat mapping, global conservation status assessments

3. Commercial Data Portals. There are arguably two large prominent commercial data portals: Esri's Living Atlas and Google Earth Engine. Both are commercial platforms that operate in different ways. Esri's Living Atlas is a data services–focused platform that allows users of their products to do "live" links to existing data on their servers. GEE, on the other hand, has built extensive interfaces tied to programming scripting languages like Python, which allows university and not-for-profit researchers to use the site for free. Esri's Living Atlas also allows many users to download shape files and geodatabases for free. Some data are only found behind a paywall.
4. University-Supported Data Portals. Although compared to government and commercially supported data portals, university-supported data portals are smaller and also tend to focus on highly specialized datasets that store and distribute data that are the focus of research at academic centers and institutes. Two examples: ICPSR (Inter-University Consortium for Political and Social Research) is based at the University of Michigan's Institute for Social Research and provides a wealth of data on economics, demographics, and health data. IPUMS at the University of Minnesota distributes harmonized census and survey microdata. Non-United States-based university data portals include the Freshwater Information Platform (www.freshwaterplatform.eu), co-managed by several European universities; and the Terrestrial Ecosystem Research Network (TERN) (www.tern.org.au) of the University of Queensland, which supports the distribution of a wealth of ecological data for Australia, described as a national ecosystem observatory.

12.12 OPEN DATA SHARING FRAMEWORKS

There are several very important **open data sharing arrangements** that have been developed over the last few decades. Open data use falls across three broad areas: rules for its use, rules for attribution, and rules for how derivative products must be reshared. Theoretically built on the Creative Commons license and related content use frameworks, the **Open Data Commons (ODC)** license framework relates to the permissible use of data across all three use areas. The **Creative Commons** 0 (or as **CC0**) provides the most open use, typically using data for any purpose, which is to provide content (data and, viz., maps one makes) without asking, attributing, or without share-alike. **CC BY-Attribution** allows for use (by not-for-profit and commercial) and redistribution so long as the data are properly attributed. **CC BY-SA** (share-alike) refers to Creative Commons that states that one can use the data, modify it, but the derivative products that are created need to be distributed through the same Creative Commons open data agreement. **CC BY-NC** (NonCommercial) and **CC BY-ND** (NoDerivatives) blocks commercial use and forbids adaption and modification, respectively. **ODC** frameworks are, for the most part, aligned to CC usage rules, with subtle differences. ODC **Public Domain Dedication Library** (PDDL) is similar to CC0, ODC-BY is similar to CC-BY, and **ODbl** is analogous to **CC BY-SA**, but all ODC relates to data whereas CC relates to a variety of types of content.

12.13 SUMMARY

The technical basics of spatial data models and the elements of maps are key concepts that anyone in spatial ecology and/or GIScience needs to master. Understanding the structure of spatial data is also important when spatial analysts use nonspatial tools like R and Python to analyze data. In general, spatial data models come in two forms: vector and raster spatial data models with geometry and attribute data stored in ways that link location and values needed for analysis and/or display. Spatial data joins and relates are also common temporary linkages that are made by spatial analysts.

Map elements follow a few standard rules, including the need to include titles, map frames, legends, and scale bars for all maps. North arrows are common map elements. Some maps may also include map text for explanations, inset maps, logos, contact information, citation information, and a neat line.

Finally, there are several forms of data sharing that users of spatial data should become aware of if use is to be within the rules that govern use and distribution.

12.14 DISCUSSION QUESTIONS

1. Discuss what you think are the benefits of storing all data for a project in one geodatabase as opposed to having shape files as separate files.
2. Describe the importance of metadata to data management. Using an internet search, what information should go into a metadata file for spatial data?
3. Search for some data and then present and discuss the use rules associated with what you found online. Are the data sharing agreements clear and understandable?

13. GISCIENCE AND SPATIAL STATISTICS

Highlights

Aims. Spatial statistics is a specialized field that has some core foundational concepts, which are summarized here.

Learning objectives. Students should be able to:

- Describe spatial autocorrelation and how it relates to means and variances of nonspatial data
- Know the basics of calculating Moran's *I*
- Describe the basics of spatial interpolation
- Describe the main components of a semivariogram
- Describe the Getis-Ord G* statistic and how it could be use in spatial ecology

13.1 INTRODUCTION TO STATISTICS

Statistics is a discipline concerned with the collection, presentation, analysis, and interpretation of numerical data (Croxton & Cowden, 1955). It has traditionally been associated with the field of mathematics, considered by some of the early statisticians as a "branch of applied mathematics" (Box, 1976; Fisher, 1925). Box (1976) argued that statistics is fundamentally designed to help support the scientific discovery process. It is not surprising to learn that most spatial ecologists develop strong skill sets in statistics as this helps to support not only their own discovery process but also the analysis, interpretation, and presentation of data used for research and/or natural resource management.

There are three fundamental measures in statistics, and these are applied to just about every subfield of statistics (e.g., spatial statistics, forensic statistics). These fundamental measures include (1) those of **central tendency**, often called the mean or average, (2) measures of **dispersion**, typically called the variability, and (3) measures of the distribution of data, which often describe the shape of the **distribution** of all values.

Averages or **means** are calculated by the simple equation of adding up all of the values you have for, viz., variable *x*, and then dividing that by the number of values—called the sample size—that you have. The equation and notation for average is:

$$\bar{x} = \frac{1}{n}\sum_{i=1}^{n} x_i$$

where n is the sample size, x_i is the value for the ith sample, and $\bar{x}$ is the mean. If we had the following data, with the first through third samples listed as $i_1 = 2$, $i_2 = 4$, and $i_3 = 6$, the average would be 12/3 = 4.0.

Variance is the measure of the difference of all data values from the mean. The equation and notation for variance is this:

$$\sigma^2 = \frac{1}{n} \sum_{i=1}^{n} (x_i - \bar{x})^2$$

which is a more complex equation than that for the mean. Let us consider the steps in making the calculation for variance. The first step is to calculate the mean, $\bar{x}$. The second step, which is placed within the parentheses, is to calculate the difference that each value, x_i, is from the mean. This difference in statistics is called the **deviation.** The third step is to square this difference (i.e., calculate the **squared deviation**). The fourth step is to do this for all of the x_i values. The fifth step is to sum all of these squared differences, which is called the "**sums of squares.**" The sixth and final step is to divide by the sample size, n, which gives the variance. The short summary of the calculation for the variance is that *variance is the average of the squared differences from the mean.* To extend our example above for the three values for x, the variance would be calculated as 4 + 0 + 4 = 8 (to calculate the differences of each value from the mean and then sum these, which is steps 1–5 above). Since there are three values, the variable is 8/3 = 2.67.

Predictive statistics is the use of data to build a model that can be used to predict values that are not observed yet. One of the simplest predictive statistical models is estimating one value given another variable. In other words, one of the most fundamental statistical models is quantifying how two variables go up or go down together; this is called **covariance.** When two variables covary so that they go up together, then this is a positive covariance. In contrast, if one variable goes up and the other goes down, then that represents a negative covariance. The type of statistic that this represents is called **correlations.**

13.2 SPATIAL STATISTICS

Spatial statistics is a specialized area of statistics that focuses on creating descriptive and predictive statistical measures and tools for all types of spatial data (Cressie, 1993). Spatial statistics focuses on quantifying spatial clustering, determining spatial dependence patterns, and quantifying how two or more variables covary together in space. There is a subset of spatial statistics called geostatistics. **Geostatistics** focuses on building and using statistical tools such as **variograms** and applying a process called **kriging** to predict values and to characterize uncertainty in estimates across space. Geostatistical tools were developed by geologists, especially those in the fields of mining and oil and gas extraction (Matheron, 1963; see also Isaaks & Srivastava, 1989).

The field of spatial statistics has had an interesting history. It began by developing statistical predictive models that used all of the data to quantify spatial patterns in data. These forms of spatial statistical models are referred to as **global** models. Later, it became evident that these global models would hide **local** patterns (e.g., hotspots, important outliers) useful for spatial analysis. So, a large body of new statistical approaches called **local indicators of spatial associations** or **LISA** (Anselin, 1995) became a focus. To date, many algorithms first developed as global models have been reconfigured for use as local model versions. One of the major differences between global and local models is that global models provide single values, such as estimates of error, which are for the entire dataset. In other words, you often get one value. Local models tend to output values, such as estimates of error, by location. So, one often gets a map of output values, such as a map of estimated error.

There are many spatial statistical models that are often applied as global models and of course, there are many that have been developed and applied as local models. Here, we present some of the classic global and local models.

13.3 SPATIAL COVARIANCE—MORAN'S *I*

There are special cases of covariance that relate to spatial data; the most fundamental one is called Moran's *I*. **Moran's *I*** quantifies how different the values are for the same variable as one considers its neighbors. This measure is called **autocorrelation** as it is the spatial measure of correlation of a variable with itself, as it occurs across space. It specifically considers whether nearby locations have similar values (positive *I*) or dissimilar values (negative *I*). Positive *I* values occur when there is a clustering of high values in some areas and also of low values in other areas. Negative *I* values means there is a checkerboard pattern of high and low values. If there is no spatial pattern of values (i.e., data values seem to occur at random across space), then *I* is zero. This statistic was developed by Moran (Moran, 1950), who referred to the statistic as the measure *I*. Other geographers (e.g., Ord) began to call this Moran's *I* in later years. This measure is one of the foundational tools in spatial statistics and GIScience (Cliff & Ord, 1973, 1981).

The equation for Moran's *I* follows from the equations above for mean and variance, but placed in the context of space:

$$\mathrm{I} = \frac{n}{W} * \frac{\sum_{i=1}^{n} \sum_{j=1}^{n} w_{ij}(x_i - \bar{x})(x_j - \bar{x})}{\sum_{i=1}^{n} (x_i - \bar{x})^2}$$

which has several components, explained in Figure 13.1. Note that the mean of all values for variable x (Figure 13.1A) are used in three locations of this equation with labels A, B, and E in Figure 13.1. On the nominator part of the equation (i.e., top part of the fraction), the difference for x at position j and that of the average is calculated (Figure 13.1B) and the difference for x at position i and that of the average is also calculated (Figure 13.1C). A spatial weight is also introduced (Figure 13.1D), which can take on several forms. For binary adjacencies, W_{ij} is 1.0 for neighbors next to one another and 0.0 for i,j pairs that are not adjacent. The W_{ij} weight can also be calculated as an absolute distance between the pairs, done so as a function of the inverse of the distance. A third approach is to place distances into distance bands, called lags, which are assigned 1 if they are in the same band and 0 if not. The sigmas (i.e., Σ) represent the mathematical symbol for "the sum of" and so these symbols state that there should be a summing of the "cross product" (e.g., the values in each of the parentheses in the numerator are multiplied) of the i and j location differences from the mean. The W represents all of the possible pairings of i and j, and n is the number of values (i.e., sample size) of variable x.

There is also a local version of Moran's *I*, called LISA Moran's *I*. It estimates *I* for each location, i, using a slightly different form of the original equation for *I*:

$$I_i = ((x_i - \bar{x})/S^2) \sum_{j=1}^{n} w_{ij}(x_j - \bar{x})$$

where S^2 is calculated as

$$S^2 = (1/(N-1)) \sum_{k=1}^{n} (x_k - \bar{x})^2$$

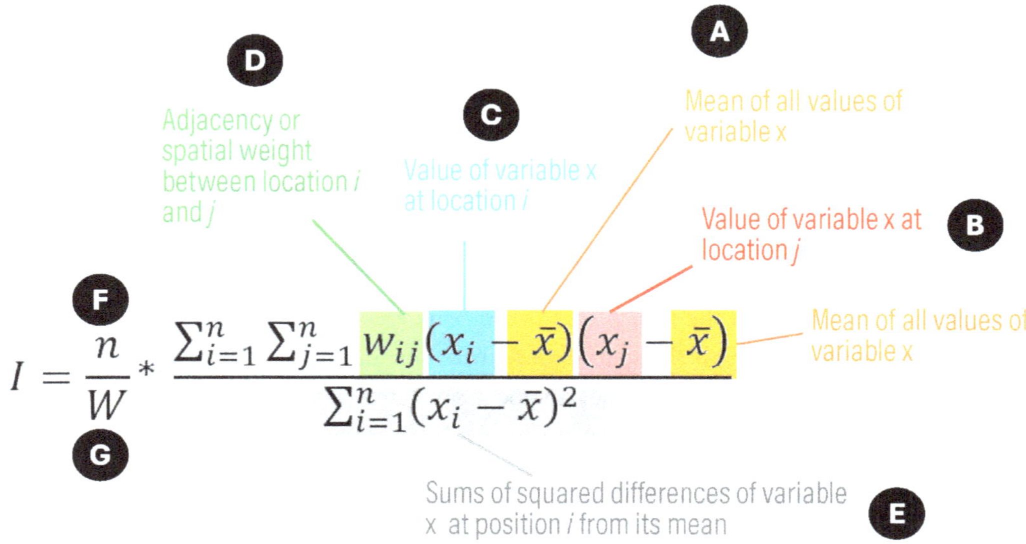

$$I = \frac{n}{W} * \frac{\sum_{i=1}^{n}\sum_{j=1}^{n} w_{ij}(x_i - \bar{x})(x_j - \bar{x})}{\sum_{i=1}^{n}(x_i - \bar{x})^2}$$

FIGURE 13.1. Major components of Moran's *I*, illustrating the important parts A–G that are described in the text.

where *k* is an indexing value, like *I* and *j* for the global Moran's *I*, used to represent all values (i.e., *I* and *j* are used for local indexing).

13.4 SIMPLE CONCEPTUALIZATION OF SPATIAL INTERPOLATION

Point data are often mapped continuously across a region using GIS-based tools called spatial interpolation. Interpolation is the process of estimating unknown values from known values that are distributed in space or time. Figure 13.2 illustrates the concept of **interpolation**. Let us assume that we have point counts for single species abundances as shown in Figure 13.2A. Note that the values in these counts range from 300 to 900 individuals per "trap." It is often necessary to determine what the population sizes would be at other locations across the area, such as those labeled "a" through "e" (Figure 13.2B). The estimated value of "a" is an easy mental estimate, as it is clearly a value between 300 and 500 in the map; it is easily determined to be 400 (note there are several easy ways using simple arithmetic to determine the halfway value, which is to subtract 300 from 500 to get 200 and then divide by 2 to get 100, which is then added to 300, OR add 300 and 500 and then divide by 2). This same approach can be used in up-down or right-left directions or for diagonals to also estimate values in between known values (Figure 13.2C).

13.5 INTERPOLATION MODELS—IDW AND SEMIVARIANCE

There are a variety of methods of interpolation to estimate values such as those indicated as locations "a" through "e" in Figure 13.2A. Let us consider two forms of spatial interpolation models (Myers, 1994):

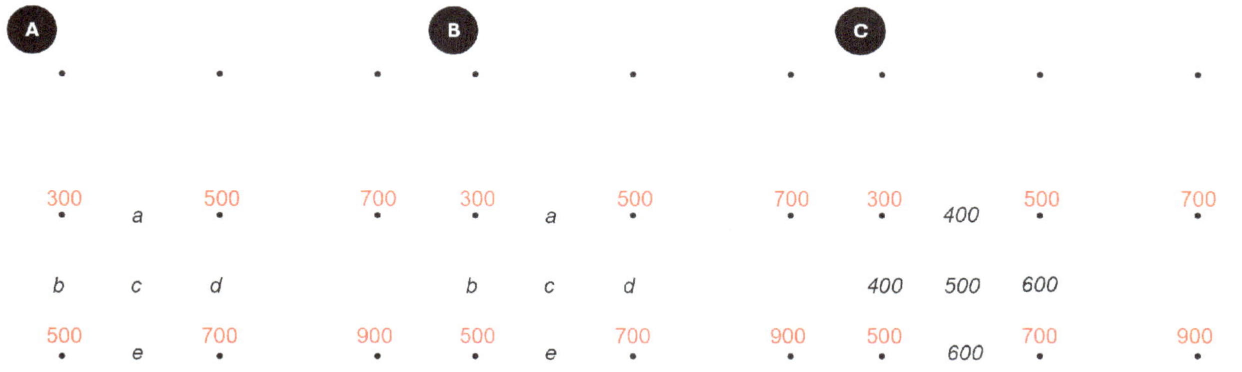

FIGURE 13.2. Illustration of the process of interpolation using known values and locations (A) to estimate unknown values (B), which are calculated using nearest neighbor known values (C). See text for further explanation.

A. DETERMINISTIC OR NONSTATISTICAL APPROACHES

This interpolation set of methods relies on mathematical models that quantify how values are estimated across space using assumptions such as these: (1) Values that are closer are more likely to be similar than more distant values (known as **Tobler's law**); (2) trends across space fit general, mathematical models. The most common of these is **inverse distance weighted** (**IDW**) models that estimate spatial values based on numerous locational data and distances between points. The equation for IDW is:

$$\hat{Z}(x_0) = \frac{\Sigma_{i=1}^{n} W_i Z_i}{\Sigma_{i=1}^{n} W_i}, \quad W_i = \frac{1}{d(x_0, x_i)^p}$$

where $\hat{Z}(x_0)$ is the predicted value at location x_0, z_i is the observed value at sample number i, $d(x_0,x_i)$ is the distance of x_0 to sample number i, w_i is the weight, n is the number of neighbors used, and p is the power parameter. For the purpose of learning about this method, the most important parts of this equation are p and n. p is often set to 2 which means that there is a distance "decay" effect that is related to the square (i.e., 2) of distance; and n is often set to 12, which is the number of nearest neighbors used to estimate $\hat{Z}(x_0)$. In courses of spatial statistics, other forms of deterministic spatial interpolation may be covered such as the use of splines and polynomial regression models (see Myers, 1994 for an excellent summary of these).

B. GEOSTATISTICAL APPROACHES

A more common approach has been the use of spatial statistical models to estimate values across space given the location of known values. This approach is called either a spatial statistical or geostatistical approach. This approach, as well as the first one listed above, uses the concept of spatial autocorrelation, which is that values close by are more similar than those far away. Some fundamental concepts are needed to understand this approach to interpolation. The first is that a group of locations and values for these locations are placed into distance classes called **lags**. The second important aspect of this approach is that a special form of variance is

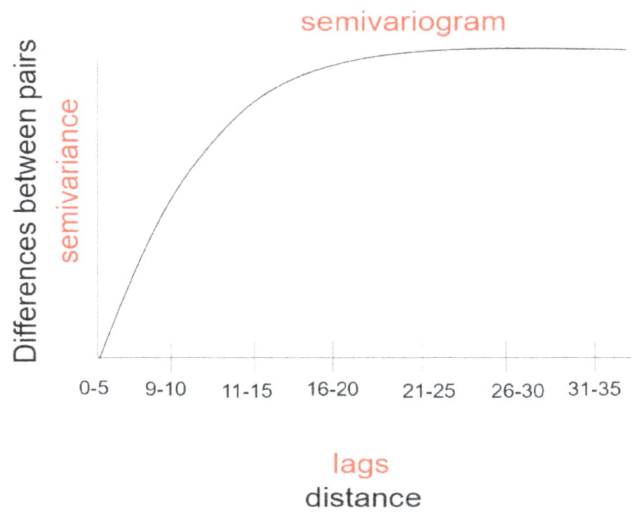

FIGURE 13.3. A plot of lags verses semivariance and the resulting best fit curve, called a semivariogram.

used to create a distance (as lags) versus variance. Figure 13.3 shows a general plot of lags versus a form of variance called **semivariance.** The curve that is fitted with all of the data from the calculation of the semivariance is called a **semivariogram.**

The equation for semivariance is as follows:

$$\hat{\gamma}(h) = \frac{1}{2N(h)} \sum_{i=1}^{N(h)} [Z(x_i) - Z(x_i + h)]^2$$

where

$\hat{\gamma}(h)$ = Empirical semivariance at lag distance h
$N(h)$ = Number of data pairs separated by distance h
$Z(x_i)$ = Measured value at location x_i
$Z(x_i + h)$ = Measured value at a paired location separated by lag h

What is also important, and not well known, is that the equation for semivariance can have two forms: (1) experimental, otherwise known as empirical, and (2) theoretical or fitted. The one above is the empirical form of the equation, which is used in software like ArcGIS Pro Spatial Analyst. It is used to calculate the semivariance based on real data.

A typical semivariogram model produces several useful values from the best fitted curve (Figure 13.4). The first is the **sill,** which is the maximum value of semivariance achieved. The value along the x-axis (i.e., the lag) is the distance at which the maximum semivariance is obtained, called the **range.** The other useful value is the **nugget,** which is the value along the y-axis that falls when the x-axis is exactly 0.0. The nugget is interpreted as the measurement error. If nugget values are large, then it could mean that the spacing of spongy moths traps is not close enough to estimate the means and variances across a range of spatial extents.

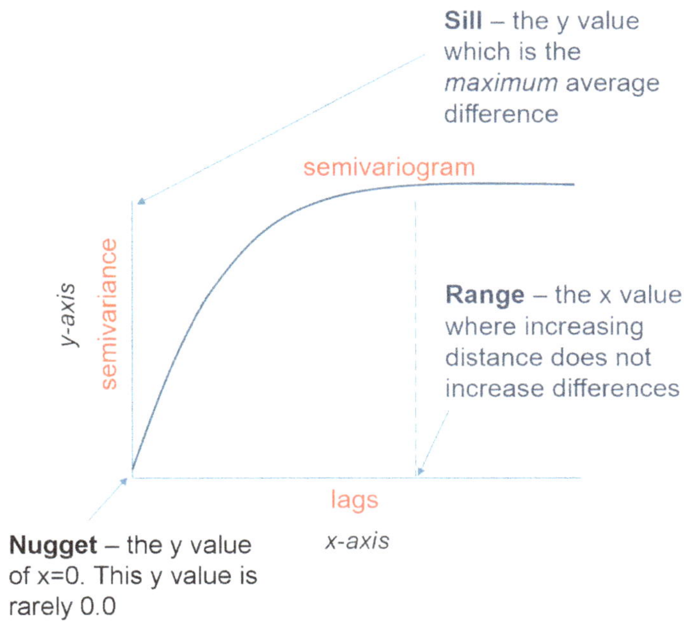

FIGURE 13.4. The main graphical parameters that are generated from a semivariogram, including sill, range, and nugget.

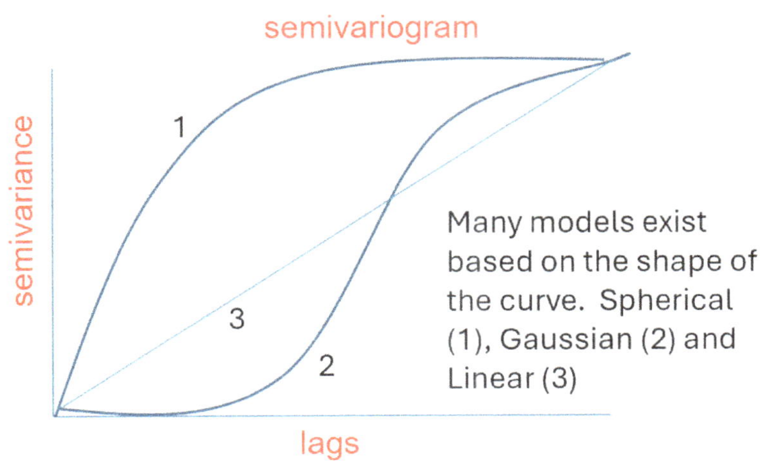

FIGURE 13.5. Comparisons of semivariogram curve shapes that are typical for environmental data arranged in space.

The shape of the theoretical semivariogram curve is known to exist in several forms. The curve shown in Figure 13.4 is one of the most common; it is referred to as **spherical.** Other forms (Figure 13.5) include: **exponential** (smoother approach to the sill than spherical), **Gaussian** (S-shaped), **circular** (similar to exponential), and **linear** (which does not have a sill or range).

Interpolation using geostatistical methods is called **kriging.** Kriging has many forms (Table 13.1), and these are reviewed briefly here. The first set of forms relates to how statistical models estimate means. **Ordinary**

TABLE 13.1. *Main dimensions of kriging and the forms that are used*

DIMENSION	FORMS	DESCRIPTION
Mean	Ordinary vs. Universal	Two approaches to calculating averages
Neighborhood	Global vs. Local	Whether all data values or those nearby are used
Geometry	Isotropic vs. Anisotropic	Quantifies directional effects that should be factored into the model

kriging (**OK**) estimates a general mean across all data points. This is contrasted with **universal kriging** (**UK**), which uses a spatially variable approach to calculating a mean using coordinates or a covariate (e.g., slope or elevation).

Kriging can also incorporate directional effects. When direction (north, south, east, and west) is not used, this form of kriging is referred to as **isotropic kriging** (nondirectional). Incorporating a directional effect is called **anisotropic kriging**. Anisotropic kriging might apply to data that are distributed over large regions where temperatures in the north would be lower than those to the southern extent, making a north-south temperature gradient an important factor that could influence the statistical model.

There is also a special version of kriging called **co-kriging**. Co-kriging is the use of another variable that is known to covary with the values that are being estimated. Co-kriging most often requires the covariate, for example, soil type or elevation, to be a map that is continuous across the area being interpolated. In other words, these models require there to be covariate values between locations being estimated with the target variable.

13.6 GETIS-ORD STATISTIC

The Getis-Ord statistic, often denoted as Gi*, was developed to quantify the spatial association of high and low values within a specified distance (Getis & Ord, 1992; Ord & Getis, 1995). This statistic essentially characterizes the concentration, or clustering, of high values and/or low values. It has become a very useful set of tools for calculating biodiversity **hotspots** (e.g., Sundaram et al., 2019; Vincent et al., 2022) and **coldspots** (see Kareiva & Marvier, 2003, for coldspot biodiversity concept and Lan et al., 2024, for its implementation to biodiversity coldspot locations). It differs from Moran's *I* in that it focuses on clusters of high and low **values** and *I* focuses on quantifying spatial autocorrelation of a variable. There are several ways to implement the Getis-Ord statistic in Esri's ArcGIS Pro. This is explored more in Exercise 15.

13.7 SUMMARY

Spatial statistics is a very technical area of spatial ecology and GIScience that has grown considerably over the last three decades. Built from basic statistics and configured for a spatial context, these tools have now been advanced in many ways and also are included in a set of tools in ArcGIS Pro. Some of the most relevant spatial statistics include Moran's *I* and Getis-Ord, which are often applied to spatial ecology analyses.

13.8 DISCUSSION QUESTIONS

1. Describe at least two situations where Moran's *I* should be used in conservation science applications.
2. Getis-Ord can provide both hotspot and coldspot assessments. Describe how you might consider doing hotspot and then coldspot analysis.

14. GISCIENCE AND CARTOGRAPHIC PRINCIPLES (WITH MAP ELEMENTS)

Highlights

Aims. The fundamental principles of cartography are summarized here. This includes a presentation of the brief history of cartography. The main elements of color theory are also presented along with a summary of the different digital formats for graphics files and some of the general rules of thumb to follow in producing digital and paper versions of maps.

Learning objectives. Students should be able to:

- Know what cartography is and its importance to GIScience
- Explain some of the core elements of color theory
- Know some of the visual color deficiencies and how one produces maps that are accessible
- Follow the general rules of thumb for producing digital and paper maps

14.1 WHAT IS CARTOGRAPHY?

The term **cartography** is translated from its Latin roots "carto-," which is "document, paper, or map," and "-graphy" for "drawing, writing or describing." Thus, the term is meant to translate as "a map that is drawn on paper." More modern definitions of the term emphasize that cartography is "the art, science and technique of making maps" (Raisz, 1948). The International Cartographic Association (ICA) defined cartography as "the art, science, and technology of making maps, together with their study as scientific documents and works of art" (ICA, 1966). Finally, Board (1967) emphasized that cartography is "primarily concerned with the communication of spatial information," emphasizing that maps are "visual models." What is important from all of these definitions is that cartography draws from the fields of science *and* art, and that maps are a form of communication that requires an understanding of the psychological interpretation of colors, shapes, and textual information that is placed on a map. Cartography also draws strongly upon an area of visualization called **color theory** and, in most recent times, on how colors in maps are perceived by people who have color vision deficiencies or CVD. As we have significantly moved into the digital information era, understanding **digital**

representation (e.g., bit depth, resolution) of how to store and represent maps is an increasingly important component of cartography. Also, cartography has its own terminology, especially as it relates to **map elements**. Terms like "neatlines" and "graticulates" are used to describe specific features of maps and how these are introduced into the map also requires great care and attention to some of the cartographic rules that have been developed over the last century by cartographers. Finally, a large movement in recent years in **participatory mapping** and issues surrounding **bias and power in map production** (sometimes called **critical geography**) emphasizes the need to devote attention to how users of maps are incorporated into their production and use. As you can see, cartography is a broad and exciting area of GIScience. Indeed, there are entire courses in traditional departments of geography that spend considerable time presenting the fundamentals of cartography, describing details of color theory, digital representation of maps, psychology of visual perceptions, map elements, and design "best practices," as well as how to properly produce maps that can be accessible to the broadest number of individuals.

This chapter will review the fundamentals of modern cartography by summarizing (1) color theory; (2) digital dimensions of producing and storing maps; (3) map elements; (4) participatory mapping approaches; and (5) issues related to accessibility of maps to a broad community.

14.2 COLOR THEORY

Many of the fundamentals of color theory were developed by the artist Joseph Albers (1963) in his book *Interaction of Color*. In this high-impact book, he outlined three core areas of what are now the fundamental principles of color theory: (1) **color wheel**—the structural organization of hues and complementary relationships; (2) **color harmony**—how colors combine aesthetically and perceptually; and (3) **color context**—how the appearance of a color depends on its surroundings as well as how color is matched to its meaning (e.g., blue is the color of water). Also presented here are complementary topics of **color symbolism** and **Brewer's color palettes**.

A. COLOR WHEEL

Hue is the physical property of a color. It is based on the dominant wavelength of light that is sensed by the eye and then interpreted as a color, such as "red," "yellow," or "blue." It is independent of the intensity or of its boldness or lightness. The variation of hues is often presented as a wheel or a 360° circle, hence the term color wheel (Itten, 1961). For example, 0° might be displayed with the color red, 120° would show yellow, and 240° might show blue, with colors between those inserted at 60° (orange), 150° (green), and 300° (violet). In this 6-component color wheel, the colors red, yellow, and blue are considered primary colors whereas orange, green, and purple are secondary colors. Hue is one of three major components of color appearance, and it is the name associated with the color. The other two components are **saturation**, which is how pure or gray a color is, and **value**, which is how dark or light the color is perceived to be. Figure 14.1A shows a standard 6-color wheel displaying the primary and secondary colors, and the three foundational components of colors of hue, saturation, and value (illustrated here as brightness) are illustrated in Figure 14.1B.

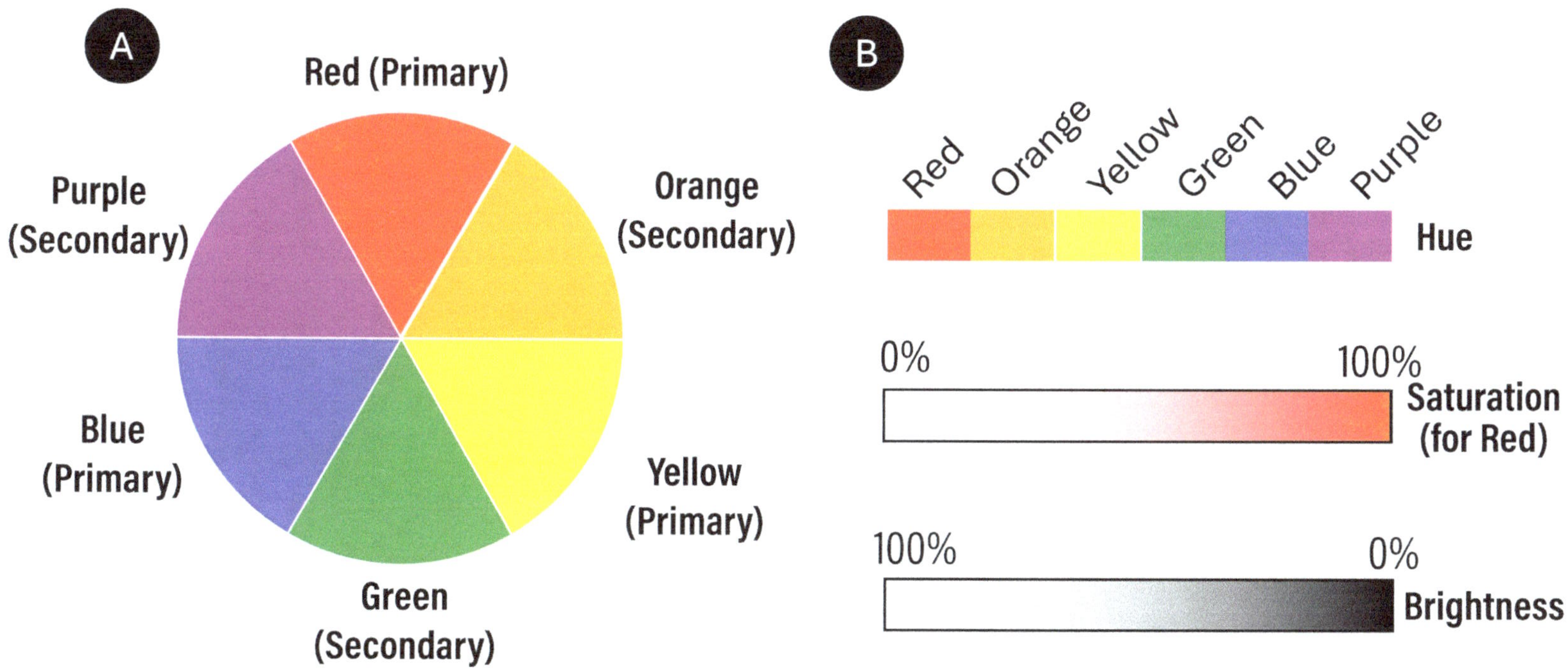

FIGURE 14.1. A 6-component color wheel showing primary and secondary colors (A) and the three fundamental properties of color (B): hue, saturation (for red), and value (scaled by brightness values).

B. COLOR HARMONY

Color harmony (Itten, 1961) is the science and art of arranging colors based on combinations of HSVs; specific arrangements of hue, saturation, and brightness can create aesthetically satisfying (the "art" part), structured (the science part) color patterns that can be described as complementary, analogous, or triadic in their organizational schemes. Itten (1961) argued that color harmony ultimately leads to a balanced, holistic, and/or "harmonious" presentation of color content. Cartographers have adopted this in the process of mapmaking with the goal of creating maps with harmonious colors. Although it was first presented as a more structured, objective approach to arranging colors by Itten (1961), others, such as Albers (1963), suggested that color harmony may also need to be a bit subjective as individuals do not precisely perceive colors as they physically exist (e.g., not everyone will perceive a pure red color wavelength as "pure red").

Itten (1961) outlined what is now referred to as the big five color harmonies. More recently, Ware (2013) describes how these color harmony schemes result in certain visual outcomes, with some excellent advice on the use and misuse of these schemes. Using a 12-component color wheel, Itten (1961, 1970) used the following terms to describe how colors can "work together":

1. *Complementary Harmony.* This is pairing two colors that are directly opposite each other (at 180°) on a 12-component color wheel. **Complementary harmony schemes** would be the pairing of blue with orange, yellow with violet, and so on. These give the visual effect of maximizing contrast, creating a sense of vibration. These color combinations, also called **direct complements**, ultimately create a binary contrast look. However, relying a lot on these color schemes poses the risk that viewers could get eye fatigue, given the strong contrast that these color pairings offer.
2. *Split-Complementary Harmony.* These **split-complementary harmony color schemes** are built by starting with one color and then using the colors on either side of its direct complement. For example, considering the base color blue, the direct complement would be orange. The colors on either side of orange, which are yellow-orange and red-orange, provide the strong contrast, but it is less harsh and thus less of a "binary" opposite. Split-complementary schemes also increase color choices that work together.
3. *Triadic Harmony.* **Triadic harmony** color schemes are ones where colors are selected based on their position around the color wheel such that three that are used are located at 120° from one another. Some of the classic triads are red-yellow-blue and orange-green violet. The aesthetic effect is to have three colors that are "balanced" but vibrant (Itten, 1961). It is often recommended to not use 100% brightness as the effect might be to produce products that appear "cartoonish."
4. *Tetradic or Double-Complementary Harmony.* **Tetradic harmony** color schemes are created by selecting two sets of "opposing" colors on the wheel, effectively creating a rectangular arrangement of colors situated around the color wheel. An example of a tetradic set of pairings would be red-green and blue-orange. It is recommended (Ware, 2013) that one use one of the pairings for most of the map and then use the second pairing for accents. It is also recommended (Ware, 2013) that saturation and value should not be at 100% as this would create a very visually busy product.
5. *Analogous Harmony.* **Analogous harmony** color schemes pair colors that are located next to each other on the color wheel, typically from a 12-component color wheel. Example color pairings in this scheme would be yellow, yellow/green, and green. Color pairings could be three or four, each one separated by 30°. The visual effect is to create a smooth set of colors that work in a unified way created from a structured color gradient.

C. COLOR CONTEXT

The third component to the traditional color theory concepts of Albers (1963) is color context. **Color context** refers to how a color is perceived based on the features of the other colors surrounding it. Background colors, adjacent colors, size and shape of the object, and its color all affect how it is perceived. The crux of this idea is that any one color can look different depending on the "context" in which it is seen. Albers (1963) famously called this the "interaction of color."

D. COLOR SYMBOLISM

Although part of the broad body of what cartographers and artists call color theory, color symbolism is another important part of how cartographers select certain colors to represent a feature on a map. This notion of **color symbolism** leverages what we know about cultural, psychological, and semantic meanings of colors. For example, Itten (1961) described yellow as "bright, expansive, and cheerful," blue as "spiritual and deep," and red as "viral and strong." Table 14.1 lists major hues, their symbolic associations, typical cartographic uses in, viz., land use/cover maps, and a citation that explores these color associations.

14.3 VISION COLOR DEFICIENCIES

There are several well-known **vision color deficiencies** (VCDs). These deficiencies are related to whether there is a problem (absent or altered) with one or more of the cones of the eye. The human eye has three cones: S-cones (peak in the blue range), M-cones (peak in the green range), and L-cones (peak in the red range). There are four known vision color deficiencies, summarized in Table 14.2. ArcGIS Pro contains tools in the layout menus that simulate the appearance of each of the four VCDs listed in Table 14.2.

14.4 BREWER COLOR PALETTES

Using the color theory components described above, combined with empirically evaluated color schemes that examine VCD, Brewer et al. (2003; see also Brewer, 1994, and Harrower and Brewer, 2003) produced a set of color scheme **palettes** that they argue are especially "tuned" for map production. These color scheme palettes fall into three broad categories (with the number per palette type):

<u>Sequential palettes (16)</u>. These are for data that are ordered scaling from low to high. They use single hue or multi-hued progressions so that light is low and dark is high.

<u>Divergent palettes (9)</u>. These are for data that have a midpoint that is meaningful with a neutral color for the center point.

<u>Qualitative palettes (7)</u>. For nominal data (described as different kinds, like forest, urban, wetlands), these palettes are designed to add enough contrast between features that are next to one another so that they can be well distinguished from one another.

TABLE 14.1. *Colors and their psychological associations and uses in maps. See citations for further information for these associations and applications in maps.*

COLOR	COMMON PSYCHOLOGICAL/ SYMBOLIC ASSOCIATIONS	TYPICAL CARTOGRAPHIC USES	SUPPORTING SOURCES
Red	Arousal, energy, danger, warning, passion; strong approach–avoidance signal; performance anxiety in achievement contexts	Urban/built-up areas, hazards, high-risk zones, fires, "hotspots," negative change	Itten, 1970; Fehrman & Fehrman, 2004; Elliot & Maier, 2014
Blue	Calm, trust, stability, coolness, competence; often rated as "favorite" color; associated with relaxation and reliability	Water bodies, oceans, lakes, rivers; cool temperature classes; "safe" or baseline categories	Fehrman & Fehrman, 2004; Elliot & Maier, 2014; Arnheim, 1954
Green	Nature, growth, renewal, safety, "go," environmental quality; restorative, low-arousal positive affect	Forests and vegetation; protected areas; "ecological" or sustainable choices; positive environmental indicators	Itten, 1970; Fehrman & Fehrman, 2004; Elliot & Maier, 2014; Brewer, 2005
Yellow	Warmth, sunlight, optimism, attention, caution; high visibility; can tip toward anxiety/irritation when very saturated	Dry fields/cropland, savannas, marginal lands; warning overlays; highlighting or callout symbols	Itten, 1970; Fehrman & Fehrman, 2004; Arnheim, 1954
Orange	Energy, warmth, enthusiasm, affordability, informality; intermediate between red's urgency and yellow's cheerfulness	Transitional land uses (suburban, exurban), moderate hazard zones, "medium" risk categories, secondary emphasis symbols	Fehrman & Fehrman, 2004; Ware, 2013
Purple/ Violet	Rarity, luxury, mystery, spirituality, "special" or exceptional status; not strongly tied to everyday natural phenomena	"Other" or special-use categories; rare classes in thematic maps; highlight of exceptional or anomalous conditions	Itten, 1970; Fehrman & Fehrman, 2004
Brown	Earth, soil, stability, ruggedness, age; low glamour, utilitarian, "grounded"; often perceived as dull but dependable	Bare ground, rock, desert, soil types, geology; traditional/heritage features; background landforms	Fehrman & Fehrman, 2004; Brewer, 2005
Black	Authority, power, formality, seriousness; in Western contexts, death and mourning; high visual weight	Outlines, text and linework; high-contrast symbols; no-data masks; occasionally for severe hazards or exclusion zones	Arnheim, 1954; Fehrman & Fehrman, 2004
White	Purity, cleanliness, emptiness, simplicity; in many Western contexts, innocence; in some cultures, mourning	Background for reference maps; snow/ice; unclassified or missing data (with care); negative space around map frames	Arnheim, 1954; Itten, 1970
Gray	Neutrality, balance, subtlety, uncertainty, ambiguity; low emotional arousal; can read as "background" or "inactive"	Neutral basemaps; secondary or de-emphasized features; infrastructure backdrops; uncertainty zones when paired with hatching	Arnheim, 1954; Ware, 2013; Brewer, 2005

TABLE 14.2. *Four VCDs, the eye cones that are affected, how the VCD affects ability to distinguish colors, and the percent of the population affected summarized here. From Simunovic (2010).*

TYPE	AFFECTED CONES	COLOR CONFUSIONS	PREVALENCE
Deuteranopia/ Deuteranomaly	Green cone	Red ↔ Green confusion	~6% of males
Protanopia/Protanomaly	Red cone	Red ↔ Brown/Green confusion	~1% of males
Tritanopia/Tritanomaly	Blue cone	Blue ↔ Green or Yellow confusion	<1% of males/females
Achromatopsia (rare)	All cones	No color vision (only light/dark)	<0.01%

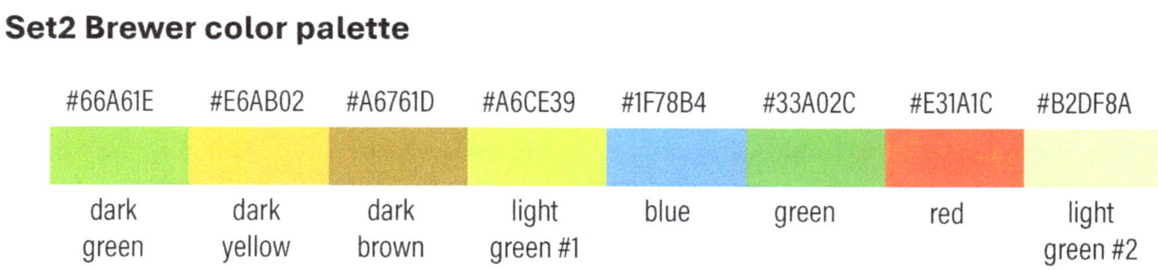

FIGURE 14.2. The set2 Brewer color palette, which is often used for land use/cover maps.

All Brewer palettes have names. For example, the set2 palette, a qualitative palette, contains eight colors (Figure 14.2). Maps that contain land use/cover classes typically use colors from the set2 palette.

14.5 DIGITAL REPRESENTATION

There are several important aspects of digital representation of images that might be viewed as "general rules of thumb" in mapmaking using computer technologies. These include (1) the **digital color model** for coding colors for computer screens and printers, (2) standards for resolution (i.e., **pixel size** and **bit depth**) for publishing images, and (3) characteristics of **file formats** that are used to store images.

A. DIGITAL COLOR MODEL

For computer screens, printers, and to store images as digital files, one of three color models is applied. These include, for computers screens, an **RGB** (red-green-blue) color model, and for printers, a **CMYK** (cyan, magenta, yellow, and black) color model. The RGB color model is referred to as an **additive color model**; here, you add light. The intensity of light for each of the three colors can range from 0 (no light) to 255 (maximum light). Each color is considered as a channel, so that an RGB coding with [0,0,0] is black, and an RGB of coding of [255,255, 255], is white. The color pure red would be coded [255,0,0] and the color for yellow would be coded [255,255,0].

The CMYK model, which is used for color printers, is a **subtractive color model**. This model refers to the physical mixing of colors (i.e., ink) to subtract colors from white (i.e., the color of paper). Thus, the color white has no ink from any of the four channels. The model also considers the fact that although black would be, in this color model, composed of all colors—C, M, and Y with values of 255—when they do mix, it produces a muddy dark color. So, black (the fourth or the K channel) is added (note that laser printers typically have four cartridges, each for the colors cyan, magenta, yellow, and black).

It should be noted that an alternative coding system for the RGB color model exists. This is a **hexadecimal** coding system, sometimes just called a hex coding system. The hexadecimal system is based on a 16-bit coding depth, which means that one can use a six alphanumeric coding system to represent values of 0 to 255 with two-character spaces for each of the three channels. For example, the hexadecimal code for pure red is #FF0000, where FF represents 255 and full light for the red channel; the second set of characters, 00, means there is no light for the blue channel; and the third set of characters, also of 00, codes for no light for the green channel. This RGB hexadecimal coding system allows for 256^3 or over 16.7 million colors, which is more than adequate variation in colors for a computer screen.

B. PIXEL RESOLUTION AND BIT DEPTH

Computer screens, including screens on mobile devices, are composed of pixels, and a printer will also create images composed of pixels, the density of which is referred to as its resolution. Often reported as **pixels per inch** or **PPI** (another term used a lot is **dots per inch** or **dpi**), a standard computer monitor has PPIs ranging from 110 to 140 PPI. High-end monitors can reach 220 PPI. Computer monitors and TVs often report the total number of pixels in terms of the number of lines vertically and horizontally. A 4K monitor, for example, has 3840 pixels horizontally and 2160 pixels vertically. Thus, the total number of pixels on a 4K monitor is over 8.2 million.

In terms of printing, printers often range in resolution from 300 dpi to around 1200 dpi. Some of the very best printers will print 2400 dpi, which is often used for high-end photo printing. It should be mentioned that glossy paper is generally used for photos and high-end graphics work and printed at 600 dpi or 1200 dpi. Handheld maps that are viewed close should be printed on high-end paper with at least 600 dpi. Scientific posters that do not have photos embedded can be printed on large-format paper at 300 dpi.

Bit depth is also an important dimension of digital information coding color. **Bit depth** is the number of bits *allocated* to store the value of a sample (e.g., a pixel or a color channel), which determines how many distinct levels you can represent. A bit can be either a 0 or 1. A bit depth of 1, termed 1-bit depth, can store two levels, 0 (black) or 1 (white). An 8-bit depth channel can store 256 different values (coded using a range of 0 to 255). Thus, with three (RGB) 8-bit depth channels, a 24-bit color palette can be created.

C. FILE FORMATS FOR IMAGES

There are many file formats for images, but, for the most part, they come in two "flavors": **bitmaps**, which are a lot like raster files, and **vectors**, which are like vector data in GIS. Bitmap formats include JPG (or sometime as JPEG for joint photographic expert group), PNG (portable network graphics), and TIFF (tagged image file format) file formats. Vector formats include SVG (support vector graphics) and EMF (enhance meta file), which store graphics as drawing features such as points, lines, and text. Vector graphics can easily scale and so tend to be a preferred format to create as output from a GIS for input to software like Microsoft PowerPoint.

Note that color JPEGs are typically stored as three-channel images that display as RGB, while color PNGs are usually either three-channel RGB or four-channel RGBA (RGB plus transparency). The JPG file format is often not preferred over PNG as JPGs store information in a lossy file format and as such there is a loss in clarity in displaying features such as thin lines and small text. The file format PNG is a compressed file format, but it is not lossy (there is no loss of information as it is compressed). Both JPG and PNG are standard formats for use on websites.

14.6 MAP ELEMENTS

We introduce here some of the more technical issues related to making proper maps using a geographic information system. These include the following (see Figure 14.3 for an example map with main elements):

Title: Most often located at the top, either centered or left justified (Figure 14.3A), this can include a title and sometimes **subtitles** (Figure 14.3B), the latter as smaller font size.

Map Frame: A map frame (Figure 14.3C) is where the spatial data are displayed, usually making up around 60–80% of the space of the entire visualization. Within the map body is the display of the vector and/or raster data and labels (e.g., names of cities, roads, or features like rivers).

Legend: As there are many spatial entities located on a map, the visual representation of these entities needs to be described, and this is done as a **legend** (Figure 14.3D) that has its own legend title and then separate labels for each entity.

Map Description: At a minimum, a description of the map (Figure 14.3E) should be included.

North Arrow: There are many styles of **north arrows** used by mapmakers. A north arrow (Figure 14.3F) helps to provide directionality for the map. Some experts in spatial data sciences argue that if a map has up as north, a north arrow is not needed. In some cases, it is convenient to create maps where the direction up is not precisely north, so including a north arrow is appropriate.

Scale Bar: A critical map element that illustrates the ratio of the true distance of the real-world entities to that of the physical map (Figure 14.3G), the **scale bar** often displays the units of distance in either meters (m), kilometers (km) for the metric system, or for the imperial system, feet (ft) or miles (mi). The metric system is highly recommended for scientific uses of the map (e.g., journal articles) and the imperial system for use in the United States for local government uses where the public is more familiar with this unit of measure.

Optional map features include:

Labels: Text that is associated with feature attributes can be added to any map. **Labels** (which are not included here) in Figure 14.3 could include the names of state or countries, or the names of the ecoregions. Labels may also have **label leader lines**, especially for point features, which visually connect the feature with the label.

Grids or Graticulates: Especially helpful in maps with large extents, grid lines (Figure 14.3H and Figure 14.3I) and associated elements represent the location of latitude (parallels) and longitude (meridians) and are referred to as grids or graticulates. Labels for grids or graticulates are often located on the edges of the map body along with ticks, to identify locations.

Neatline: A map can be surrounded by a border, called a **neatline** (Figure 14.3J), often just a bounding box with a solid line or decorative border.

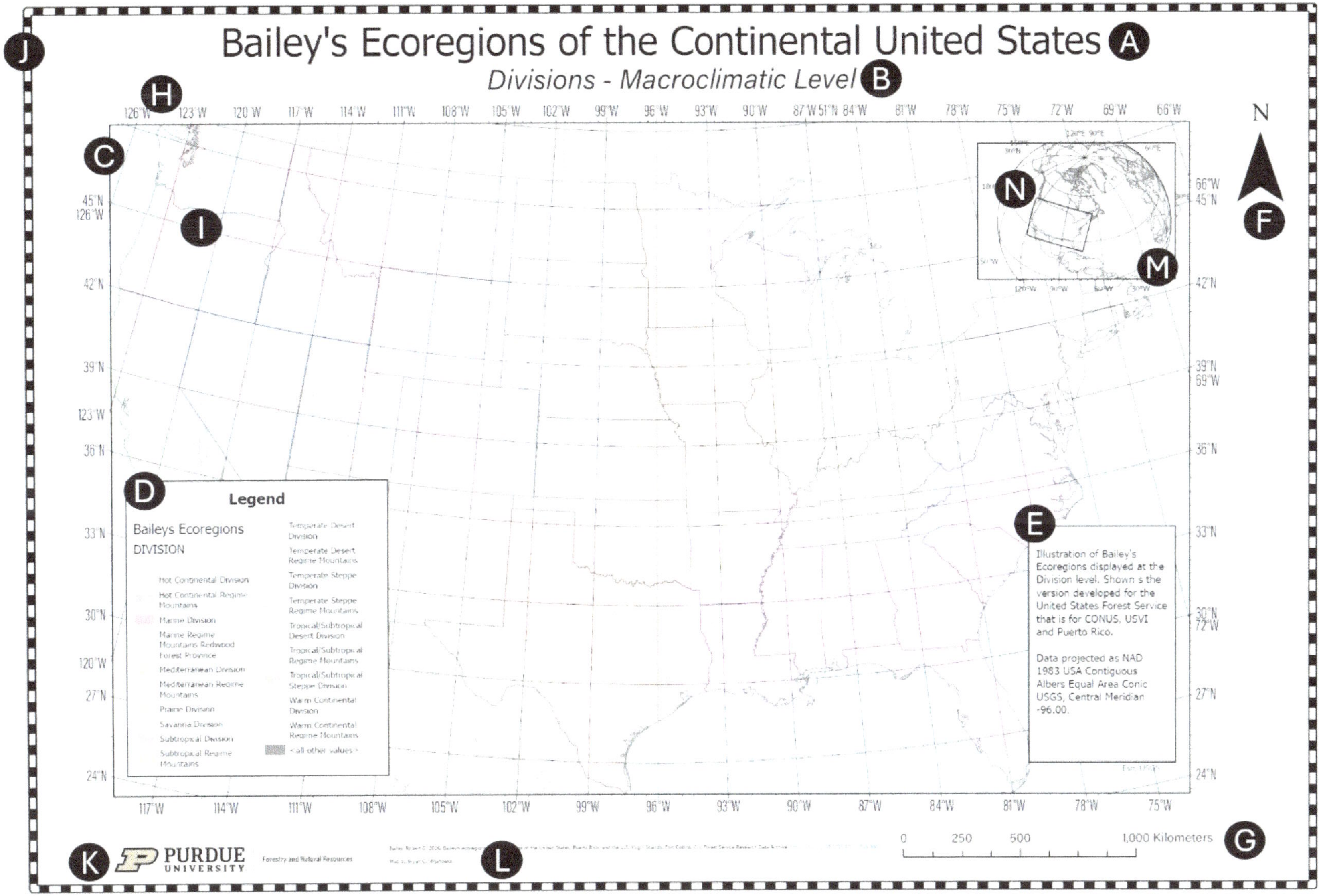

FIGURE 14.3. A sample map and map elements that can be included. See text for details.

Logo: If you, as the mapmaker, belong to an organization, it is often useful to include the organization's official logo (Figure 14.3K), or if more than one organization, all logos.

Metadata: The source of the data with any specific references published, which is often called the map metadata, should also be included as part of the required map elements. The person who developed the map could also be added here (Figure 14.3L). Sources of data (and dates), and perhaps some other important information such as contact information (mapmaker name, email address, physical address, date of map, event, etc.) are often useful to include. Citations can also be part of the map metadata and may be helpful to those viewing a science poster with a map where the map references a report or scientific paper.

Locator or Inset Map: If the location of the map in the map body is not well known to those viewing the map, it is considered best practice to include a **locator map** (Figure 14.3M) that illustrates where the smaller location of the map body belongs. Often, there is a linked map frame border (Figure 14.3N) that shows how the extents of the two maps relate. For example, if you have the main map frame illustrating the streets and land use of a city, it would be best practice to include the boundary of the state it is located in along with a border that illustrates the map body boundaries. How to make a locator, also called an inset map, will be part of Exercise 10. The lines that connect the boundaries of the inset map with the larger map are called **map leader (connecting) lines**, which, in ArcGIS Pro, are drawn using an insert line tool.

14.7 SUMMARY

Knowing the fundamental principles of cartography is an essential part of making impactful maps. The field of cartography has been developed through studies of how the human eye perceives color and how to arrange colors, shapes, text, and map elements to tell a visual story. There are several rules of thumb for producing digital and paper maps and these are summarized here briefly. Making an effective map is also important for presentations at scientific conferences and community events where the public is gathered to learn about and provide input on important decision-making events.

14.8 STUDY QUESTIONS

1. Discuss some of the new knowledge that you gained after reading and studying this chapter. How has it changed your view on how to present information in a map? How has it changed the way you judge the maps of other mapmakers?
2. Visual color deficiency knowledge and awareness has grown in recent years. How might you go about using digital tools to produce highly accessible maps for use in publications?

15. SCIENTIFIC WRITING AND LAB REPORT BASICS

Highlights

Aims. This chapter summarizes the key components of a lab report as it reflects the structure of a standard data-oriented scientific journal article. Some general rules for writing sections of these documents are provided.

Learning Objectives. Students should be able to:

- Know and describe the main components of a data-oriented paper that is organized around IMRAD sections
- Explain some of the history of scientific reporting as it relates to scientific papers
- Compare and contrast the content for each section

15.1 INTRODUCTION TO LAB REPORTS AND SCIENTIFIC PAPERS

Submitting lab reports is a common assignment in university science and engineering courses. Most lab reports follow the standard data-driven scientific journal article format, which includes an **Introduction, Materials and Methods, Results**, and **Discussion**; this format is sometimes abbreviated as **IMRAD**. Other parts of lab reports also include the title, author, and date. Some may require an abstract, conclusions, and references. In some cases, appendices might also be required. Writing strong, clear, succinct lab reports is excellent training for developing good journal article writing skills. The most universal currency for gauging the success of a scientist is the number of journal articles published in top journals and the number of citations these papers generate (note: clearly written journal articles are well cited). This chapter will summarize (1) writing fundamentals for lab reports and journal articles; (2) content for each of the IMRAD sections; and (3) tips for organizing references for research.

15.2 BRIEF HISTORY OF IMRAD AND SCIENTIFIC WRITING

Galileo Galilei (1564–1642) is often credited with elucidating how observations, experimentation, and hypotheses are used to frame the scientific discovery process. He is famous for developing powerful telescopes to make observations of the surface of the moon and to observe planetary orbital patterns in his attempt to describe the phenomenon of motion. He is also considered the first to emphasize the use of hypotheses and careful documentation of scientific hypotheses, methods, and results. Louis Pasteur (1822–1895) is often credited for using the IMRAD format for communicating scientific research. It was not until the mid-20th century (Sollaci & Pereira, 2004) that the IMRAD format became standard, with each of the four major terms Introduction, Methods, Results, and Discussion becoming headings for sections of scientific journal articles.

There are many great guides for how to write a scientific journal article, especially using the IMRAD format. One of the most comprehensive is Nair and Nair's (2014) *Scientific Writing and Communication in Agriculture and Natural Resources*, which should be consulted for details that cannot be included here. Another great resource is Peat et al.'s (2013) *Scientific Writing: Easy When You Know How.*

Style guides have been developed over the last several decades. A few of the most consulted guides are ones produced by the Council of Science Editors (CSE) titled *Scientific Style and Format: The CBE Manual for Authors, Editors and Publishers* (1994), edited by E. J. Huth et al., and *The Chicago Guide to Communicating Science* by Montgomery (2017).

15.3 CONTENTS OF THE INTRODUCTION

The **introduction** should be a brief presentation of the problem or topic; a history of the most important results to date on the topic or problem; context for the current research, which is usually more specific than a general description of the topic; and the research questions or hypotheses of the current paper. It is suggested by Nair and Nair that strong introductions should not be more than 500 words.

Avoid long sentences in the introduction, especially the first sentence. It is also common to cite "big literature" (papers that describe problems broadly and provide ideas or terms that are highly relevant). Also avoid defining well-known terms; define those terms that are unique or are used a lot in the paper.

Research questions and/or **hypotheses** should be stated in a way that is directly "answerable." Starting research questions with "Does," "Are," or "Is" allows researchers to answer questions with specific analyses; starting research questions with "How" typically focuses on open-ended answers that are not easily tested. For example, "Is the biodiversity in the Sundarbans West Wildlife Sanctuary protected area greater than outside in similar habitats?" is an answerable research question. "How does biodiversity differ between the Sundarbans West Wildlife Sanctuary protected area and areas outside the protected area?" generally leads to an endless number of analyses and is not a recommended way to pose a research question.

15.4 CONTENTS OF MATERIALS AND METHODS

This section typically has several subsections, usually described in the following order. First is a description of the **study area**. This often includes a map as well as a description of the ownership of the area, size, composition (e.g., what kinds of vegetation are most common), and the history of the location, all of these provided depending on their relevance to the study. The next section describes **observations** and data that were collected and the protocols used to collect the data. This would be followed by a description of the **approaches** used, including the analysis tools, that might be described as a workflow. Workflows commonly include a box and arrow diagram and labels for steps in the workflow. The final section typically includes **statistical procedures** used, citing the rules that are being used to frame results and conclusions. Strong papers will also include a summary of assumptions made in data collection or in the analyses. Citing methodological papers is done in this section as well as where data and code are posted for readers to download.

The materials and methods are most often written in the past tense. Writing style is very precise and concise. Scientists should use the standard units of measure as described by the **International System (SI) of Units** of measurement. The SI unit for length and area is meters (m), but kilometers (km) can also be used.

15.5 RESULTS

Results sections present the new information that was part of the study. This section, as with Methods, focuses on precise and concise writing. The order of the results should reflect the steps conducted as part of the research and also be framed around the research questions or hypotheses. This is also the section that contains tables and figures that describe the results of the study. Figure and table captions are included as well and are numbered sequentially, all as Tables and then as Figures. Plots of data should be clearly legible, especially titles of figures and elements of figures. Make sure that all axes (x-axis and y-axis) are labeled with a name and with units. Avoid presenting the same data in **Tables** and **Figures**. Long tables are best included in an appendix or posted as a digital table on a website.

The results section should also avoid interpreting the findings of the study and should be written in the past tense.

15.6 DISCUSSION

The **discussion** section should try to accomplish several objectives. First, a presentation of how the study has helped to answer each of the research questions or hypotheses should be summarized. A good structure is to devote one paragraph to each research question or in the case of a lab report, to address each of the assignment questions. Second, each of the results should be placed into the context of current knowledge. Describe how the work either supports or refutes current knowledge. For the lab reports that are part of the hands-on guide, show how the results are consistent with expectations that are in the suggested readings list. You may want to do a Google Scholar search of more recent papers to determine if the historical work that is often listed in this section of the hands-on guide is still relevant. Third, shortcomings of the current study could be mentioned. Finally, ideas for future work could be included at the end of the section.

15.7 TITLE

Concise **titles** are most often preferred by journals although if longer ones are informative, then often a title that has two parts to it is separated by a colon. A succinct, informative title might be *Principles of Soundscape Ecology: Discovering Our Sonic World.* This title reflects an article that is about sound and the principles needed to conduct research. Titles should avoid acronyms, verbs, or highly technical terms and should be short.

15.8 ABSTRACT

The **abstract** contains the highlights of the paper written in such a way that all readers can determine if they should read and study the rest of the paper. Abstracts are thus "triage" parts of the paper. The abstract should contain a unique statement of objectives, methods, a summary of the most important results, and a statement about the study's conclusion(s). Most abstracts are from 100 to 300 words long.

Abstracts should not include abbreviations, acronyms, or references.

15.9 CONCLUSIONS

If this section is allowed or required, concisely state the study's conclusions, not repeating them as they might be presented in the discussion.

15.10 REFERENCES

It is a good idea to use a reference manager such as EndNote or Zotero to help manage the references cited in the paper. There are several reference citation styles. Probably the most used in the sciences is the APA (American Psychological Association) format, which uses a last name of first author order, followed by year of publication, title of the article, journal name, and volume and page numbers.

Do not include references that are not cited in the article.

15.11 SUGGESTED FORMAT FOR HANDS-ON GUIDE SUBMISSIONS

The number of paragraphs or words for each of the IMRAD lab reports are summarized in Table 15.1. Course instructors can provide alternative suggestions for lab reports.

TABLE 15.1. *IMRAD Lab Report Sections and Suggested Lengths*

SECTION	LENGTH
Title	<15 words
Introduction	3–5 paragraphs
Methods	3 paragraphs, one study area map, and one flow diagram
Results	3 paragraphs and a mixture of figures and tables (see assignment details if there are required products needed)
Discussion	3 paragraphs that include answering research questions and stating conclusions
References	Most assignments request a minimum of 5 references to be cited (likely for introduction, methods, or discussion)

15.12 SUMMARY

The IMRAD format for reporting on scientific work has become the standard for journal articles and also for lab reports in university courses in science and engineering. This chapter summarized some of the most important considerations for the major sections and some of the other sections that are common to scientific journal articles. It also made recommendations to students for other more detailed references to consider for writing scientific articles and reports.

15.13 DISCUSSION QUESTIONS

1. Discuss at least three papers that have been published in tier 1 journals in spatial ecology that follow the data-oriented paper format and how they each have developed their introduction. What are some of the common elements, for example, of all papers in the first paragraph in the introduction? In the last paragraph of the introduction?
2. Using these same papers that were part of discussion question #1, what are some of the common elements of the methods and results section in terms of (1) use of tense; (2) level of brevity; (3) the quality of the figures and tables and how these impacted your understanding of the study?

EXERCISES

This section contains exercises designed to be completed in a 2- to 3-hour computer lab period. Each exercise contains some brief background information necessary to understand the exercise, a summary, in most instances, of data that are typically used for an exercise such as the one to be completed, a set of ArcGIS Pro, Microsoft Excel, or other software, interfaces that are marked up with narratives that attempt to describe a typical workflow. Each exercise focuses on a case study, most of which are from the author's long-term research program. A workflow diagram is also provided for each exercise toward the end to assist students in understanding how tools are combined to conduct analyses. Two sets of suggested exercise questions or tasks are listed: those for intermediate (first-year) students and then those for advanced students (e.g., graduate students). The latter are often framed in a way that requires a report or additional work. Each exercise has a stated aim, a set of learning objectives, and also a list of the data and/or software resources required to complete the assignments. Words that are in bold match those in the software interface that is described in the navigation sections of the exercises. Words that are in `Courier New` are specific commands used in ArcGIS Pro.

EXERCISE 1: MAKING YOUR FIRST MAP

Highlights

Aims. This exercise is designed to introduce you to the basic interface components of ArcGIS Pro so that you can make your first map. This chapter builds from many of those in the GISci fundamental chapters described above.

Learning objectives. After completing this exercise in about a 2-hour period, students should:

- Know how to create a Map project using Map Views and then insert these into a Layout View
- Know how to control zoom, insert data layers, create and modify major map elements such as legends, scale bars, and north arrows
- Be able to produce a digital version of their map

Project Data Resources. A sample set of vector and raster data for the Muskegon River Watershed and the state of Michigan.

1.1. INTRODUCTION

Making maps in GIS used to be a couple days' effort. Now, the interfaces to many GIS packages look like software packages that we work with on a more frequent basis, such as Microsoft PowerPoint and Adobe Photoshop. Adding many elements to a map is also a simple, drag and drop task, making composing a map easy. That said, a good map requires quite a bit of skill and patience. Maps are important to make for scientific publications and reports, for use in scientific posters at conferences, as well as for posters that are used for community events.

Most good maps still require a few hours to make even if you have had years of experience. Getting the right colors, nice balance, and a clean and readable set of map elements takes time. Creating the final product, which is to send a map to the printer or to create a publication quality (generally 300 dpi or greater) map means that the entire mapmaking task can still take several hours. Great maps are often posted on walls and hallways so taking that extra time to make an effective and visually appealing map is important.

This exercise will introduce you to the main map features that will be used repeatedly to make effective maps. In general, if this is your first introduction to any GIS or to Esri products, you should be able to follow these simple instructions to start ArcGIS Pro and eventually produce a digital file that can be used to show your work.

1.2. NAVIGATING ARCGIS PRO

The interface for ArcGIS Pro is composed of several different kinds of user interface components. At the very top of the interface are tabs. There are nine **standard tabs** (Project, Map, Insert, Analysis, View, Edit, Imagery, Share, and Help). Selecting a tab activates a **ribbon**, which organizes many of the tools users will need: **views**, which contain main visualizations including maps, layouts, tables, and catalog (i.e., files and directories); and **panes**, which tend to be "dockable" (located at specific locations on the screen). A host of drop-down and pop-out **menus** are also part of the interface and are accessed through ribbons, views, and panes.

ArcGIS Pro also contains several kinds of projects (Figure Ex1.1A). These include **map** projects (i.e., making maps and analyzing spatial data, location 1), **catalog** (i.e., used for managing data), **global scene** (i.e., visualizing large-scale data in 3D, and **local scene** (i.e., visualizing data in 3D but at a local scale and typically with buildings as a major feature of the visualization).

There are several domain-specific extensions, which typically are add-ons to the base ArcGIS Pro package. These include **Spatial Analyst**, **Geostatistical Analyst**, **3D Analyst**, Image Analyst, Network Analyst, Data Interoperability, and Business Analyst, along with domain-focused extensions such as Aviation, Maritime, and Production Mapping (those listed in bold are ones that will be used in this textbook's hands-on exercises).

It should be noted that Esri also has several **contextual tabs**. These will appear to the right of the standard tabs when different kinds of data (e.g., raster data) or other features are turned on.

1.3. RULES FOR WORKING WITH ARCGIS PRO

After working with many GIS software packages over the years, the author has learned several lessons. These lessons are based on the fact that (1) ArcGIS Pro contains an interface to hundreds to potentially thousands of commands, so this is an enterprise software platform; (2) it likely takes a few years to be competent using many features of this software; (3) there have been several significant "reboots" of the interface over the last several decades, so understanding the fundamental design features and terminology of spatial analysis is essential to being able to transition to a new, major GIS design. Based on these general facts, the following best practices are good to follow:

1. <u>There Is Usually More Than One Set of Tools to Do an Analysis.</u> Given that there are so many commands and tools, it is often common that there is more than one workflow and set of commands that can produce the same outcome.
2. <u>"Left Click" on Properties.</u> Access to some of the most important features that you will need is by pulling up the Properties menus, usually through left clicks on menu items.
3. <u>"Right Click" on Item.</u> To gain access to some of the most "active" lists of items for a component of an interface, right clicking on that item will bring up some of the most common tasks that are related to that item (e.g., right clicking in a data item will bring up a list of potential tasks such as Symbology that will allow users to change a host of colors, patterns, etc. related to features).
4. <u>Learn by Exploring.</u> As an early learner of such a large, enterprise-level software package, it is always helpful to stop and explore buttons or menus that are not used to determine how you might use them in the future. Approaching this enterprise software package by thinking "I wonder what that icon and term does?" is one of the best ways to discover the breadth and depth of this powerful software package.

5. Seek Out Help. The Esri Help (?) button, located on the top quick access toolbar, will take you to their online help where there is a massive, and always up-to-date, set of explanations for all of the tools and commands contained in the package. In addition, there are a host of video tutorials that are short lessons on how to use the software for a specific purpose. In short, the online documentation is outstanding and is helpful in learning new tools or approaches to working with GIS.
6. Don't Forget the Hamburger. In many menus, there is a symbol some refer to as the hamburger (☰). Right clicking on this icon will often bring up important menus. The hamburger helps to keep the interface clean and not so cluttered, but it also may hide some of the software features that a user might need.
7. Stay Organized. Database management requires strong organizational skills. Make sure that you organize and document the data you use; placing all of your data in one folder is not considered good database management.
8. Save Work Often. Once within ArcGIS Pro, you are frequently using it for long periods of time. It is recommended to frequently save your work using the save project icon located on the top of the interface (on the left).
9. Don't Forget to Apply. Often, when you change an option for any of the menus, these are not applied until the Apply button is selected.
10. Know Mouse Movements. Placing your cursor over an icon (i.e., hovering the mouse) or some parts of the menu will often bring up more detailed information that is helpful as you navigate through the interface. The scrolling wheel on a mouse is often used in active maps for zooming in and out.
11. Legacies Exist. There are a handful of legacies from the 1980s that still carry into the current software platform. Probably the most important is that the name restrictions for the Esri grid format will still maintain a limit to the number of characters that can be used for filenames. Currently, that limit is 13 characters, although it is advisable to limit the number of characters for grid files to 11 characters or less.

1.4. NAVIGATING ARCGIS PRO FOR MAKING YOUR FIRST MAP

A. OVERVIEW OF RIBBONS AND MENUS FOR MAKING YOUR FIRST MAP

Figure Ex1.1B–E contains screen grabs for many of the ribbon menus that will be needed for Exercise 1. One of the most often used on the **Map** ribbon is the button for **Add Data** (Figure Ex1.1B, box 2). One can also change the standard Esri basemap to one of nearly 30 preset basemaps that have common features (e.g., roads, places) and labels (e.g., place names).

For the **Insert** ribbon, two often used tools are **New Map** (Figure Ex1.1C, box 1) and **New Layout** (Figure Ex1.1C, box 2). A new map insertion is done when a new collection of data is to be added to the project. If, as a spatial analyst, it is to add a map inset to a layout, then a second map would need to be created with new data, possibly with data that are for larger spatial extents. For example, as we will create a land use/cover map for Michigan's Muskegon River Watershed (MRW) for this exercise, a second map that has the 48 lower states as features will be needed to show the location of the MRW with regard to the rest of the states.

The **Analysis** ribbon will be used a lot for other exercises in this textbook; the most selected icon will be for the **Toolbox** icon, which is a red toolbox (Figure Ex1.1D, box 1). Selecting this icon will bring up a fixed menu to the right of the map display frame.

Finally, depending on the need to share data and products from your work (e.g., maps), the **Share** ribbon provides users with a host of options that allow one to share output. After we create a map in the layout frame,

FIGURE EX1.1. The opening ArcGIS Pro menu (A) and the main tabs (B–E) that display the ribbons for access to many of the ArcGIS Pro features needed to make your first map. See text for details.

we want to produce a high-quality product in one of the several standard graphics file formats (see Chapter 14) using the **Export Map** icon (Figure Ex1.1E, box 1).

B. MAP VIEW ELEMENTS

Figure Ex1.2 shows a very common arrangement of major map view elements along with the ArcGIS Pro user interface. At the top, the Insert ribbon (Figure Ex1.2, box 1) is active, showing all of the options for this broad set of insert features. To the right and below are is the **Contents** menu (Figure Ex1.2, box 2). The view of the map is provided in the central window (Figure Ex1.2, box 3), and if selected, a view of the tabular data (Figure Ex1.2, box 4) is displayed below the view of the map. One of the more common map options windows is that for Symbology, shown in Figure Ex1.2, box 5.

Perhaps one of the most used menus is the one that is brought up by right-clicking on a map item (e.g., lu98_level1_new that is listed in Figure Ex1.2, box 2). Right clicking this map item brings up menu with a list of important options (Figure Ex1.2, box 6). Most often used options from this menu are: **Attribute Table** (which creates the display as shown in Figure Ex1.2, box 4), **Zoom to Layer** (which, for the selected dataset, is used to zoom to this extent in the view in Figure Ex1.2, box 3), **Label** (which sets on the field selected to be the **Label** viewed in Figure Ex1.2, box 3; the example here is county names as labels), **Symbology** (which brings up the Symbology menu to the right as shown in Figure Ex1.2, box 5), **Data** (which allows users to export the data to another file), and **Properties** (which will show important information; arguably the most significant is the coordinate system of the data).

To change the background color of the map (here it is blue, which matches the color of lakes), users would do this in the Map frame, selecting the top item, Map; right clicking Map to show the Options menu; selecting Properties; and then changing Background from no color to the color shown here.

C. LAYOUT VIEW ELEMENTS

Creating and inserting a **Layout** into the project is necessary to create a professional-looking map that has all of the main map elements (see Chapter 14). On the **Insert** ribbon, selecting **New Layout** (Figure Ex1.3A, box 1) will result in a presentation of a large menu of layout formats (Figure Ex1.3A, box 2). Cartographers are very familiar with all of the sizes of paper used to create maps, and readers of this textbook will eventually find several that are most useful. Printing a map using standard, ANSI format paper sizes such as letter size (8.5 in by 11 in), would be good for maps that are part of a standard report. The **Architectural B, Landscape Orientation** (Figure Ex1.3A, box 3) is one of the best options for use in a computer display or presentation package such as Microsoft PowerPoint.

Once the layout has been created (Figure Ex1.3B), users may consider changing the extent of the map through the interactive features toggled on and off with the **Activate** map frame button (Figure Ex1.3B, box 1). Also note that in the view frame (Figure Ex1.3B, box 2), there are tabs to get users to a **Map** and also to a **Layout**. In many projects, users have more than one map and sometimes even more than one layout.

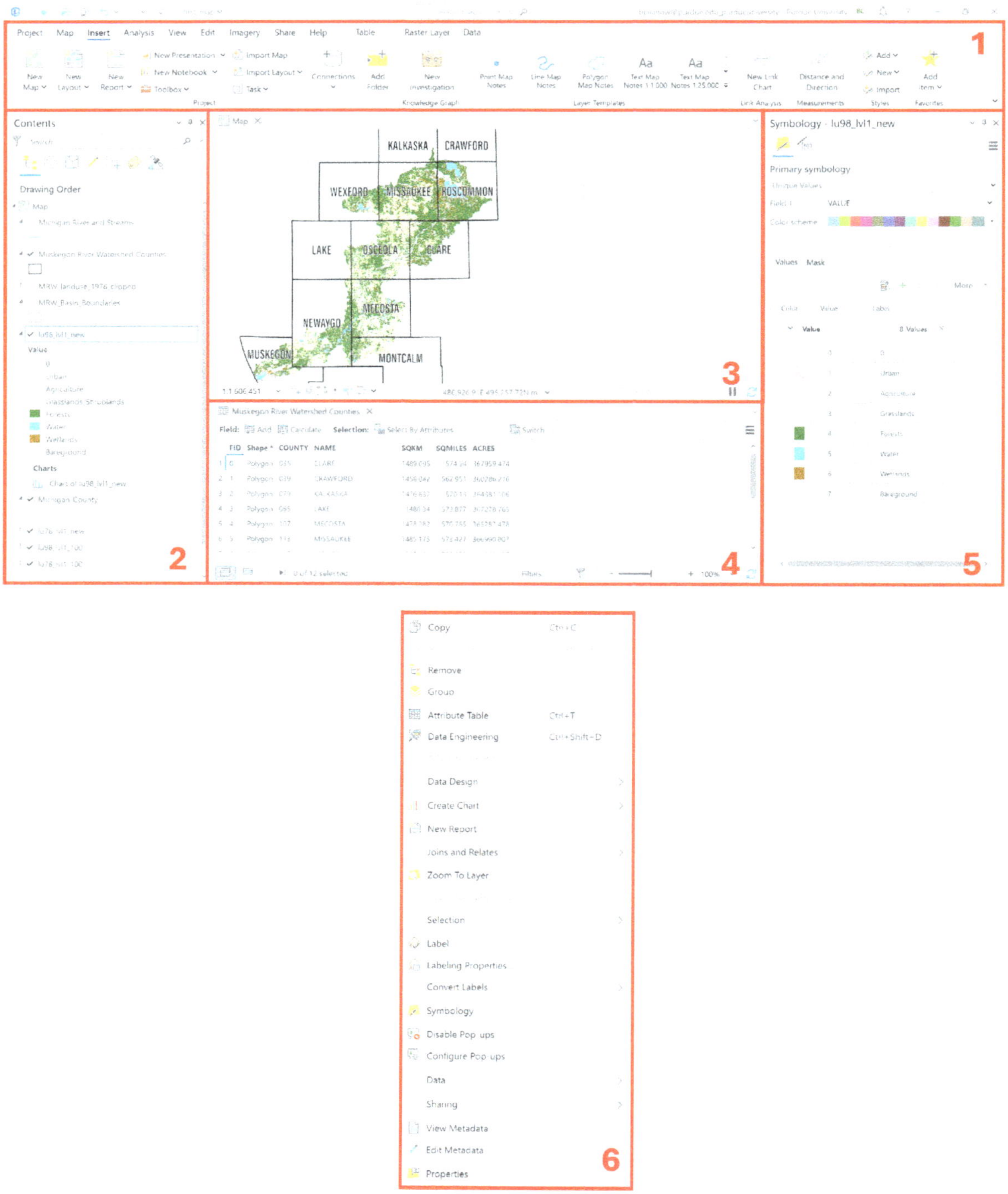

FIGURE EX1.2. Map View windows that aid in the management of tools in ribbons, data layers, maps, tables, and options for how features are controlled. Also shown are the Options menus, activated by left clicking a data layer in the Contents pane. See text for more details.

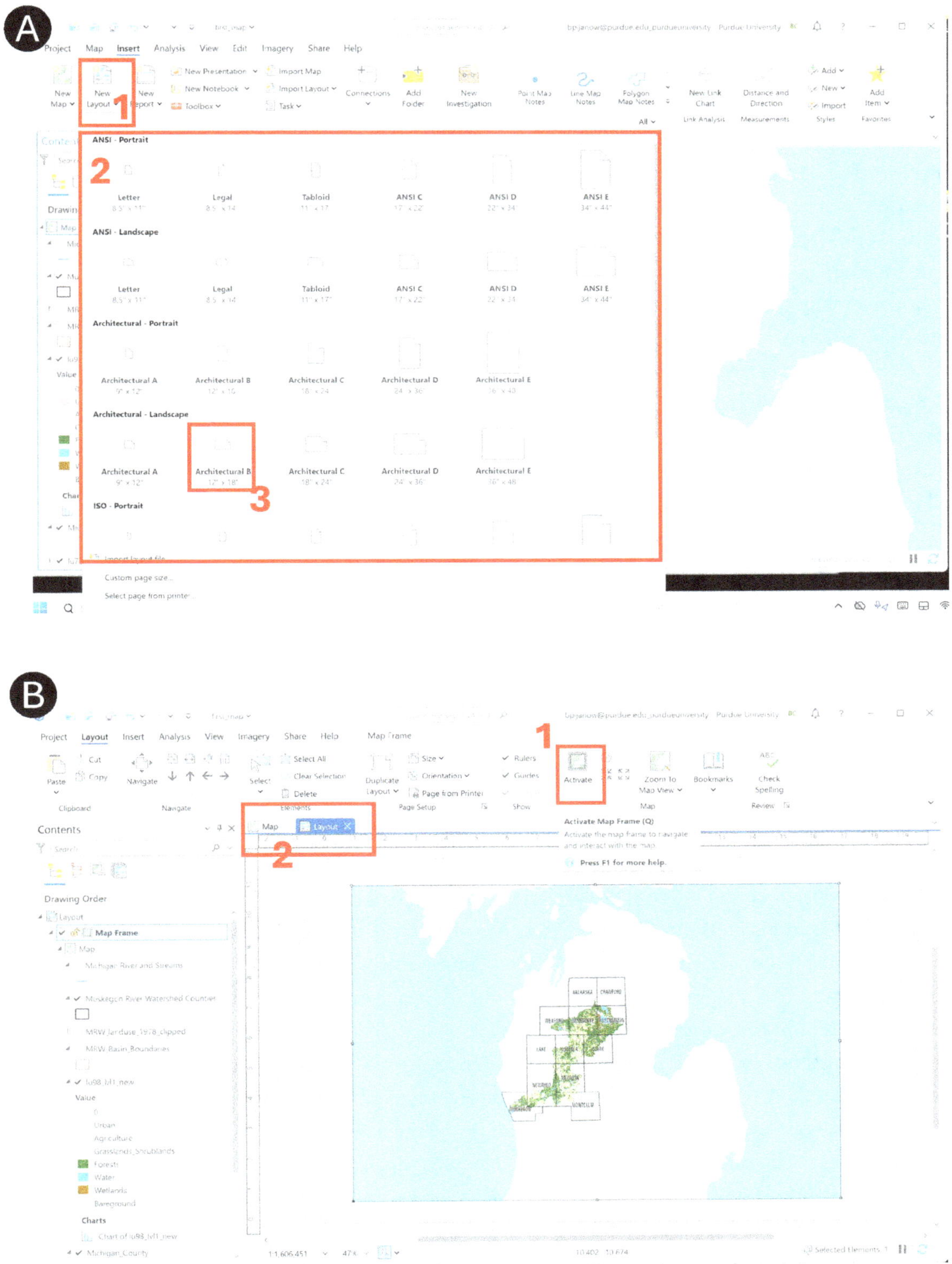

FIGURE EX1.3. Steps to create a Layout View (A) and then a resultant Layout View for the first map (B).

D. ADDING MAP ELEMENTS TO A LAYOUT

The user interfaces to the layout are organized around the same designs as those for creating maps; that is, with tabs, ribbons, and panels that contain a wealth of inputs for users to select. Let us consider a layout and all of the controls offered to users (Figure Ex1.4A). At this point in the map layout making process, the user has just inserted a **Legend** (Figure Ex1.4A, box 1). Once that legend is inserted, the Legend contextual tab is activated and can be selected to control some of the main legend features. At this moment, however, the user has selected the Insert tab (Figure Ex1.4A, box 2) which shows all of the options for inserting. Of important note are the main map elements that are shown within box 3 of Figure Ex1.4A. Selecting any of these icons will bring down a menu of options. For example, the upper left icon is for north arrow. Selecting the north arrow will result in a presentation of many north arrow styles. After selecting one, the user is then provided with an active selection box to insert the north arrow anywhere on the layout frame (Figure Ex1.4A, box 4). The **Element** menu (Figure Ex1.4A, box 5) shows many options, toggled through four buttons that operate like a tab. The one selected here shows options for the user to change the map element's Border, Background, and to add an optional drop shadow. Note that the user here has added a white background to the **Legend** box and then added some gaps in the x and y directions to create a buffer space between the edge of the legend box and the text. The size of the text and its font were modified using the **Legend** contextual tab.

There are a couple of other useful features to this layout interface. One includes the ability to precisely scale the map to a percentage or to a specific map scale (Figure Ex1.4, box 6). A second feature is that there can be bottom tabs that allow users to toggle between menus on the right side (Figure Ex1.4A, box 7).

Note that this map has most of the necessary map elements added but they are in draft form. These map elements include the map, map frame, north arrow, scale bar, legend, and map title.

If we want to add an inset map (see Chapter 14) to our layout, the best way to do that is to create a new Map frame (Figure Ex1.4B), set the map to the desired Coordinate Systems (Albers Equal Area), and then insert that Map frame into the layout (Figure Ex1.4B, box 1). Note that when a user selects the Map frame, they are presented with the ones that are available with a thumbnail view of that map (Figure Ex1.4B, box 2).

Pushing forward to creating that first, final map will require work in the following areas if you have a draft that is like what is in Figure Ex1.4A. Adding a raster grid to the inset map can be accomplished by selecting Grid in the Map Frame area of the Insert ribbon (Figure Ex1.5, box 1). This will be added to the inset map (Figure Ex1.5, box 2). To change any appearance in the map, one of the most straightforward approaches is to right click the item in the Contents panel (Figure Ex1.5, box 3) and then Properties to bring up the options to change any of those elements. The default appearances for the legend (Figure Ex1.5, box 4) often need to be edited. The best way to do this is to right click on the selected text and then select **Convert to Graphics**. Next, right click again and then **Ungroup**. Then, all of the text can be edited. Labels (Figure Ex1.5, item 5) to the Muskegon River Counties can be changed by selecting that layer in the Contents panel and then properties. Inserting a title and subtitle (Figure Ex1.5, box 6) can be accomplished using the **Graphics and Text** area of the Insert ribbon. Text properties provide an interface to change font size, font type, font color, and font style (regular, bold, italics).

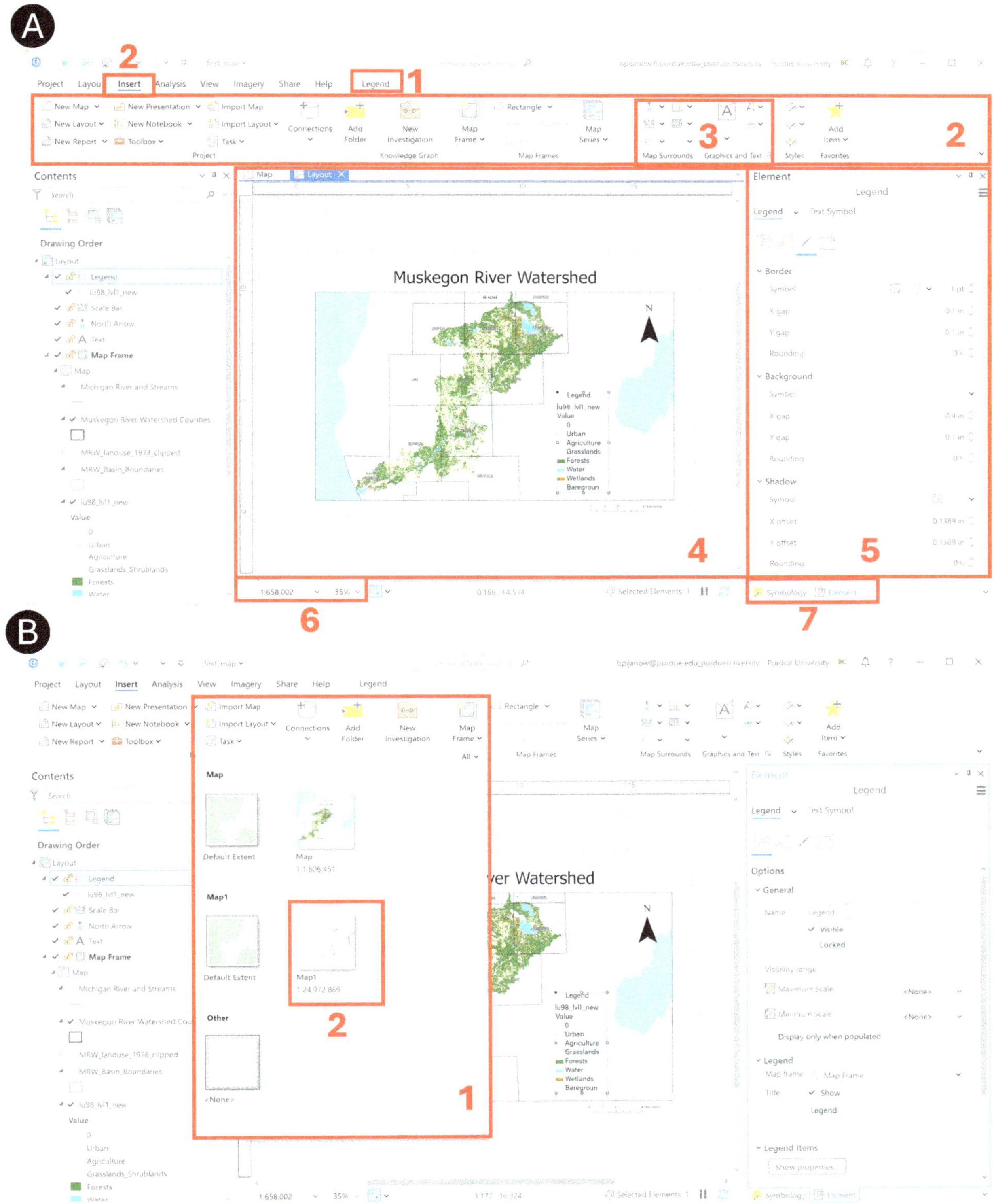

FIGURE EX1.4. Layout View windows that allow access to the mapmaking features from the Insert tab and its associated ribbon (A) and the controls for inserting a Map into a Layout (B). See text for descriptions of the user interface components.

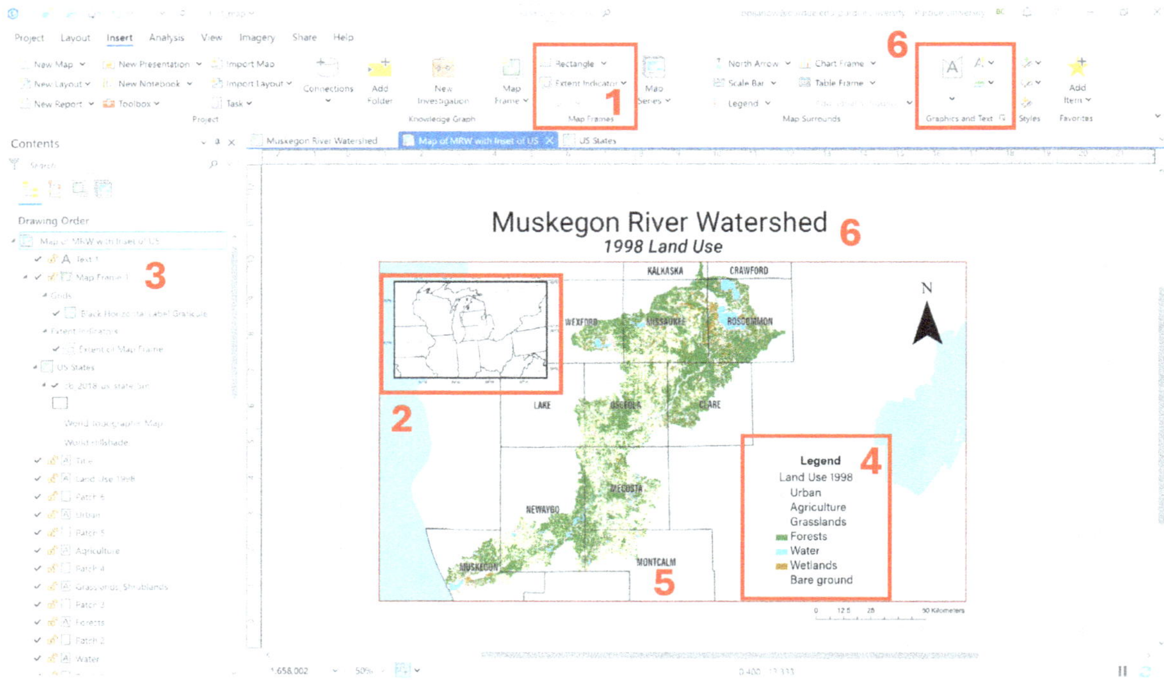

FIGURE EX1.5 Layout View showing all of the map elements for the near final map. Also note how the Contents pane provides access to all map elements (right click will bring up Properties). See text for further details.

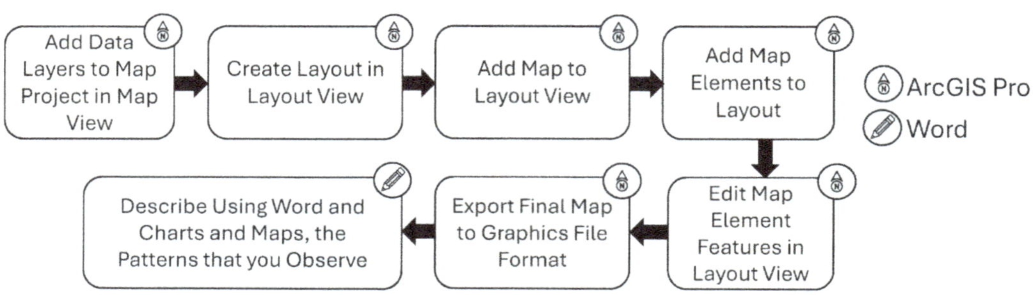

FIGURE EX1.6. High-level workflow for this exercise. These are at this level a very simple workflow.

1.5. WORKFLOW FOR THE EXERCISE

The workflow for this lab (Figure Ex1.6) is straightforward but to take advantage of being in ArcGIS Pro's interface, perhaps for the first time, some effort should be made to explore the interface, especially for those features that affect map production.

1.6. ASSIGNMENTS

A. INTERMEDIATE STUDENTS

If you can produce a map with the following, you've cleared the first hurdle. I would like a map of the Muskegon River Watershed with land use (level 1) displayed with colors of your choosing for each land use. Make sure that the borders are off on the land use shape file, otherwise that is all you see on the map. I would also like to have a map border, a scale bar (using miles is fine), a north arrow, and a legend that is readable. Create a Word document that includes the map described above. In no less than 200 words, describe the patterns you see for the major (level 1) land use classes of urban, agriculture, forest, and wetlands.

B. ADVANCED STUDENTS

Using data layers from more than one Map view (for example, one Map View that is for the entire Muskegon River Watershed and then for one county in the watershed), create a Layout that has more than one map. In Layout mode, in the Insert ribbon, use the **Extent Indicator** in the Map Frames portion of the ribbon to configure a layout that shows how the one smaller map extent is framed in the larger map extent. For example, find the **Leader** option to select the **Style Callout to Edges**. Also include scale bars for all maps, aligning them in a way that is not confusing to the viewer of the map. Create a Word document that includes the map described above. In no less than 200 words, describe the patterns you see for the major (level 1) land use classes of urban, agriculture, forest, and wetlands.

EXERCISE 2: LAND CHANGE ANALYSIS

Highlights

Aims. This lab is designed to introduce you to land cover/use databases and land change analysis tools in ArcGIS Pro that can be imported into Microsoft Excel for further spatial-temporal analysis. Land use/cover change is one of the most common spatial analyses conducted by natural resource analysts.

Learning objectives. After completing this exercise in about a 2.5-hour time period, students should know:

- What sources of remote sensing can be used to develop land use/cover data and the scale at which these are collected
- How to create a land change transition table using ArcGIS Pro
- How to use Microsoft Excel to compare and contrast changes in land cover/use over time in nearby regions in a watershed

Project Data Resources. Students will be provided with land use/cover databases from Michigan's Muskegon River Watershed for 1978 and 1998 to complete several analyses. These data were originally created using heads-up digitization of aerial photographs from these time periods. The data were developed for a Fisheries Trust project that aimed to understand how changes in land use/cover alter stream health for cool water stream fish and macroinvertebrates. Students are encouraged to read the Project Brief for the Muskegon River Watershed that describes the demographic and economic history of this large watershed in the Laurentian Great Lakes Region.

2.1 OVERVIEW

In spatial ecology, one of the most common types of spatial databases are those that contain maps of land cover and land use. Indeed, there are a plethora of land cover/land use maps for the world and many that are for the United States or Europe. This exercise reviews some of the most common sources of land cover/use data and then provides an overview of how one can analyze land cover/use change, sometimes referred to simply as land change.

2.2 COMMON REMOTE SENSING SOURCES OF LAND USE/COVER DATA

There have been several major sources of land use/cover data for use in mapping the land surface and for performing land change analysis since the late 1930s. A description of land use/cover products that have been developed by each of the remote sensing platforms follows. It should be noted that each of these remote sensing sources for land use/cover have a different spatial resolution, temporal resolution (also called repeat interval), and spectral resolution (number of spectral bands). A summary of the most common land use/cover products from each of these sensor platforms is given here.

1. Aerial Photography. In the late 1930s, a company called Abrams Aircraft Corporation in Lansing, Michigan, built the first airplane that was designed to operate a camera for the specific purpose of producing **aerial photographs**. Since that time, many states in the conterminous United States have supported the creation of statewide databases. Many of the land grant universities archived these in libraries as paper databases. In the 1990s, many of these were converted to digital formats and offered to users online. Around 2003, a national program was initiated called the National Agricultural Imagery Program (NAIP), funded and organized by the USDA's Farm Services Agency located in Salt Lake City, Utah. NAIP is planned to be active every five years, providing local government (cities and counties) with high-resolution imagery and tools for analysis.
2. Landsat. One of the most widely used sets of land use/cover databases is derived from the long-term Landsat satellite program. To date (i.e., 2026), nine satellites have been launched (note: Landsat 6 failed to reach orbit and Landsat 7 has a key technology not functioning). Landsat 5, which operated for nearly 30 years, is considered one of the most successful Earth observing satellites ever built. Landsat 7, 8, and 9 are currently in operation. NASA and the USGS have plans for a triplet satellite constellation called Landsat Next, which would increase the temporal resolution from 16-day intervals to 6-day intervals and increase the number of spectral bands from 11 (latest versions of Landsat 8 and 9) to 26 while also reducing the spatial resolution to 10 m x 10 m (from 30 m to 30 m for Landsat 5–9).
3. AVHRR. Known as the Advanced Very High-Resolution Radiometer (AVHRR), this land surface mapping program has been supported by NOAA and, more recently, the European Space Agency. Compared to more recent satellites, it is not advanced or high resolution, but the program has maintained a consistent set of products that have mapped the surface since 1981. Originally designed to track atmospheric and thermal changes, it proved to be useful to map land cover in broad classes of vegetation (e.g., forests, grasses, shrublands) and in nonvegetation classes (urban, barren, and ice).
4. MODIS. Also known at the MODerate Resolution Imaging Spectroradiometer), this is a NASA pair of satellites, one called Terra and the other Aqua, that have moderate spatial resolution (e.g., 250 m x 250 m or 1 km x 1 km) with high temporal resolution (every one to two days). Land use/cover maps are generated from these sensors, but their most common applications are products generated by NASA that map fires, oil spills, and other natural hazards (e.g., floods, mudslides, earthquakes). The sensor can also be used to map sea surface temperature, algal blooms in aquatic systems, and quantify the impact of droughts on vegetation.
5. CBERS. A less well-known set of satellites, the CBERS (China-Brazil Earth Resources Satellite) program has been used to track changes in land use/cover at regional to continental scales, used most often for monitoring changes in the Amazonian Forest.
6. Sentinel-1 and Sentinel-2. These two European Space Agency satellites use different sensing technologies (Sentinel-1 = synthetic aperture radar, or SAR; Sentinel-2 = optical) that are combined to create land use/cover maps at 10 m x 10 m resolution. Many products are global and Esri has partnered with several groups to offer these data as part

of the data accessible to ArcGIS Pro users. The common land use/cover product is also available through Esri's LivingAtlas.

7. SkySat. Imagery from nearly two dozen SkySat satellite-based sensors from the commercial firm Planet Labs can produce sub 1 m resolution land surface data for specific, targeted regions.

2.3 SOURCES OF LAND COVER DATABASES

Land use/cover databases have been developed for the world, with many countries and groups of countries coordinating the development of regional (continent to subcontinent extent) databases. Reviewed here are land use/cover databases for the world and then some that are continent focused.

A. GLOBAL LAND USE DATABASES

Several databases now exist for global land use/cover, each developed for specific uses. Likely one of the most widely used is the **ESA World Cover** database that is produced at 10 m x 10 m resolution. It uses Landsat and Sentinel 1–2 to create LCCS-based land cover classes. It is released as large tiles in a GeoTIFF format (Zanaga et al., 2022). This two-year (2020 and 2021) set of maps was developed with different algorithms, and so they vary based on real change in the landscape but also due to outcomes from different methodologies. The ESA World Cover database has nine land cover classes: built, crops, trees, water, rangeland, flooded vegetation, snow/ice, bare ground, and clouds.

The **Moderate Resolution Imaging Spectroradiometer (MODIS)** 500 m x 500 m MCD12Q1 product (Friedl et al., 2010) contains five separate classification schemes; most of them have more than a dozen classes. The classification schemes focus on classes using Earth System models, a vegetation-based set of classes, another set that contains ecological parameters such as leaf area index, a broad set of biomes (nine), and another that relates to about a dozen plant function types (PTFs). The latter classification schemes relate to leaf and plant structural types (e.g., broadleaf or evergreen needleleaf). These are released as annual products from NASA and the USGS.

A third common global land use/cover database is the European Space Agency's (ESA) **CCI Land Cover database**. This is a 300 m x 300 m database specifically designed to interface to global climate models. The database was developed using several remote sensing platforms, including MERIS (2002–2012), SPOT (1998–2014), and AVHRR (1992–1999). The database contains 22 global land cover classes and fractional cover (percentage vegetation, snow, and bare ground). Data are also released annually.

The **GLC_FCS30 database** from Tsinghua University in China (Zhang et al., 2021, 2024) uses Landsat 8, Sentinel-2, MODIS (NDVI), and SRTM DEM to create maps of 10 main land cover classes along with finer subclasses (Level 2), totaling 30 global land cover classes. The data are released as 30 m x 30 m products and are being processed for five-year time intervals starting with the 2015 "epoch."

A long-term historical database of land covers is contained in the **HYDE 3.3 database** (Klein Goldewijk et al., 2011). This database, at 10 km x 10 km resolution, contains a long-term 10,000 BCE to current that is often used for backcasting of land-climate interaction modeling.

B. NATIONAL/CONTINENTAL LAND COVER/USE DATABASES

The United States **National Land Cover Database** (**NLCD**) is developed by the USGS, NASA, USDA, NOAA, and EPA at 30 m x 30 m resolution using Landsat 5–9 data. It is released in five-year increments starting in 1992. It contains 16 primary land cover classes (Homer et al., 2020). This database is perhaps one of the most widely used databases at national to local levels. Those that are interested in historical land use/cover change might consider examining the **Land Change Monitoring, Assessment and Projection** (LC-MAP) dataset (Brown et al., 2020) of the USGS, which has been extended back to 1985 using historical Landsat imagery. There have also been recent efforts to extend the maps to Canada and Mexico by harmonizing the NLCD to two other national land cover maps and classification systems in a program called the **North American Land Change Monitoring System** (NALCMS).

The FAO (2002a) in the early 2000s developed a 100 m to 300 m resolution database for Africa called **Africover**. Initially developed for East and Central Africa, it was expanded to include the entire continent. This database contained about a dozen major classes that are further subdivided into 80+ land cover classes to level 3.

There is a European land cover database (EEA 2000) called **CORINE** (Coordination of Information on the Environment) that was developed using SPOT, Landsat, and later Sentinel-2 sensors. The data were first developed in 1985 and then reproduced every five years from 1990 to 2020. The database contains about 44 land use cover classes (five main level 1 classes).

Finally, a few not as well-known continental land cover databases exist including the Asia-Pacific Land Cover (APLC-30) database (Asia Pacific 2000–2020, Chen et al., 2019), MapBiomas (Souza et al., 2020) for South America (1985 to present), and the Dynamic Land Cover Database (DLCD) for Australia.

C. STATEWIDE AND LOCAL DATABASES

Historically, there have been a few very well used statewide land use/cover databases developed. One of the most "classic" statewide databases was the one for Michigan (MDNR, 1981), called the **Michigan Resource Information System** or **MiRIS**. It was created in the late 1970s through to the early 1990s to map land use/cover. It was developed using heads-up digitizing instruments (planimeters and stereoscopes) creating polygons of similar land use/covers based on aerial photographs. Land use/cover classes had multiple levels for a total of about 40 classes. These data were also created along with river and road networks to create a large-scale, comprehensive spatial database of natural and built resources. An interesting note is that in the early years of MiRIS, the data were stored in a CAD/CAM system (software used for engineers) called Intergraph Interactive Graphics Design System (IDGS), as it predated most GIS software.

D. HISTORICAL LAND COVER MAPS

In the Great Lakes region of the United States, the states of Michigan, Wisconsin, Minnesota, Illinois, and Ohio used paper records from the U.S. General Land Office from the early 1800s that documented the land cover in about 10 classes of forest and grasslands. The Michigan effort focused on creating maps for about 40 vegetation classes that were documented during the surveying of the state just after statehood (Comer et al., 1993). Many of these European pre-settlement maps were produced on paper first (e.g., Minnesota's was produced by Marschner, 1930) and then converted to digital form in the 1990s. Currently, all five states maintain these historic land cover databases, and they are used for restoration efforts by the state Department of Natural Resources.

2.4 LAND CHANGE ANALYSIS APPROACH

Let us consider one of the most commonly used tools in land change spatial analysis, the cross-tabulation matrix or table. Figure Ex2.1 shows two small, simple maps of land cover separated by 20 years. Note that in 1978 (Figure Ex2.1A), the first cell (i.e., upper left corner) is forest. In 1998 (Figure Ex2.1B), that forest cell is now an urban cell. As we move one position to the right, that cell is an agricultural cell (value = 2) in 1978 but is then an urban cell (value = 1) in 1998. In moving another position to the right in the map, that cell is forest (value = 4) in 1978, but note that in 1998, it remains a forest cell (value = 4).

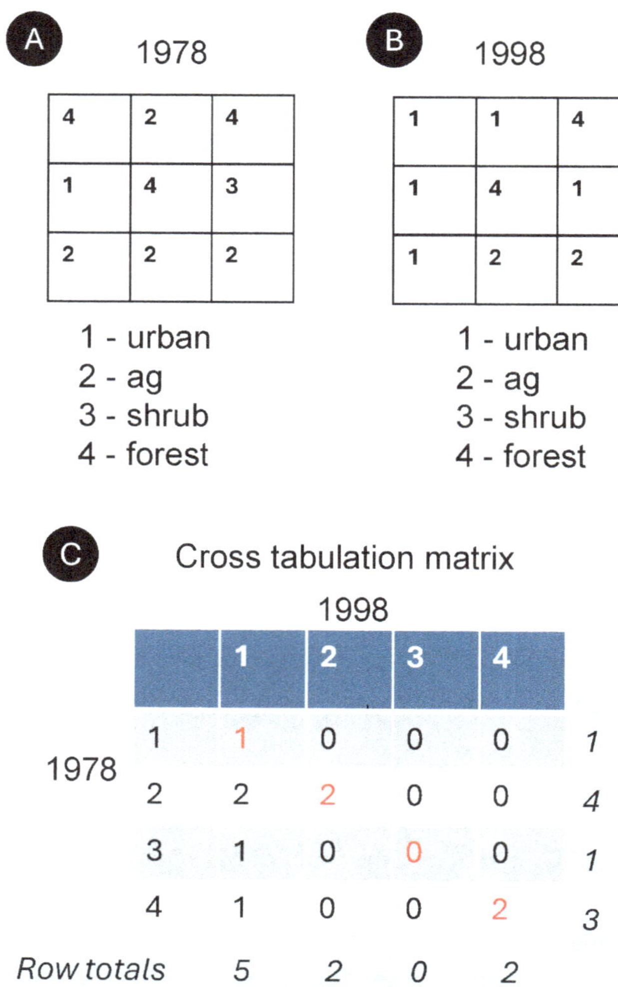

FIGURE EX2.1. Illustration of how land cover/use maps (A and B) are represented as cross-tabulation tables or matrices (C). See text for details.

A cross-tabulation table keeps track of all possible combinations of change (first two examples listed where they both transitioned to urban) and no change (the last example just provided where forest remained forest). A cross-tabulation table contains four main elements: year, land cover code, the number of cells that are counted for each of the transition outcomes, and row/column totals. If we compare the maps (Figures Ex2.1A and Ex2.1B), we can see that in 1978, there was one cell for urban but there are five cells of urban in 1998. In 1978, there were four cells of agriculture, which then decreased to two cells in 1998. We can see from these two maps that urban was a "gain" class and agriculture was a "loss" class. The cross-tabulation table provides not only those quantifying changes, but also the kinds of transitions between 1978 and 1998. For example, in 1998, one cell remained urban between the two time periods, two cells of agriculture changed to urban, one cell of shrubland changed to urban, and one cell of forest changed to urban. The row total is one cell for urban (right side column, top value) for 1978 and we have five cells of urban (see row total) for the 1998 time period. Also note that the diagonals of this cross-tabulation matrix represent the number of cells of "no change"; adding up the diagonals gives the total number of static cells (i.e., 5) which means that 5/9 (i.e., 55.6%) of the map did not change.

2.5 NAVIGATING IN ArcGIS PRO

For this lab, we will process land use/cover data using a variety of Spatial Analyst tools. As an introduction to these tools, we will first load level 2 land use/cover raster data layers into our Map view in the **Contents** pane (Figure Ex2.2A, box 1 and label 2). By selecting the **Analysis** tab, a host of analysis tools will be shown in the **Analysis** ribbon. One of the most frequently selected icons you will likely use in ArcGIS Pro is the **Tools** icon, which is red and looks like a toolbox (Figure Ex2.2A, label 3). By selecting that icon, you will activate the **Geoprocessing** pane to the right (Figure Ex2.2A, box 4). Note that it defaults to the **Favorites** tab. Selecting the **Toolboxes** tab brings up a very long list of tools, including one toward the bottom called **Spatial Analyst Tools** (Figure Ex2.2A, box 5). Spatial Analyst tools are designed mostly to manipulate, integrate, analyze, and export raster data layers and tables. Note that the tools that are presented in this textbook, but not associated with this lab, are contained in orange-colored boxes. The two sets of tools we will use in this lab are organized under **Reclassify** (Figure Ex2.2A, box 6) and **Zonal** (Figure Ex2.2A, box 7).

Let us explore how we would need to use the **Reclassify by ASCII** Tool. Due to the many ways that one could reclassify a raster table, there are several major options. When you have a lot of different values from the first raster data layer to classify, the best option is to use a reclassify table which is a plain text, "look-up table." Figure Ex2.2B shows a reclassify table that was created in **NotePad**. Here, the file was named **level2tolevel1.txt** (Figure Ex2.2B, box 1), using the *.txt extension which is one of the defaults Windows uses to launch the file in Microsoft Notepad. In this example, we included several lines of notes (Figure Ex2.2B, box 2), which is a good idea when you have data around that needs to be frequently reclassed. We also include notes here on what the various file sections mean (Figure Ex2.2B, boxes 3 and 4). The reclass table for level 1 was included since some of the aerial photography interpretation that was done in 2002 could only get to a level 1 class for shrubland; this was agreed to by the scientific members of the team to ensure that the data could be created in enough time for use in modeling and analysis. Also note that the reclass table should be a **Plain Text** file (Figure Ex2.2B, label 5) and the form of the text file is **UTF-8** (Figure Ex2.2B, label 6).

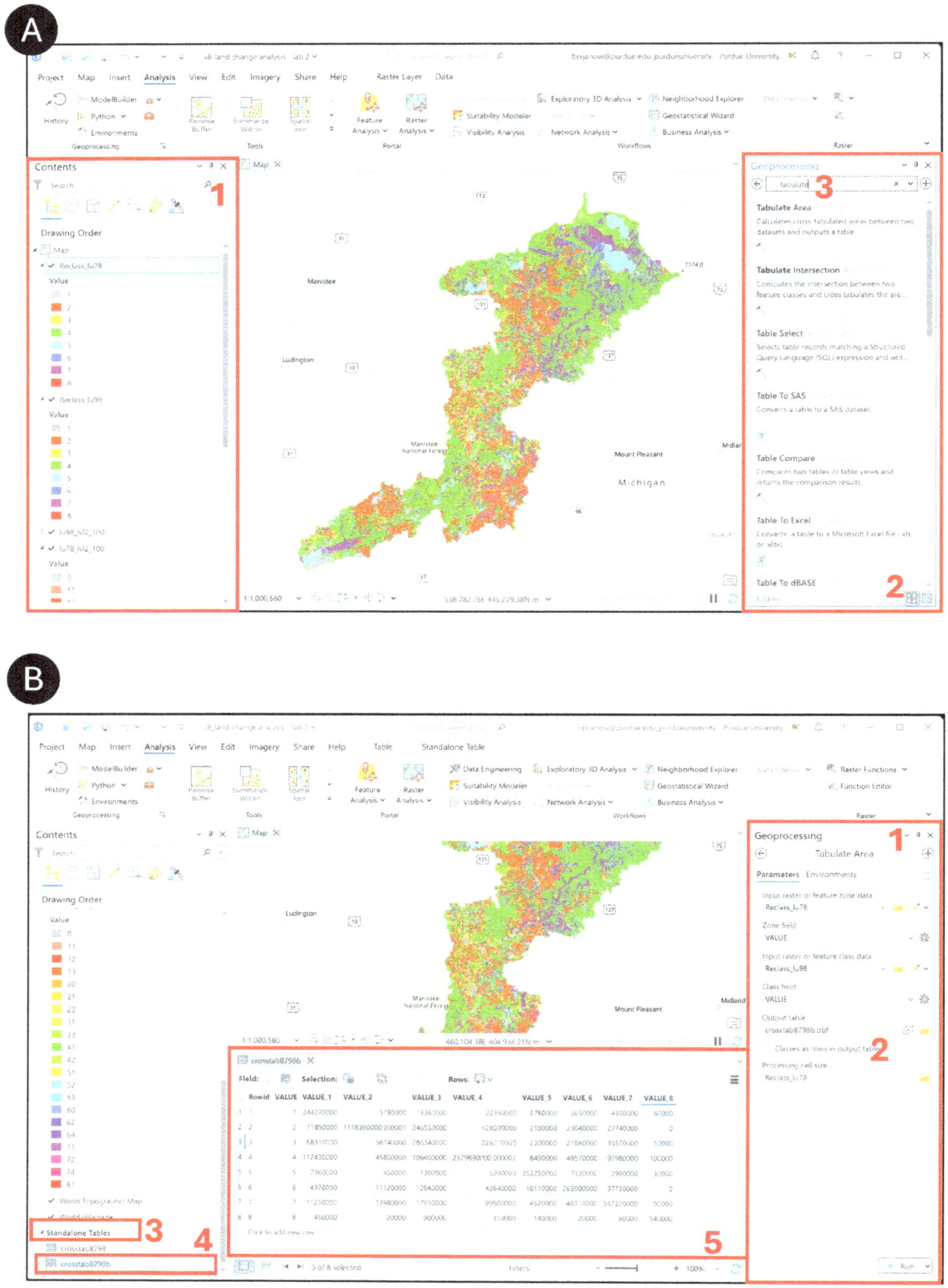

FIGURE EX2.2. Main components of (A) Map View with the Contents pane and the Geoprocessing Tools and (B) a reclass table that is coded as a text file.

Another very useful tool for conducting land change analysis is the **Tabulate Area Geoprocessing Tool**. This tool uses two raster tables and computes an area of cross tabulation. For the land change analysis, two raster data layers, one for 1978 and the other for 1998, will be used as input (Figure Ex2.3A, box 1). Rather than searching through all of the toolboxes, another way to find a tool if you know any parts of the tool name is to type in that name and ArcGIS Pro will bring up a set of tools (Figure Ex2.3A, box 2) that might be of interest. Here, type in **Tabulate** (Figure Ex2.3A, label 3), which gives a list of potential tools and short descriptions of the tool that could be related to this keyword.

The interface to the **Tabular Area Geoprocessing** Tool (Figure Ex2.3B, box 1) provides areas for the first year of the land use files (or, more broadly, as a zonal file, which could be something like a map of counties with county codes used for the categorical values in the map). Select Value and then enter in the second map to perform the cross-tabulation operation. Make sure that **Classes as rows in output table** are selected off (Figure Ex2.3B, label 2). After running this tool (by hitting the **Run** button), the output table will appear in the list of **Standalone Tables** (Figure Ex2.3B, box 3) with the specific table listed below it (Figure Ex2.4B, box 4). Right clicking this table and then selecting **Open** in the Options menu presents the raster table view (Figure Ex2.3B, box 5). The file should be available in the folder that was designated in the **Tabulate Area** tool pane. Note that since the cross-tabulation file is to be brought into Microsoft Excel, it is a good idea to add the extension *.dbf to the file.

Bringing the cross-tabulation file into Microsoft Excel is simple. Drag it from Windows Explorer into a blank Microsoft Excel file. Doing so gives the matrix that is shown in Figure Ex2.4. Note that summing rows and columns (Figure Ex2.4A, boxes 1 and 2) gives the total number of cells that are in each of the level 1 land use classes for each of the two years (note that since the row total for class 1, which is urban, is less than the column total for level 1, then the row totals are for 1978 whereas the column totals are for 1998). You can also use the cells to determine the number of cells undergoing specific changes. For example, the position that is the row for urban (1) and then agriculture (2) shows that over 71 million cells underwent change between 1978 and 1998 (Figure Ex2.4A, box 3). As explained above, the diagonals of the cross-tabulation table indicate no change. For forests (value 4 for both rows and columns; Figure Ex2.4A, box 4), there are over 2.5 billion cells that were coded as forest in both years. Tables such as these are rich with information about how two maps differ.

Analyses can be also conducted within subsets of larger regions, which we will do with this lab. The Muskegon River Watershed is divided into three zones, an Upper, Middle, and Lower Zone, by the Muskegon River Watershed Assembly. One of the ways to subset a raster file is to clip it using a polygon boundary from a shape file. There are three shaped files for the three watershed zones. The **Clip Raster Geoprocessing** Tool (Figure Ex2.4B) contains interfaces to the inputs (Figure Ex2.4B, label 2) and also important parameters for processing of the raster file. Note that there are two checkboxes here; both appear in other **Geoprocessing Tools**. The checkbox for **Use Input Features for Clipping Geometry** (Figure Ex2.4B, label 3) should, in most cases, be set on. This requests ArcGIS Pro to use the boundary of the shape file to process the clipping; otherwise, it clips using a minimum bounding rectangle. The second option, **Maintain Clipping Extent B** (Figure Ex2.4B, label 4), is needed if you desire to keep the cells exactly at the positions in the original raster data layer (i.e., having it off means that it adjusts slightly to the boundaries of the clipping extent).

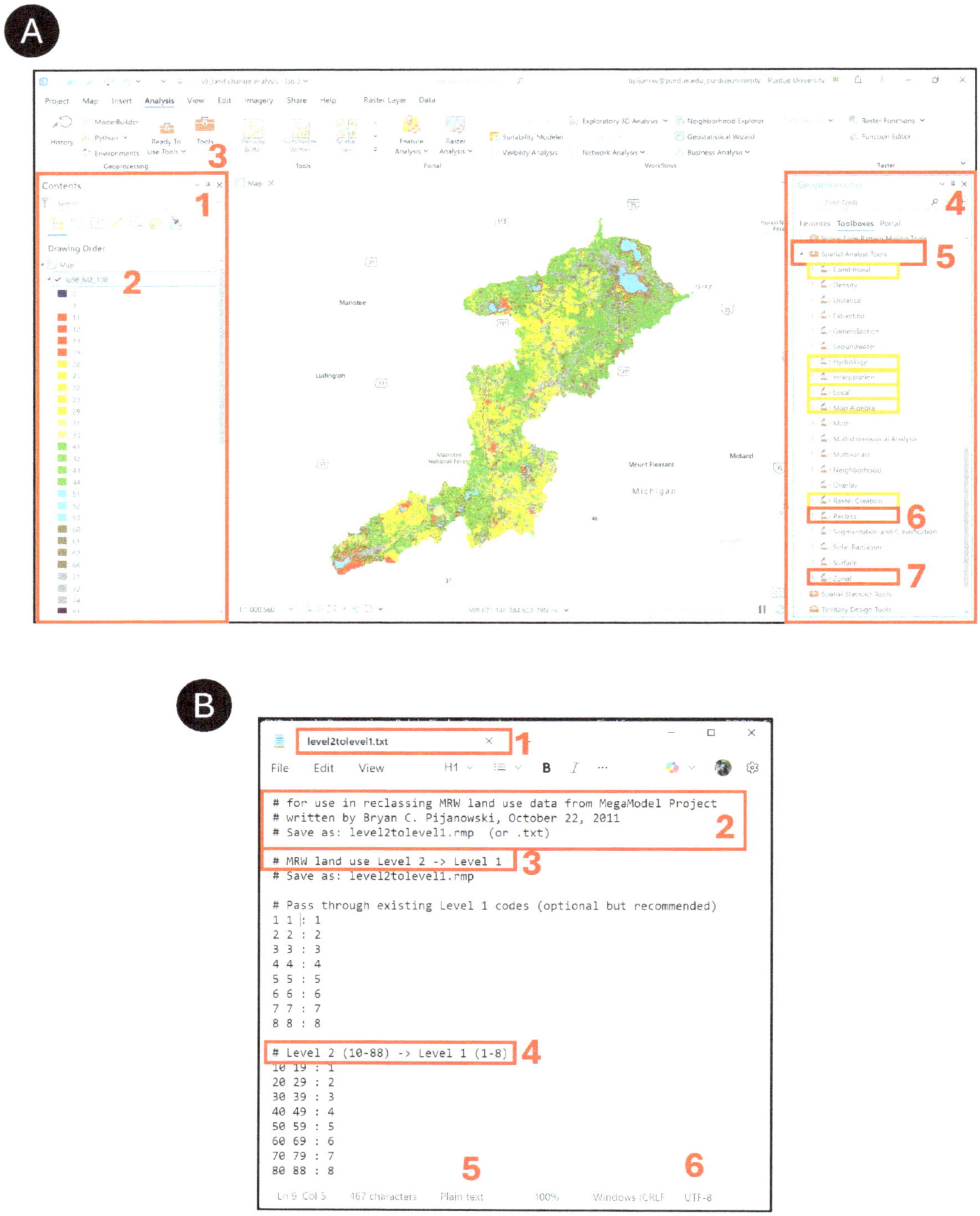

FIGURE EX2.3. Main components of (A) Map View with the Contents pane for the two land use raster data layers and keyword search for Geoprocessing Tools and (B) output of the TabulateArea tool showing the table view.

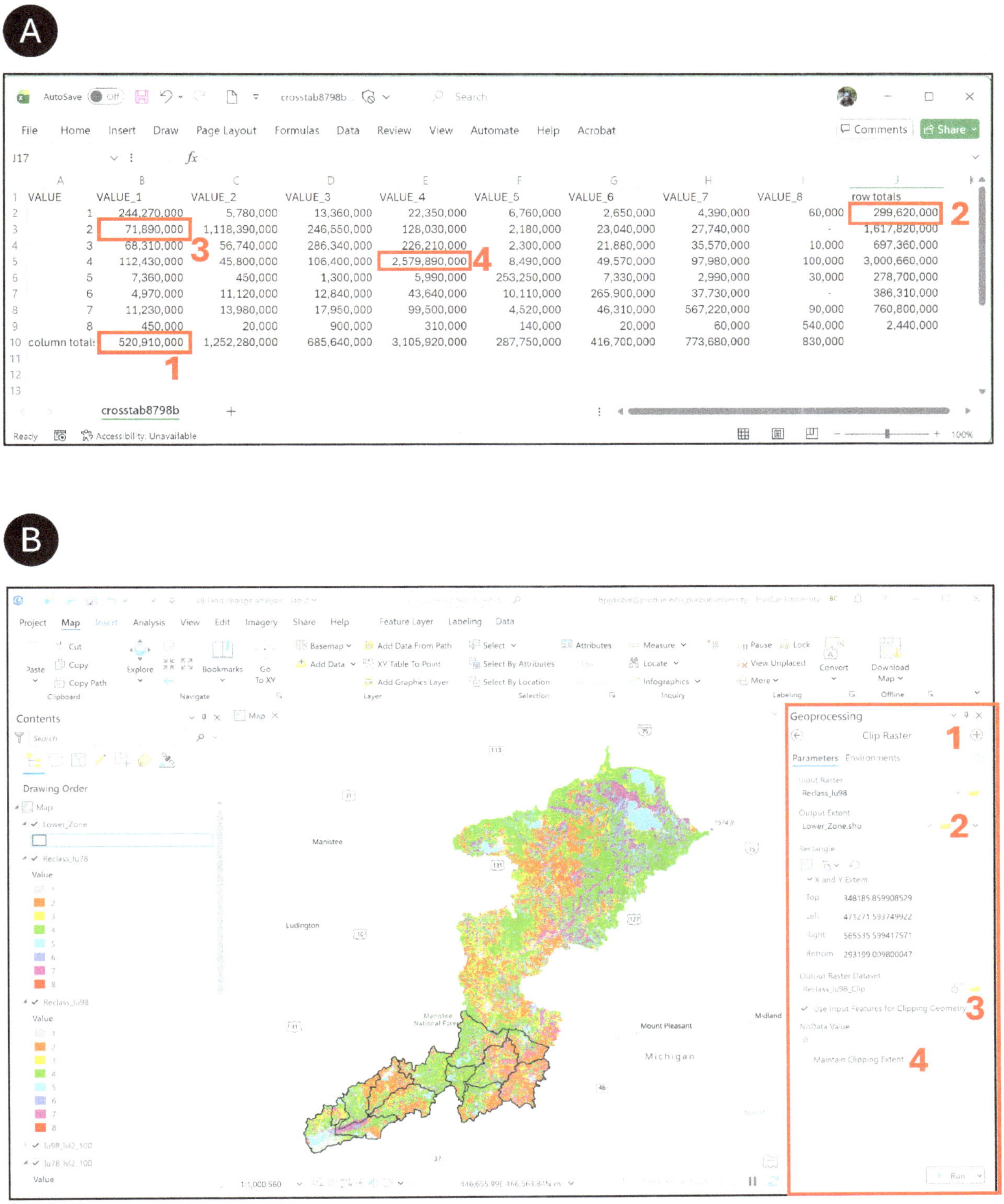

VALUE	VALUE_1	VALUE_2	VALUE_3	VALUE_4	VALUE_5	VALUE_6	VALUE_7	VALUE_8	row totals
1	244,270,000	5,780,000	13,360,000	22,350,000	6,760,000	2,650,000	4,390,000	60,000	299,620,000
2	71,890,000	1,118,390,000	246,550,000	128,030,000	2,180,000	23,040,000	27,740,000	-	1,617,820,000
3	68,310,000	56,740,000	286,340,000	226,210,000	2,300,000	21,880,000	35,570,000	10,000	697,360,000
4	112,430,000	45,800,000	106,400,000	2,579,890,000	8,490,000	49,570,000	97,980,000	100,000	3,000,660,000
5	7,360,000	450,000	1,300,000	5,990,000	253,250,000	7,330,000	2,990,000	30,000	278,700,000
6	4,970,000	11,120,000	12,840,000	43,640,000	10,110,000	265,900,000	37,730,000	-	386,310,000
7	11,230,000	13,980,000	17,950,000	99,500,000	4,520,000	46,310,000	567,220,000	90,000	760,800,000
8	450,000	20,000	900,000	310,000	140,000	20,000	60,000	540,000	2,440,000
column totals	520,910,000	1,252,280,000	685,640,000	3,105,920,000	287,750,000	416,700,000	773,680,000	830,000	

FIGURE EX2.4. (A) Cross-tabulation table shown in Excel and (B) ArcGIS Pro interface for clip raster.

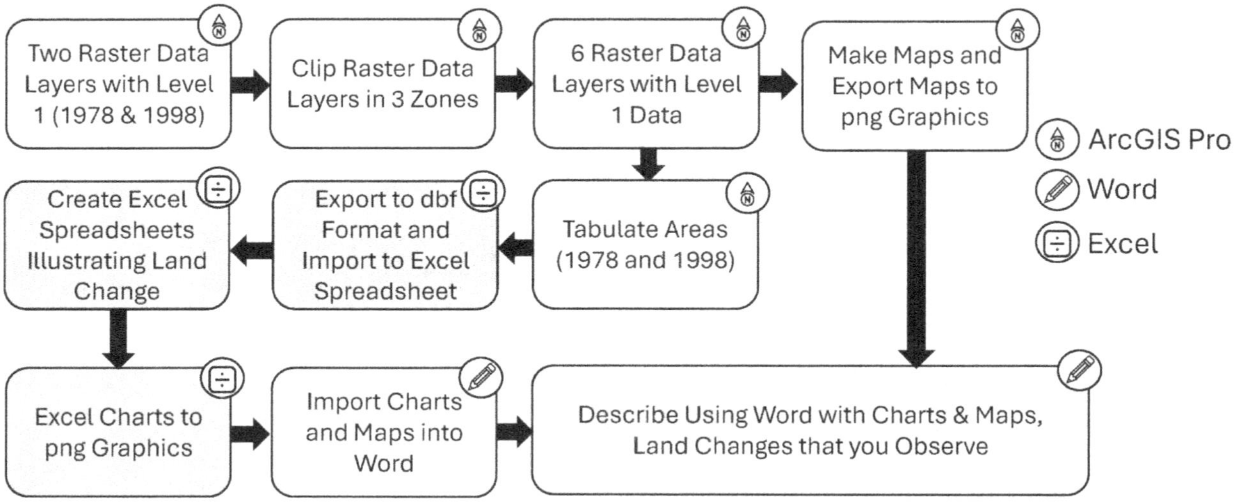

FIGURE EX2.5. Workflow for Exercise #2—Land Change Analysis.

2.6 WORKFLOW FOR THE EXERCISE

We have two different years (1978 and 1998) of land use/cover for the Muskegon River Watershed of Michigan. A general workflow for this exercise is provided in Figure Ex2.5. We will use the **Tabulate Area** command in ArcGIS Pro to determine how much new urban area has been added during this 20-year period. A cross-tabulation table will be produced, which must then be imported to MS Excel so that you can determine the land use/cover types that were converted to urban. Key data for use today are in the Muskegon River Watershed Data Products folder. Land use data from 1978 are from the MiRIS database; those from 1998 were developed using the same methods as for the 1978 land use maps, but GIS technicians used 1998 aerial photography, and 1978 polygons were modified based on changes in land use/cover during that 20-year period. Files here are therefore land use (1978 and 1998), digital elevation models (DEMs), streams, political boundaries, subwatersheds, fish stocking sites, and so on, so this is a rich database!

2.7 ASSIGNMENTS

A. INTERMEDIATE STUDENTS

For those of you that are undergraduates, produce a document that contains the following. Note that there is no one right way to do this. Use your own ideas for how to present this to, viz., a watershed planning community interested in the landscape dynamics of this important watershed.

- A map or set of maps that illustrate changes in land use/cover
- At least two figures that illustrate these changes in a quantitative way for each of the three zones in this large watershed in Michigan

- In at least 150 words, describe these changes that are presented in both the maps and the figures. Make sure you label the figures appropriately; the x- and y-axes should have values, units, and a short description of the variable.

B. ADVANCED STUDENTS

Write a report that includes the standard IMRAD format that addresses the following research questions:

- What are the major "gainers" and "losers" in land use/cover across the watershed during this 20-year period?
- How do the subwatersheds differ in the trends of land change?
- What are the most common types of transitions involving urbanization and what might be the causes of these transitions?

In your report, cite at least five peer-reviewed journal articles that also describe land change theories, global/regional/local trends, or alternative approaches to spatial analysis.

EXERCISE 3: LANDSCAPE PATTERN ANALYSIS USING FRAGSTATS

Highlights

Aims. Today's lab is designed to introduce you to categorical pattern analysis and the popular landscape ecology package called FRAGSTATS. This exercise will also introduce you to workflows that require at least three software packages where understanding the various spatial data models is critical to moving data between these software packages.

Learning objectives. After completing this exercise in about a 3-hour time period, students should:

- Know how to export a raster data layer to a common format for other software packages to use as inputs
- Know and be able to select appropriate spatial pattern statistics using FRAGSTATS
- Be able to create a variety of outputs from FRAGSTATS (e.g., at patch and landscape levels) that can be used for analysis of land change in a large regional watershed

Project Data Resources. Students will be provided with land use/cover databases from Michigan's Muskegon River Watershed for 1978 and 1998 to complete several analyses of spatial patter statistics.

3.1 OVERVIEW

Many maps in spatial ecology are categorical. That is, a pixel or cell within a raster map is coded with a value, generally an integer value. One type of the classic categorical maps that spatial ecologists work with is the land use/cover map. **Land cover classification systems (LCCS)** have been developed that have established standards for ways to code these maps for use across a variety of disciplines and applications.

There are a variety of ways to work within a GIS to conduct categorical map analysis. Exercise 2, for example, introduces the concept of land change analysis and the development of transition tables to understand how specific land use/cover classes change through time. Exercise 4 will focus on how categorical maps of crop types change over a long time period using time series tools. This lab will examine how patterns of patches, edges, and distribution of patches of the same class change through time and also how these changes vary in space. A "classic" tool developed by landscape ecologists is the pattern analysis tool called FRAGSTATS (McGarigal & Marks, 1995). FRAGSTATS was developed in the early 1990s by McGarigal and colleagues. It has been updated several times over the years (e.g., McGarigal et al., 2012) and has been applied to characterize spatial patterns represented in categorical maps (e.g., McGarigal et al., 2009). It has been distributed to the broad community of spatial ecologists and GISci analysts for a few decades now with a distribution site at www.fragstats.org.

3.2 ABOUT PATTERN STATISTICS

There are easily over 100 pattern statistics built into FRAGSTATS. Riiters et al. (1995), in an often-cited paper, found that many of the pattern metrics developed up to that time strongly covaried and may be redundant, based on how they are constructed (i.e., are "near duplicates"). So, the careful rule for applying pattern metrics is to (1) select a few; (2) make sure that they are relevant to the study; (3) read and compare the descriptions of the metrics to ensure that they are measuring something different. Riiters et al. (1995) list, in their Table 2, statistics that they found to be fairly similar to one another, and they also recommend ones that should be preferred where there is a large number of similar metrics.

It should also be noted that the online documentation (at www.fragstats.org) for the pattern metrics is extensive with many explanations of the metric, also citing the original paper that described the metric.

3.3 NAVIGATING ArcGIS PRO: PREPARING CATEGORICAL MAPS

Preparing spatial data for use in FRAGSTATS is straightforward. Let us take, for example, a raster map of land use from the Muskegon River Watershed developed as part of the historical MiRIS database from 1978. The data shown here (Figure Ex3.1) have been reclassified from a level 3 LCCS (i.e., dozens of classes) to level 1 (i.e., seven classes). Right clicking on the raster data layer will launch the actions menu (Figure Ex3.1, label 1). Selecting **Data** (Figure Ex3.1, label 2) and then **Export Raster** (Figure Ex3.1, label 3) launches to the **Export Raster** menu (Figure Ex3.1, label 4). Here, the user needs to type in the name of the output file (Figure Ex3.1, label 5); select the pixel type, preferably 4-bit (Figure Ex3.1, label 6); and then hit **Export** (Figure Ex3.1, label 7).

3.4 NAVIGATING FRAGSTATS FOR SPATIAL PATTERN ANALYSIS

After launching FRAGSTATS, a user is provided with a minimalist interface (Figure Ex3.2A) with two main options, to create a **New** (Figure Ex3.2A, box 1) or to open an existing FRAGSTATS project. Selecting **New** starts the more familiar work menus that are provided to users (Figure Ex3.2B). Note that all major control options are now available (Figure Ex3.2B, box 1), but also the input file menu (Figure Ex3.2B, box 2), the menu to select the pattern statistics (Figure Ex3.2B, box 3), which can be expanded through a set of menu tabs (Figure Ex3.2B, box 4), and side toggle tabs (Figure Ex3.2B, box 5). For this lab exercise, we will be primarily interested in **Class** and **Landscape** Metrics and selecting a few of the pattern metrics that are toggled on and off through the checkboxes located in box 3 of Figure Ex3.2B). Another important area of these main menus is the **Activity log** (Figure Ex3.2B, box 6).

Importing data requires using the Input Layers menu. Selecting **Add Layer** (Figure Ex3.3A, label 1) brings up the **Data** type selection menu (Figure Ex3.3A, label 2). Selecting GDAL *.img option, which was created using the ArcGIS Pro **Export Raster** feature earlier, will launch the **File Selection** menu (Figure Ex3.3A, label 3).

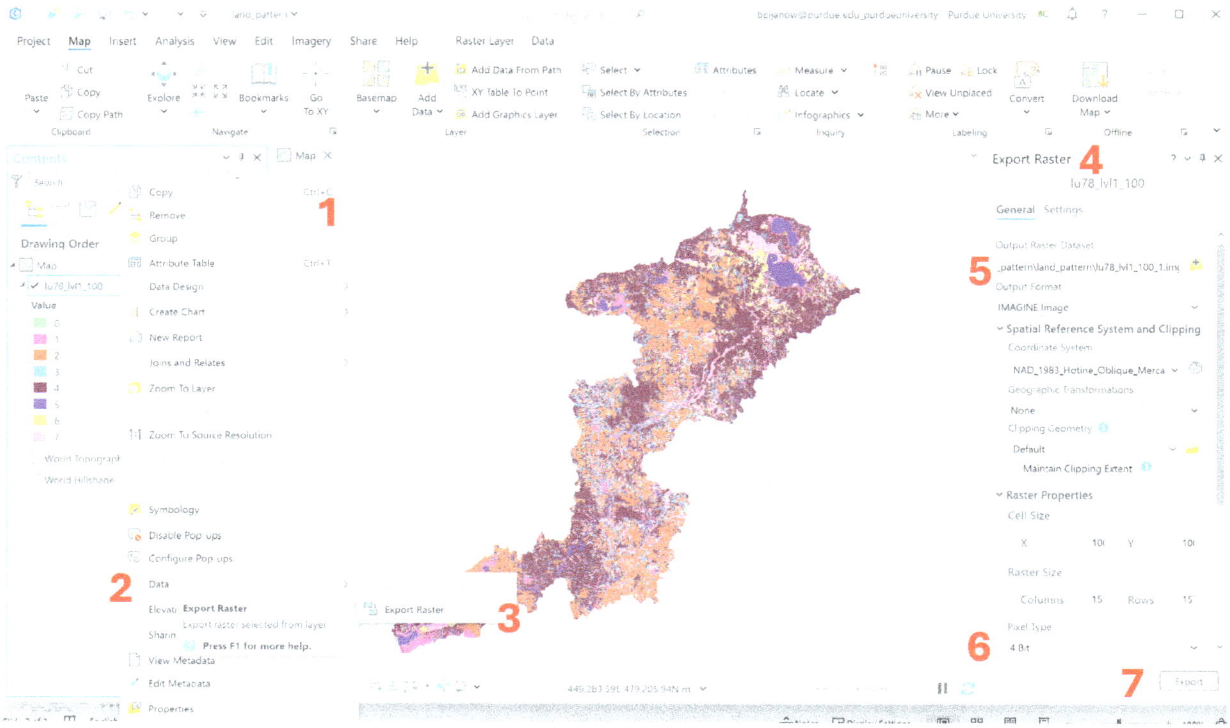

FIGURE EX3.1. Steps needed to process raster data layers for export to an img file for use in FRAGSTATS.

Once the input file is selected, and all spatial patterns are togged on, AND the **Analysis Parameters** (Figure Ex3.3B, label 1), and also selected for output saving (Figure Ex3.3B, label 2), the user can then select **Run** (Figure Ex3.3B, label 3) to have the **Run Properties** menu presented (Figure Ex3.3B, label 4); **Proceed** needs to be selected to run the program (Figure Ex3.3B, label 5). After running the program, an Activity log will present the status of the steps for processing, most importantly, documenting the file names and directories (Figure Ex3.4A). Note that in the file, there is a header followed by lines of data. It should be noted that the user names the base file name for FRAGSTATS output and FRAGSTATS appends the following names onto the base name file name:

.patch. This is output that has patch statistics for each patch in the raster file
.class. This is csv output that has class statistics for each class in the raster file
.landscapes. This is a csv output that has landscape statistics for patch in the raster file
.adj. This is a csv output file that has adjacency statistics; this has to be selected as a checkbox

A way to analyze the output data from FRAGSTATS is to import the csv file into Excel. The best way to do that is to start Microsoft Excel (Figure Ex3.4C, box 1) and then use the **Data** (Figure Ex3.4C, box 2) tab and insert text data (to the far left), which brings up the **Import File** menu (Figure Ex3.4C, Box 3). Navigate to the folder where the FRAGSTATS output is located, and then **Select** the file (Figure Ex3.4C, Label 4) for

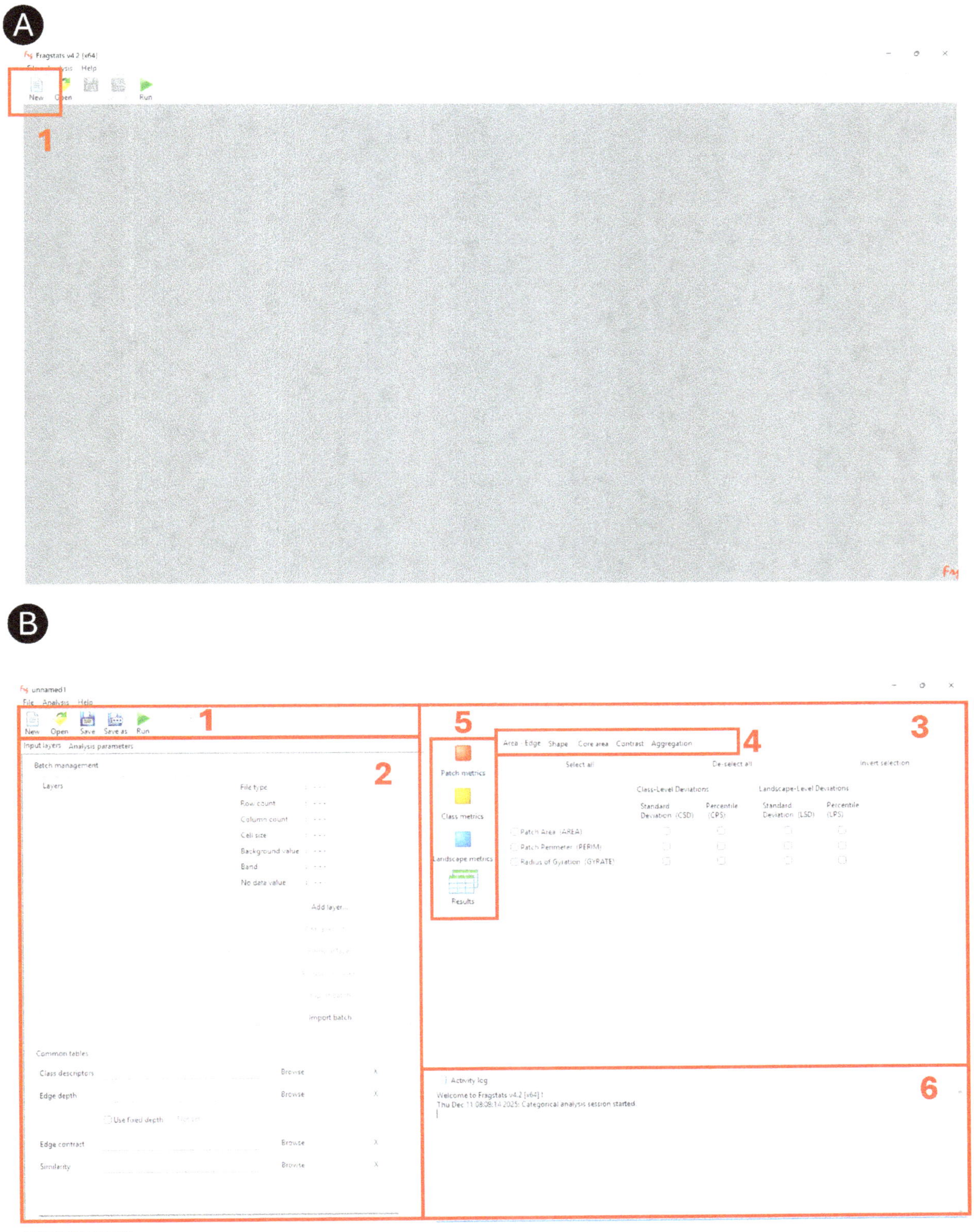

FIGURE EX3.2. (A) The main interface after startup for FRAGSTATS and (B) with a New project started that allows the user to have access to input layers, settings for parameters, and an activity log.

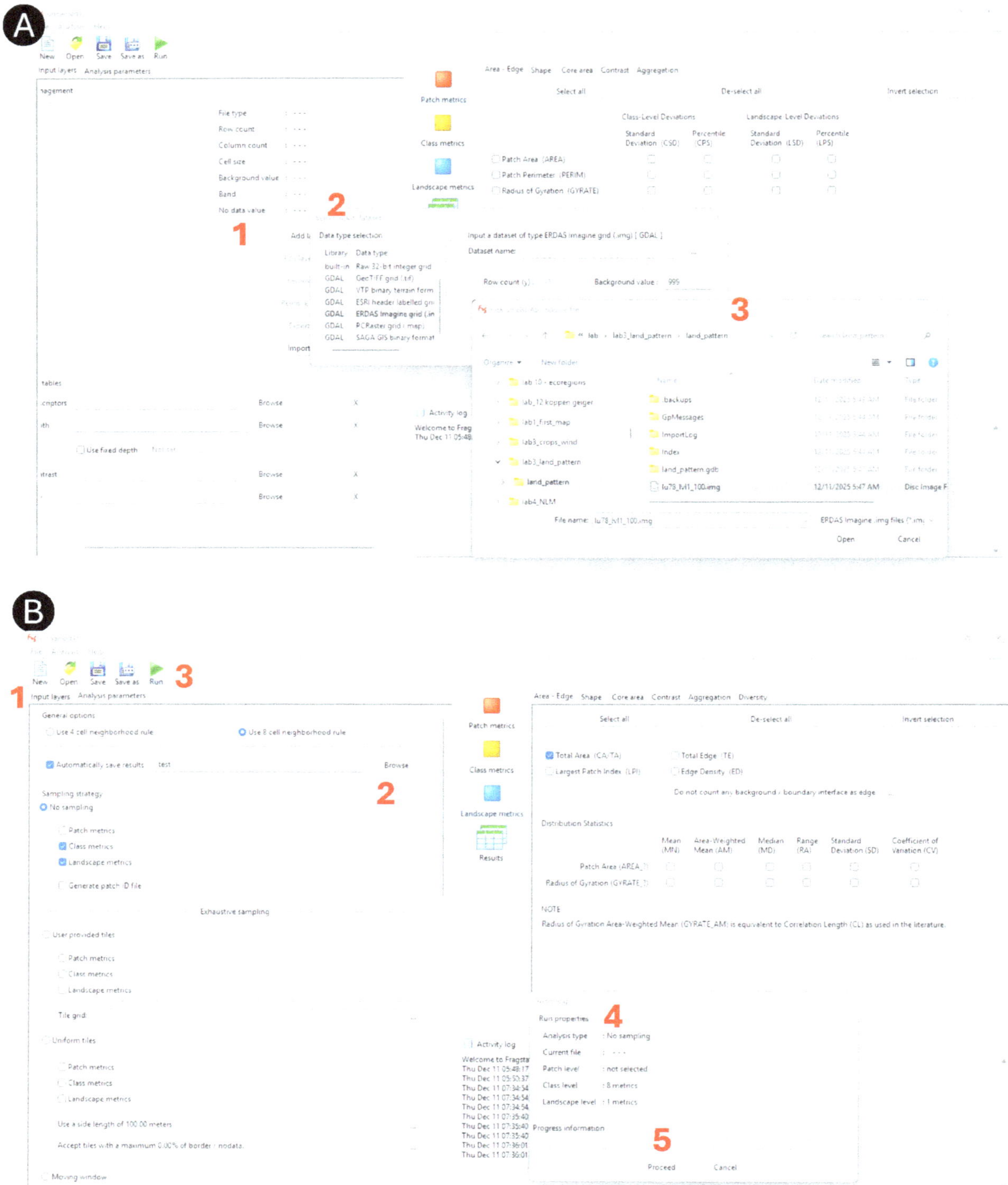

FIGURE EX3.3. (A) The main interface importing data and (B) execution of the tool to calculate the spatial statistics.

FIGURE EX3.4. (A) An example activity log, (B) format of the output for a class statistic file, (C) how a file can be imported into Microsoft Excel, and (D) what the spatial statistics file looks like in Microsoft Excel.

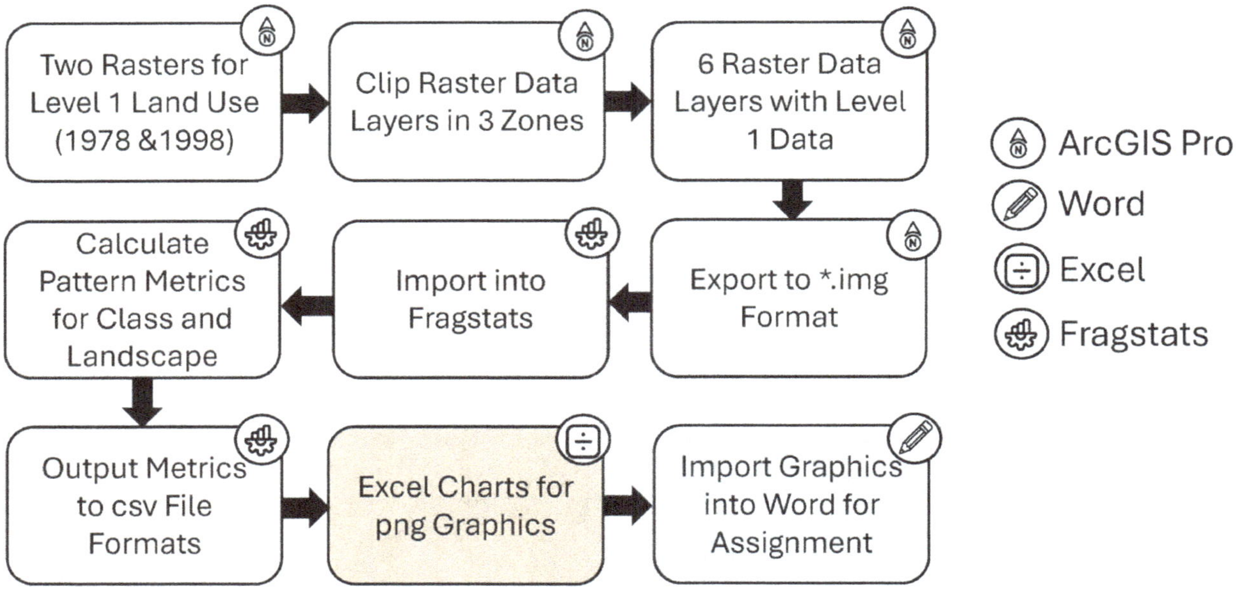

FIGURE EX3.5. Workflow for Exercise #3—Pattern Analysis.

input (Figure Ex3.4C, Label 5). Finish this by selecting the **Import** (Figure Ex3.4C, Label 6) button. Once imported to Excel (Figure Ex3.4D), output can be analyzed. Note here that the column to the far left lists the file name and path (Figure Ex3.4D, label 1); the next column, for this .class file, is the class name (Figure Ex3.4D, label 2); and then the rest of the columns represent pattern statistics (Figure Ex3.4D, label 3).

3.5 WORKFLOW FOR THE EXERCISE

The workflow to accomplish an analysis of landscape patterns across space and time is shown in Figure Ex3.5. Note that raster data are processed using ArcGIS Pro to create 6 raster maps—years 1978 and 1998 and then for Lower, Middle, and Upper portions of the Muskegon River Watershed. The six img files will then be input into Fragstats, with analysis in Microsoft Excel. Maps and figures can be imported into Microsoft Word for a lab report.

3.6 ASSIGNMENTS

A. INTERMEDIATE STUDENTS

For those of you that are undergraduates, produce a document that contains the following. Note that there is no one right way to do this; use your own ideas for how to present this to, say, a watershed planning community interested in the landscape dynamics of this important watershed.

- Figures (bar charts) of three pattern metrics (two for class-level and one for landscape-level) for each of the six maps
- In at least 150 words, describe these changes that are presented in these six charts, emphasizing how changes (1) occurred during the 20 years and (2) how these changes differ across the three zones of this watershed.

B. ADVANCED STUDENTS

Write a report that includes the standard IMRAD format that addresses the following research questions:

- Conduct the same work as that for the undergraduates above except select four class-level and two landscape-level pattern metrics.
- Using the IMRAD format, describe the research in more specific terms (specify research questions), and also include references for the indices you used (trying to determine the original source) and for how others have used these metrics for similar studies.

In your report, cite at least five peer-reviewed journal articles that also describe land change theories, global/regional/local trends, or alternative approaches to spatial analysis.

EXERCISE 4: SIMULATING NEUTRAL LANDSCAPE MODELS WITH GIS

Highlights

Aims. This lab is designed to introduce you to the concept of neutral landscape models and the use of simple tools in GIS that can create random spatial data layers that represent a neutral landscape model. This lab also introduces you to zonal statistics and other zonal commands that can make patch calculations as you would in FRAGSTATS.

Learning objectives. After completing this exercise in about a 3-hour time period, students should:

- Know how to generate a random value raster map
- Understand the concept of percolation
- Understand the workflow of building models, using their output, and then assessing the model through a variety of tools in ArcGIS Pro Spatial Analyst and Microsoft Excel

4.1. INTRODUCTION TO THE CONCEPT OF MODELS

In science and engineering, a **model** is a simplified representation of reality. Models are needed for a variety of reasons. Some models help scientists to understand complex processes by simplifying key elements in equations or diagrams. Other models are used to generate predictions. Yet other models are used to test ideas or hypotheses. For example, an ecologist might ask: If half of the vegetative habitat were to be removed, what population level of species A would I expect to have? Finally, some models are representations of the real world using data, and for the most part, these are known as statistical models.

A common phrase used to describe models is that "all models are wrong, but some are useful." By this we mean that all models simplify the complex real world so that we might understand this world better, but because they are oversimplified, they are likely to not have all of the elements necessary to characterize the complexity inherent in the system under study.

Spatial models are developed and used all the time. In fact, most environmental models have some form of spatial representation. And nearly all spatial models are based on a grid or raster data model.

Overall, scientists have identified three kinds of models and a fourth special form of model (see Keane et al., 2019). The three common forms of models include **conceptual** models, which generally consist of box and arrow diagrams that illustrate key interactions; **mechanistic** models, which attempt to simulate the way the world works using mathematical representations of processes; and **empirical** models, which are models that describe data and relationships between key variables. The fourth form of model, a **neutral** model, uses a variety of random number generators to create hypothetical outcomes that are useful for comparisons, often to the real world (i.e., if your model cannot do better than a random model, then is it useful?). This model overlaps with mechanistic and empirical models but has important distinctions.

This lab will introduce you to one spatial model called a neutral landscape model. It was developed in the late 1980s and 1990s to test some key ideas about cause-and-effect relationships in landscapes. The key question posed by the developers of these models was: Can we generate the patterns that are observed in our landscapes using a random number generator? If so, what does that tell us about understanding the causes of spatial patterns and the ways that we need to assess landscape change?

4.2. KEY PRINCIPLES AND TERMS OF NEUTRAL LANDSCAPE MODELS

A. FUNDAMENTALS OF A NEUTRAL LANDSCAPE MODEL (NLM)

The classic neutral landscape model (NLM) is composed of a grid of m columns and m rows (i.e., it is square). The size, A, of this grid is

$$A = m^2$$

If we allocate values in the grid such that 0 = no habitat and 1 = habitat, and do so such that the occurrence of 0 or 1 is random with a probability of p (i.e., if we were to use a coin flip to determine the value of 0 or 1, the probability of either would be 0.50, or $p = 0.5$), the number of cells in the grid that have a value of 1 becomes:

$$N = p * m^2$$

where N is the number of cells that have habitat in the grid. If the size of m is 10, and $p = 0.7$, then the number of cells that have habitat in the grid is 70 (note that the size of the grid would be 100).

Let us now consider generating a random, binary map just like the landscape ecologists did who developed the first neutral landscape model. If we were to use a random number generator to create a value for each cell, this would be done by the computer's random number generator, which typically assigns a value between 0.0 and 1.0. Most GIS software can do this and create the random value map shown in Figure Ex4.1.

If we were to threshold the map using a value of p, it would create a binary map of 0s and 1s where 0 = no habitat and 1 = habitat. Let's consider how this can be done using three threshold values of p in the maps shown in Figure Ex4.2.

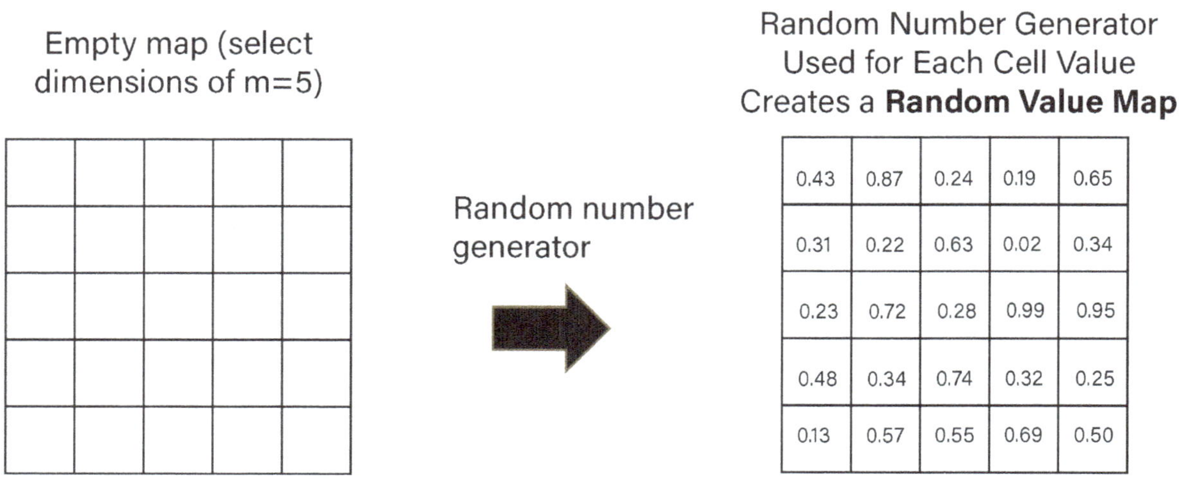

FIGURE EX4.1. How a GIS populates a random value map. An empty map is first created and then a random number generator assigns values between 0.0 and 1.0 as shown on the right.

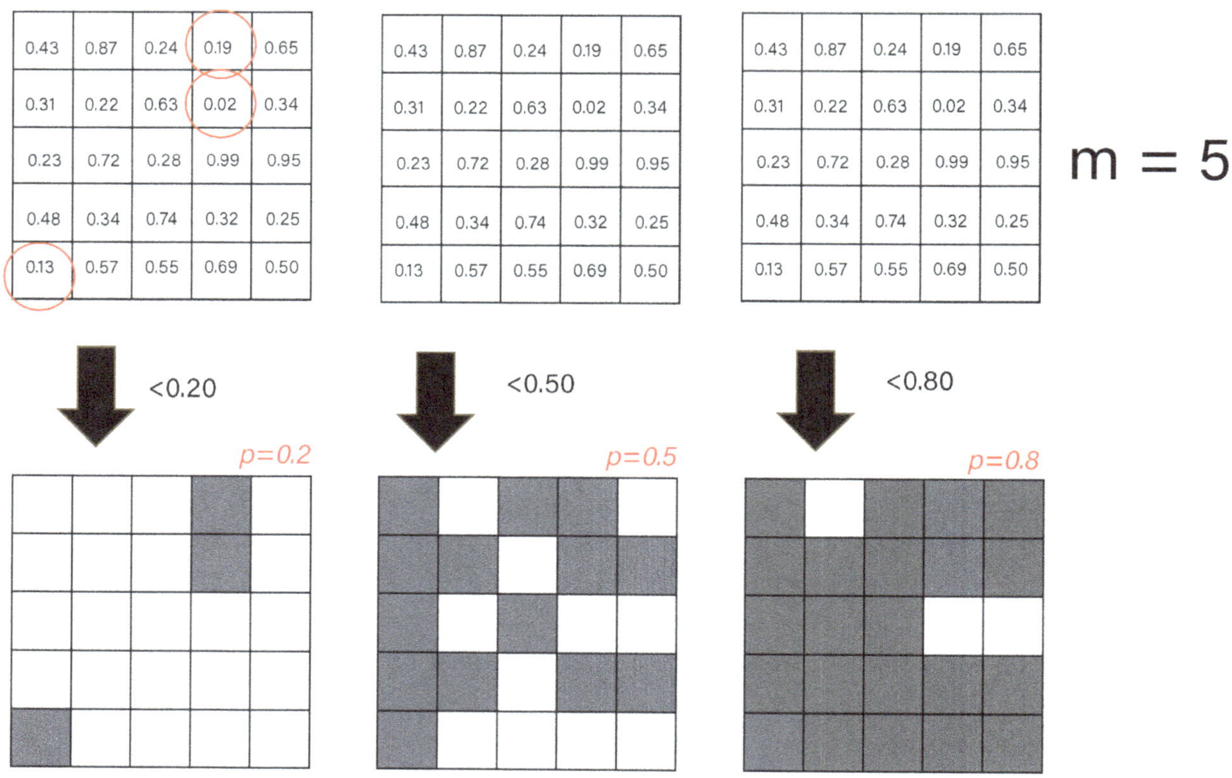

FIGURE EX4.2. A simplification of how a landscape ecologist can use a random value map to create maps of habitat that are simulated by a random number generator. Note how the value of p is used to threshold the random value map to create the presence (1)/absence (0) map.

B. NLM ANALYSIS AND PERCOLATION

Let us now consider how we might analyze NLM maps. Note that in Figure Ex4.2, the map on the left has two patches; the one in the middle has anywhere from one to four, depending on whether you use the 4-cell or 8-cell adjacency rule; and the one on the right has one patch (regardless of the adjacency rule).

Note that if you are managing your landscape to help support animal movement through the landscape, the landscape on the left would be difficult to move through if the animal requires habitat to support the movement (or, if the no-habitat cells are barriers to movement such as being a wetland or urban cover/use). The middle landscape, if the animal can traverse the diagonals, would allow it to traverse from one side to any other side. When this pattern occurs in the landscape, it is said to **percolate** (i.e., think about this as the ability of the animal to move from one side to another). If a 4-cell rule were to be applied, then the right and middle landscapes would not percolate, but the one on the left would have the property of percolation!

C. COMPLEMENTARY SPATIAL ANALYSES

In most geographic information systems packages, a set of spatial analysis tools within the raster environment are ideal to conduct some simple analyses of the output of a neutral landscape model. Since this book and most exercises are designed around the ArcGIS Pro Spatial Analyst Extension, we will use commands and tools in this package as examples.

Four broad categories of spatial analysis are contained in this raster toolbox that can be useful for analysis of NLMs as well as a host of other spatial ecology applications. These include local, regional, focal, and regional tools, which are briefly described here:

1. Local Operations. These are a group of operations that are performed on each cell in a grid. The simplest operation might be a BOOLEAN operation that takes a grid with a range of values (e.g., between 0.0 and 1.0) and then converts these into a binary map using the BOOLEAN statement: if a value is >0.5, then set the value of the output grid to 1, otherwise convert that cell to a value of 0. BOOLEAN operations are logical operations that use terms such as IF, AND, NOT, often with the mathematical operations of >, <, or = (you can also use >=, which is "greater than or equal to"). The result of TRUE throws a value of 1 to the output grid at that cell location, and FALSE a value of 0. Multiplying raster data together is also a local operation (divide all values by, say, 10), as are complex statements that combine mathematical and Boolean operations into a single query. For many of these local operations, `rastercalculator` is the most commonly used tool in the ArcGIS Pro Spatial Analyst Extension.
2. Regional Operations. A small group of operations are often used when conducting spatial analysis across grids. The `regiongroup` command is especially useful for landscape ecologists as this command considers the nearest neighbors in assigning same-value cells to spatial groups (e.g., patches!). `Regiongroup` assigns a unique value to each unique patch (defined using either the 4-cell or 8-cell rule). In a way, this is an alternative to using FRAGSTATS to calculate patch sizes. `Regiongroup` is often a necessary step as part of a zonal operation set of analyses.
3. Zonal Operations. These sets of operations perform calculations on zones or regions. These can be very useful commands to know. For example, let's say you have two raster data layers, one that is land cover and the second that is digital elevation. You would use the zonal operations to create a grid of zones coded by land cover class and then make the calculation of the average elevation for each of the land cover classes.

4. Global Operations. There are numerous global operations that can be used by landscape ecologists in conducting spatial analysis of grid data. Some of the most useful are those that use vector input (points and lines especially) to calculate a Euclidean distance from the nearest feature. For example, you can use the distance tool in Spatial Analyst Extension to assign the vector data for input and create an output with all cell values representing the distance to the nearest feature in the input.

The steps to use a binary map as input to assign unique values to zones and then how to use this analysis are summarized in Figure Ex4.3.

4.3. NAVIGATING ARCGIS PRO FOR DEVELOPING AND ANALYZING RANDOM SPATIAL MODELS

First, open ArcGIS Pro and start a new project. Call it something like MakeRandomAnalyze. Then, using the Toolbox (Figure Ex4.4A, box 1), create a random raster (Figure Ex4.4A. label 2), naming the random map NLM (Figure Ex4.4A, label 3), with the default range (Figure Ex4.4, label 4) and grid size dimensions (Figure Ex4.4, label 5). After running this tool, a raster will appear in the Map view (Figure Ex4.4A, box 6) and then be added to the Contents pane (Figure Ex4.4, label 7). Note the range of values here in the output random map (0.999984 to nearly 0.0).

Next, create a raster data layer for each threshold value from 0.0 through 1.0. To accomplish this, use one of the most used tools in Spatial Analyst, **Raster Calculator** (Figure Ex4.4B). Raster calculator can alter values in one or more raster data layers by specifying an equation to apply. To create threshold values, we will use a Boolean operation on the random map, **NLM** in the **Raster Calculator** pane (Figure Ex4.4B, box 1). Note that there are rasters listed that are active in the **Contents** pane that can be applied to the calculator operations. Select the random raster data layer, **NLM**, and type in the equation as shown at the position in Figure Ex4.4B indicated with label 4. After hitting **Run** at the bottom of the **Geoprocessing** pane, a map of the result that applies the Boolean operation is shown (i.e., if less than 2, then true, which is 1; if not, then false, which is 0). The map is also added to the **Contents** pane (Figure Ex4.4B, label 6).

As we will only analyze the cells that are coded as 1 (i.e., these are the "forests" that will form "patches" in our hypothetical, 250 x 250 cell raster landscape map), let us set values of 0 to no data (i.e., which is also another way of setting this value to null). There is a Set Null operation (a Local Operation) that can be done to any map (Figure Ex4.5A, box 1). To accomplish this, int input has to be set (Figure Ex4.5B, label 2) as well as the Value field with values of 0 (Figure Ex4.5B, label 3). Create an output that can be quickly interpreted (Figure Ex4.5B, label 3) and managed and then hit Run ((Figure Ex4.5B, label 4). Doing so will generate one value (i.e., 1) raster data layer map (Figure Ex4.5B, box 5) with non-1 values set to no data.

Region Group (Figure Ex4.5B) in ArcGIS Pro's Spatial Analyst Toolbox is an interesting command and has some similar functions as FRAGSTATS when combined with other features that are zonal operations. Using the **Region Group** (Figure Ex4.5B, box 1), let us use as input the raster maps (Figure Ex4.5B, label 2) that have 1s and no data as input, and use the neighborhood rule (Figure Ex4.5B, label 3) to define a region (ArcGIS Pro's term for "patch"). Selecting **Run** (Figure Ex4.5B, label 4) will bring up the region map (Figure Ex4.5B, box 5) that codes each unique patch with a code, starting with 1. These region group maps (Figure Ex4.5B, label 6) are added to the **Contents** pane. Viewing the Attribute table (right-click the raster

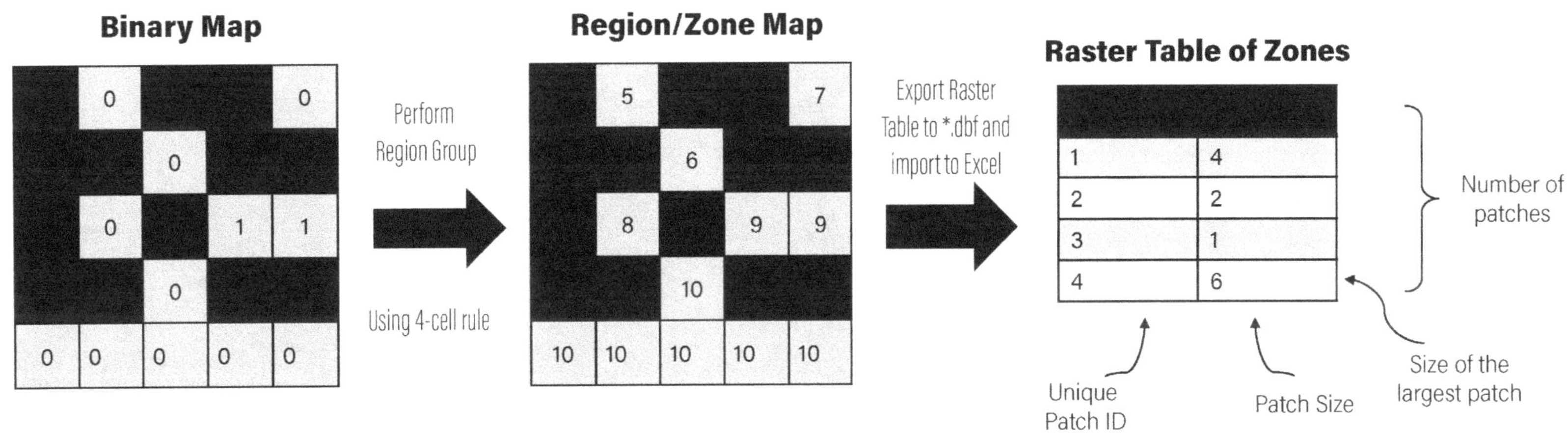

FIGURE EX4.3. Steps involved in assigning zones to a binary map (e.g., 1 = forest, 0 = nonforest) using a 4-cell rule to define unique zones (i.e., patches). Viewing the resultant table for zones would allow users to determine patch sizes and identify the largest patch as well as the number of patches.

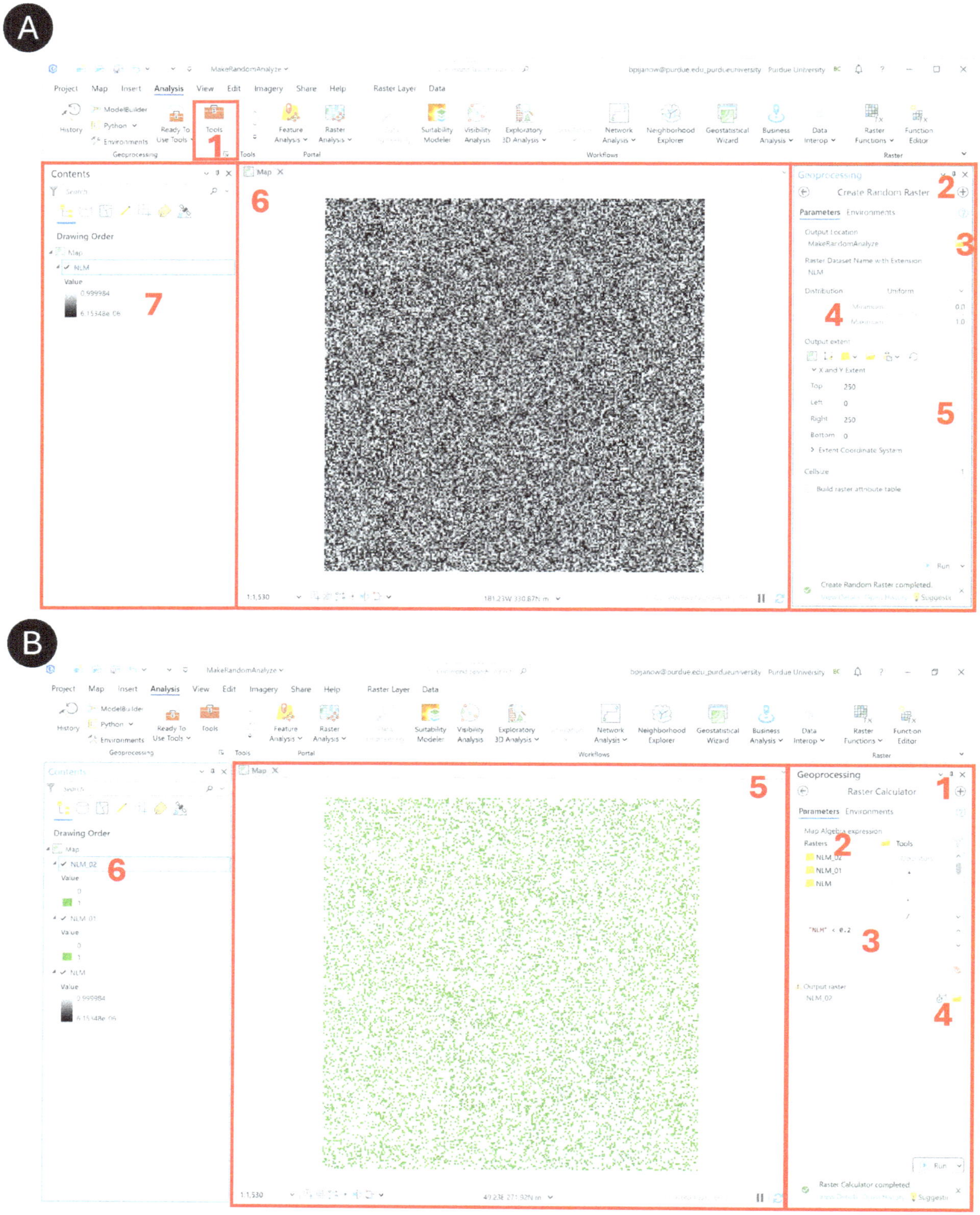

FIGURE EX4.4. ArcGIS Pro interfaces for (A) creating a random raster data layer and (B) applying the threshold p across the random raster data layer.

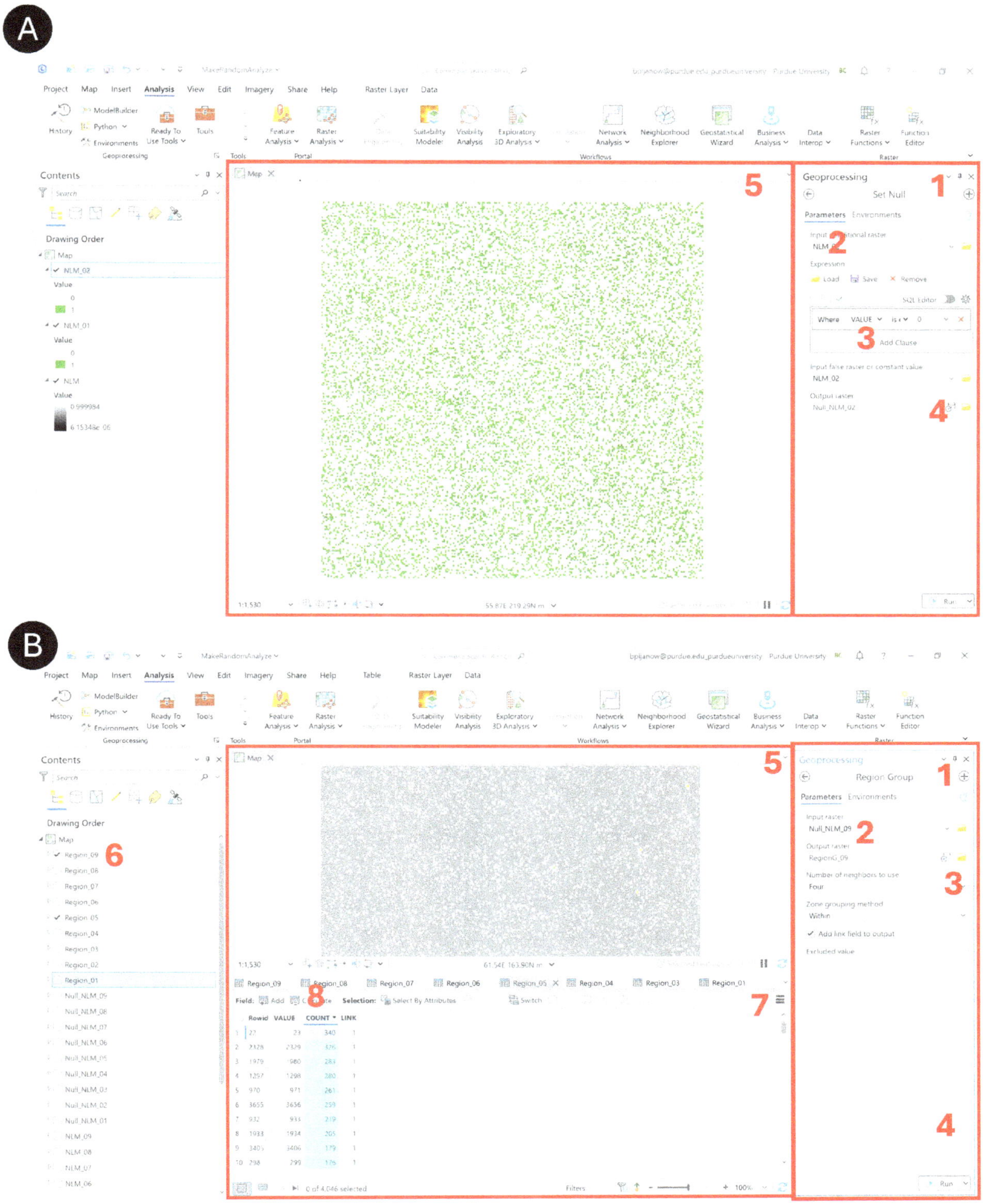

FIGURE EX4.5. ArcGIS Pro interfaces for (A) setting null values and (B) determining largest patch size and number of patches from the regional group attribute table.

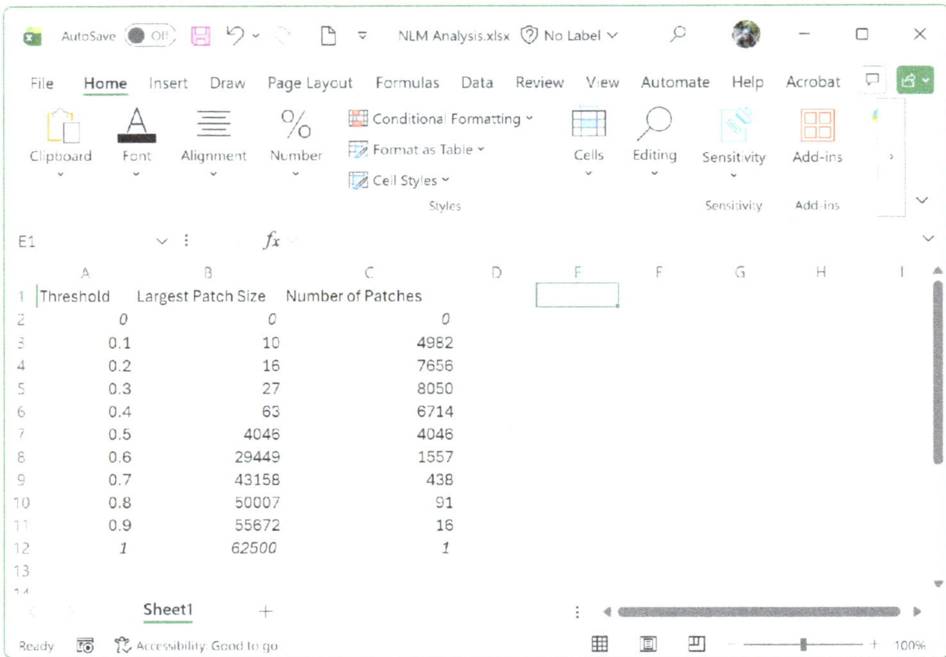

Threshold	Largest Patch Size	Number of Patches
0	0	0
0.1	10	4982
0.2	16	7656
0.3	27	8050
0.4	63	6714
0.5	4046	4046
0.6	29449	1557
0.7	43158	438
0.8	50007	91
0.9	55672	16
1	62500	1

FIGURE EX4.6. Organizing the largest patch size and number of patches by threshold value, p, in Microsoft Excel.

data layer in the Contents pane; Figure Ex4.5B, box 7) will show a table of values (Figure Ex4.5B, label 8) in the region group map; note how count is the number of cells that are in the group (i.e., patch), coded here at 23 (Figure Ex4.5B, label 8). Note also in the Contents pane how the raster data layers have been named so that they can be easily recognized by their short name.

Once all of the region group maps have been produced, two statistics should be gathered for the analysis. The first is the size of the largest patch. This value is obtained from the **Attribute Table** view and then sorted to find the largest value in the **Count** field (which is the size of the largest region or patch). Note that the number of records in the table indicate the number of unique regions (or patches). Thus, the region group map created using the threshold of 0.1 (Figure Ex4.5B, label 8) has the largest patch, composed of 340 cells (as indicated in the **Count** field), and the number of unique regions or patches is 4,046.

Saving these two values for each of the eight raster data layers and then using some simple logic to infer values for 0.0 and 1.0 thresholds, you can input these into Microsoft Excel (Figure Ex4.6). Note how each of the measures of largest patch size and the number of patches change across the value of the threshold (which from the above description of neutral landscape models is the value of *p*) used to create the binary maps.

4.4. WORKFLOW FOR THE EXERCISE

A sample workflow is shown in Figure Ex4.7. Note that this will involve using ArcGIS Pro to create a model and then conducting an analysis that attempts to replicate Gardner et al.'s elegant analysis of model output.

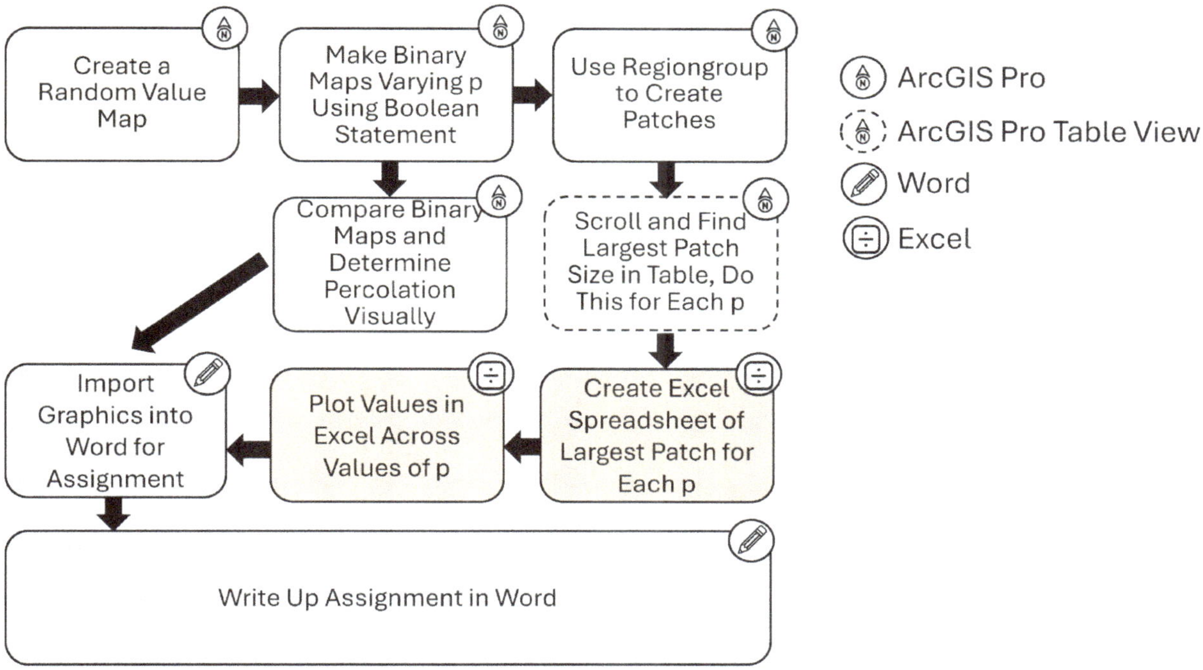

FIGURE EX4.7. Workflow for Exercise #4—Neutral Landscape Model Assessment.

4.5. ASSIGNMENTS

A. INTERMEDIATE STUDENTS

For this assignment, please complete the following:

- Provide two graphs (each replicating Gardner's work) that illustrate how varying *p* affects the number of patches and the largest patch size. Create these two graphs for the 4-cell rule and the 8-cell neighbor rule (i.e., produce four graphs!). Describe the patterns that you observe (in no fewer than 50 words).
- Develop a map that illustrates the concept of percolation. Don't bother with a north arrow or scale bar since these are theoretical simulations. Describe the pattern of percolation (in no fewer than 50 words).

B. ADVANCED STUDENTS

Write a report in the standard IMRAD format that addresses the following research questions:

- What is the effect of using the 4-cell versus 8-cell rule on percolation if you were to use $p = 0.5$? Use the maps you generate to support your answer.
- Use FRAGSTATS to calculate two other patch statistics (e.g., you might consider average patch size and patch shapes) for the raster data layers that are coded as habitat = 1 and all other cells are no data. Use some of your knowledge from Exercise 3 to complete this task.

- How does the 4-cell or 8-cell nearest neighbor rule affect the average patch size and largest patch size trends as one varies p from 0.1 to 0.9?

In your report, cite at least five peer-reviewed journal articles that also describe land change models and the use of neutral models in ecology.

EXERCISE 5: ANALYZING CROPS THROUGH TIME

Highlights

Aims. This exercise is designed to introduce you to the time series analysis tools in ArcGIS Pro and to the use of the USDA's Crop Data Layer (CDL).

Learning objectives. After completing this exercise in about a 2-hour time period, students should know how to:

- Extract categorical values from a long time series (>6) of spatial data
- Process for analysis and then extract these tables and bring them into Excel for analysis
- Work with the rich number of crops that are contained in the CDL **Project Data Resources**, USDA's CDL data for Indiana

5.1 INTRODUCTION TO FOOD PRODUCTION

This lab is designed to introduce you to the rich spatial databases that are available for analysis of crop spatial-temporal patterns and associated environmental (e.g., soils), economic (e.g. farm income), and demographic (e.g., farm size) databases that can be used to assess crop production patterns across large areas of the United States. Much of the data have been collected historically; in some cases, data are available back to the early 1900s. Agricultural data are also released for global coverage, and these are often used by global organizations (e.g., World Bank) to track food insecurity.

One of the common features of crop production data is that the data are often released annually, meaning that spatial-temporal analysis is a critique element for any assessment of patterns of change. Some of the tools in ArcGIS Pro are configured especially to conduct temporal analysis of time series data. We will explore some of the simple time series data analysis tools in the lab within the context of a U.S. Farm Bill (Lark et al., 2015; Plourde et al., 2013). You will also be presented with a brief overview of some of the most common databases that are available for federal and international organizations.

5.2 INTRODUCTION TO MAJOR GLOBAL SPATIAL AGRICULTURAL DATABASES

Globally, there are several large-scale databases that can be used for spatial analysis as it relates to food production in landscapes. These are summarized in Table Ex5.1. Note that many are from the United Nations Food and Agriculture Organization (FAO), which has been collecting these data for decades. Some of the data are produced in collaboration with research institutions (e.g., International Institute for Applied Systems Analysis in Austria) and USDA's Foreign Agricultural Service.

5.3 MAJOR AGRICULTURAL SPATIAL DATABASES FOR THE UNITED STATES

Nearly all data that are used in the United States to track the spatial distribution of agricultural land use and the statistics of agriculture (e.g., acres harvested, number of head of livestock, farm income) are from the United States Department of Agriculture (USDA). Within the USDA, several agencies (e.g., NRCS, NASS, FSA) within a department specialize in various aspects of data collection, processing, and distribution. Table 5.2 summarizes major agricultural databases distributed by the USDA; nearly all are developed as annual databases. Several, such as those from NASS, which is often called the Census of Agriculture, go back over 100 years.

TABLE EX5.1. *Major agricultural databases that are available for global spatial analysis*

DATABASE	SOURCE	DESCRIPTION
Sentinel-2 Land Cover	European Space Agency and Living Atlas at Esri	10 m TIFF images distributed by large tiles or for the entire world for each year. Two products are available, a simple 9-cover class and a 29-cover class (Sentinel2GlobalLULC). Note: pastureland use not specified, rather grasslands and shrublands are listed.
GLC-SHARE (Global Land Cover SHARE)	Food and Agriculture Organization (FAO) of the United Nations	Harmonized global land cover at 1 km
GAEZ v5 (Global Agro-Ecological Zones)	Food and Agriculture Organization (FAO) of the United Nations	Crop suitability in 6 classes, potential yield
SoilGrids (FAO–ISRIC collaboration)	Food and Agriculture Organization (FAO) of the United Nations	Soil property maps
GMIA (Global Map of Irrigation Areas)	Food and Agriculture Organization (FAO) of the United Nations	Irrigated land fractions
GLW (Gridded Livestock of the World)	Food and Agriculture Organization (FAO) of the United Nations	Livestock density maps
FAOSTAT Spatial Data	Food and Agriculture Organization (FAO) of the United Nations	Geo-referenced stats
FRA (Forest Resources Assessment spatial datasets)	Food and Agriculture Organization (FAO) of the United Nations	Forest extent, biomass

TABLE EX5.2. *Major agricultural databases available for spatial analysis in the United States*

AGENCY	DATABASE	SCALE/RESOLUTION	PRIMARY USE
NASS	CDL	30 m or 56 m	Crop mapping for about 120 different crops made each year
NRCS	SSURGO	Field scale	Soil analysis that is used for conservation planning (generally used at county levels)
NRCS	STATSGO	State/regional	Regional soil planning database (generally at state or regional spatial extents)
FSA	NAIP	1 m or smaller aerial imagery	Land cover mapping and for identifying crops
USDA	NRI	Sample-based	Conservation and land use analysis
NRCS	SCAN	Point	Soil moisture and climate monitoring
NASS	QuickStats	County Summaries	Contains statistics, by crop, in acres, of planted, harvested, the number of farms, value of the crops, number of head of livestock, farm income, government payments
USGS	NAWQA	Summary by river reach	Contains geochemical properties of streams including major nutrients

5.4 CROP DATA LAYER OVERVIEW

One of the most useful agricultural databases in the United States is the Cropland Data Layer or CDL. The CDL is produced by the USDA National Agricultural Summary Statistics (NASS); it contains slightly more than 120 crop types for locations at 30 m x 30 m pixel sizes (it was reduced to 10 m x 10 m in 2024). CDL has been released for the US CONUS region since about 1997. Data for download can be found (as of December 2025) at https://croplandcros.scinet.usda.gov/.

We would like to examine how corn planting has changed over time in Indiana. We will focus on two years prior to the 2008 Farm Bill (2006–2007) and then two years after (2009–2010). Recall from Chapter 4 that the 2008 Farm Bill contained a lot of funds to help support biofuel initiatives, especially ethanol production from corn. As a result, our work (Plourde et al., 2013) showed that corn on corn production in many areas of the country, such as in Iowa, changed drastically from the early 2000s through to 2010.

You are being provided with raster files extracted from the original geotiff files that are distributed from the USDA. The data layers are coded 0 (background) and then 1–255 for different crop types and for a variety (~10) of noncrop types (e.g., open water, urban classes, forest classes). Note that corn = 1 and soybeans = 5. Other interesting crop covers include fallow/idle (coded as 61) and pasture/grass (coded as 62). Cell size for these raster databases have already been standardized to 56 m, which is a resolution that is no longer supported as the remote sensing imagery has been reprocessed using native Landsat TM resolution of 30 m x 30 m.

5.5 NAVIGATING IN ARCGIS PRO FOR TIME SERIES ANALYSIS

The key tool in ArcGIS Pro that combines multiple raster data is **Combine.** This Spatial Analyst tool creates an output raster that has values from each input raster layer so that sequences can be analyzed using a table that can be exported to Microsoft Excel. Figure Ex5.1A shows the interface of the **Combine** tool and a sample output raster. Note that in the **Combine Geoprocessing** pane (Figure Ex5.1A, box 1), you can enter a list of raster data layers and name the output raster data layer. Executing this tool is accomplished by hitting **Run.** Using the right mouse button over the raster data layer combine_crop brings up the Options menu, and from that you can select **Attribute Table** to show the raster table view for combine_crop (Figure Ex5.1, box 2). If you sort on **Count** (Descending), you will see results like this where the largest number of cells in the combine_crop map is at the top containing data, in this case, for noncrop land covers (e.g., code 141 is for deciduous forest). Inspection of this table provides clues to how crops were rotated during this time period. Note that for Record 3 (Figure Ex5.1, box 3), we observe a crop rotation pattern of 5,1,5,1,5, which is for a sequence of years planted of soybeans, corn, soybeans, corn, and then soybeans. This represents the traditional corn-soybean rotation. Note that record 4 (Figure Ex5.1, box 4) is for 1,5,1,5,1, which is for the annual sequence of corn, soybeans, corn, soybeans and corn, the other temporal pattern that also represents the traditional corn-soybean rotation. Moving down to record 7 (Figure Ex5.1, box 5), gets us to one of the nontraditional cropping patterns of 1,1,1,1,1, which is the condition where corn is planted for all five years. Record 8 (Figure Ex5.1, box 6), lists 1,1,5,1,5, a condition where there are two years of corn that are then followed by a traditional crop rotation sequence in years 2–5 of the analysis. Also note that the total number of cells that are in traditional crop rotation

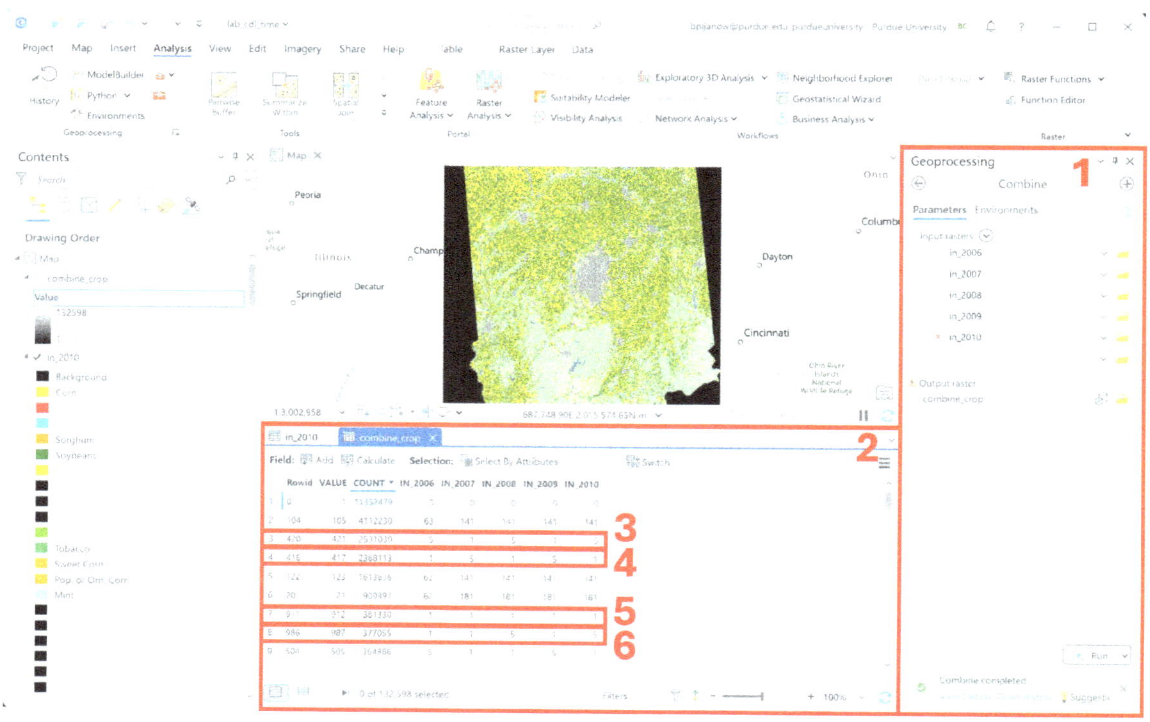

FIGURE EX5.1. Interface to the Combine Geoprocessing Tool and the output of this operation, in Table View, of the time series analysis of the crop data layer.

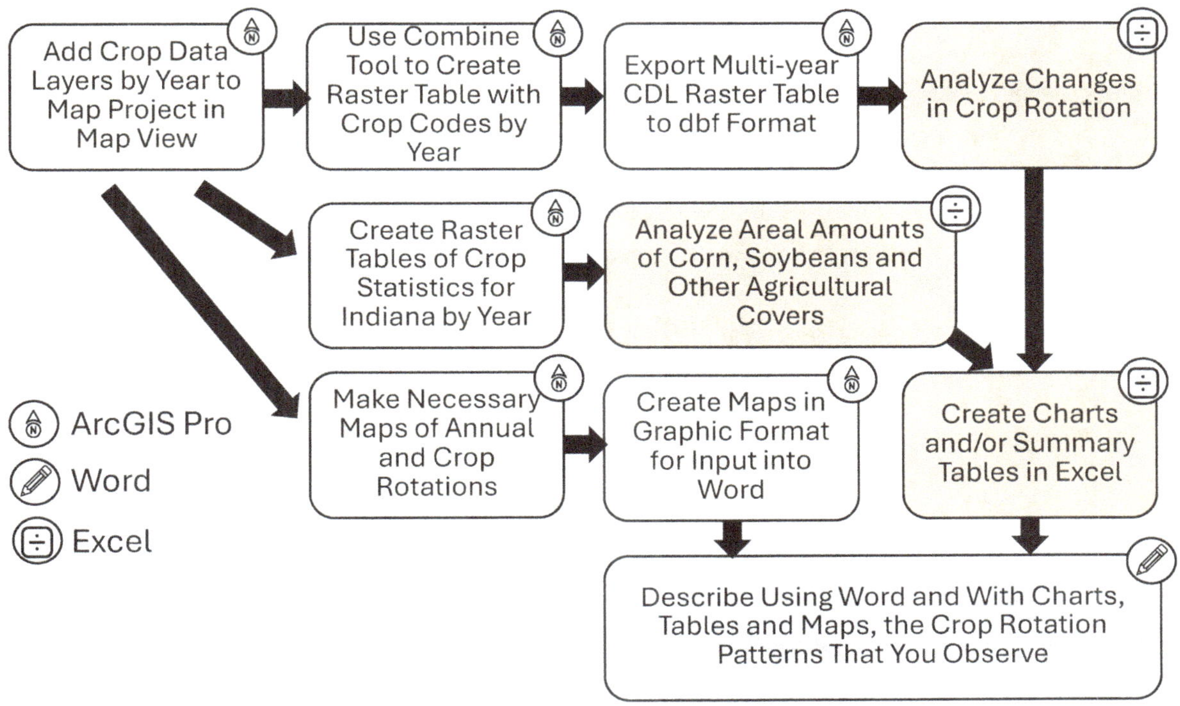

FIGURE EX5.2. Suggested workflow for the crops through time in the Indiana analysis.

(i.e., 2,531,030 + 2,368,113 = 4,899,143) is far greater than the number of cells that are in an "all years in corn" condition (i.e., 381,330). Dividing one by the other, that is, 381,330/4,899,143, means that the all-year crop rotation only represented 7.8% of the area (measured as the number of cells; to calculate true area would require us to multiply by the area of each cell) for the all-year rotation pattern when compared to the area in complete crop rotation. Keep in mind that this represents only one kind of comparison as some of the years will contain traditional rotation that may also include fallow or later years when the land is transformed into urban. Indeed, the number of unique sequences is large (>132,000) and therefore analysis of these data has to be strategic!

5.6 WORKFLOW FOR THE EXERCISE

The workflow (Figure Ex5.2) for this exercise involves an analysis of the top rotation sequences for the state of Indiana prior to, and just following, the 2008 Farm Bill. Analysis also includes creating maps that demonstrate crop rotation patterns.

5.7 ASSIGNMENTS

A. INTERMEDIATE STUDENTS

For this assignment, please answer the following questions:

- How has the *area* of corn and soybeans changed in Indiana between 2006 and 2010? You might consider plotting the total area of corn over time. Use any combination of maps, figures, and/or tables to describe these changes through time. Provide a summary of these illustrations using at least 100 words.
- Use the **Combine** tool to create rotation pattern tables. Compare *patterns of crop rotations* in 2006 and 2007 to those 2009 and 2010. Are there any differences in the total area devoted to continuous corn planting prior to the 2007 Farm Bill (i.e., years 2006 and 2007), which was released after farmers planted? Use any combination of maps, figures, and/or tables to describe these changes through time. Provide a summary of these illustrations using at least 100 words.
- How do fallow and idle land change over this time? Use any combination of maps, figures, and/or tables to describe these changes. Provide a summary of these illustrations using at least 100 words.

B. ADVANCED STUDENTS

Answer the questions above for intermediate students but place your analysis and description of cropping patterns in a standard lab report format with an introduction, methods, results, and discussion. You must cite at least five papers from the literature. You should read Plourde et al. (2013) and include *in your discussion* an assessment of crop rotation patterns as they are influenced by policy and economics. In your report, based on the Plourde et al. (2013) paper on crop rotation, determine whether Indiana is similar to, or different from, crop rotation patterns in the rest of the country. What new information is presented by Lark et al. (2015) that describes the impact of policies on cropping patterns in the United States?

EXERCISE 6: CREATING STREAM NETWORKS AND WATERSHED BOUNDARIES USING HYDROLOGIC TOOLS

Highlights

Aims. This lab is designed to introduce you to the various tools in ArcGIS Pro's Spatial Analyst Toolset that can be applied to any digital elevation model to simulate stream networks, watershed boundaries, and pour point catchments.

Learning Objectives. At the end of this exercise, students should be able to:

- Know how to combine various ArcGIS Pro's Hydrology tools in a simple workflow
- Understand how there are various approaches to coding flowdirection paths
- Understand how these tools can be used to create watershed boundaries and water quality sample point catchments for use in watershed assessments

Project Data Resources. A sample DEM for use for advanced analysis of model output along with a host of other hydrologic data (e.g., streams, well logs) to conduct analyses and to create maps of water resources.

6.1 INTRODUCTION TO HYDROLOGY TOOLS IN GIS

There is so much you can do with GIS to model water on the land surface. A subset of the broader field of digital terrain modeling, modeling the movement of water over the land surface often requires a set of carefully executed commands. ArcGIS Pro's Hydrology tools represent a collection of procedural tasks that are performed on **digital elevation model (DEM)** data to create (1) flow paths; (2) stream networks; (3) basin boundaries; and (4) snap points or pour points. Much of the details of these steps and the rationale behind the coding of each step is described in Maidment (1993).

6.2 PROCEDURAL TASKS FOR HYDROLOGIC MODELING USING DEMS

Figure Ex6.1 outlines the various steps required to process a DEM to create one of several hydrologic maps. The first set of steps involves the processing of the digital elevation model (DEM) to ensure that "blemishes" (e.g., sinks) are removed. Then, a "clean" DEM is used to create either streams, basins, or pour points. The "procedural tasks" interface to one of three major surface ArcGIS Pro commands that relate to water movement: `flowdirection`, `flowaccumulation`, and `flowlength`. How each of these are calculated is described briefly below.

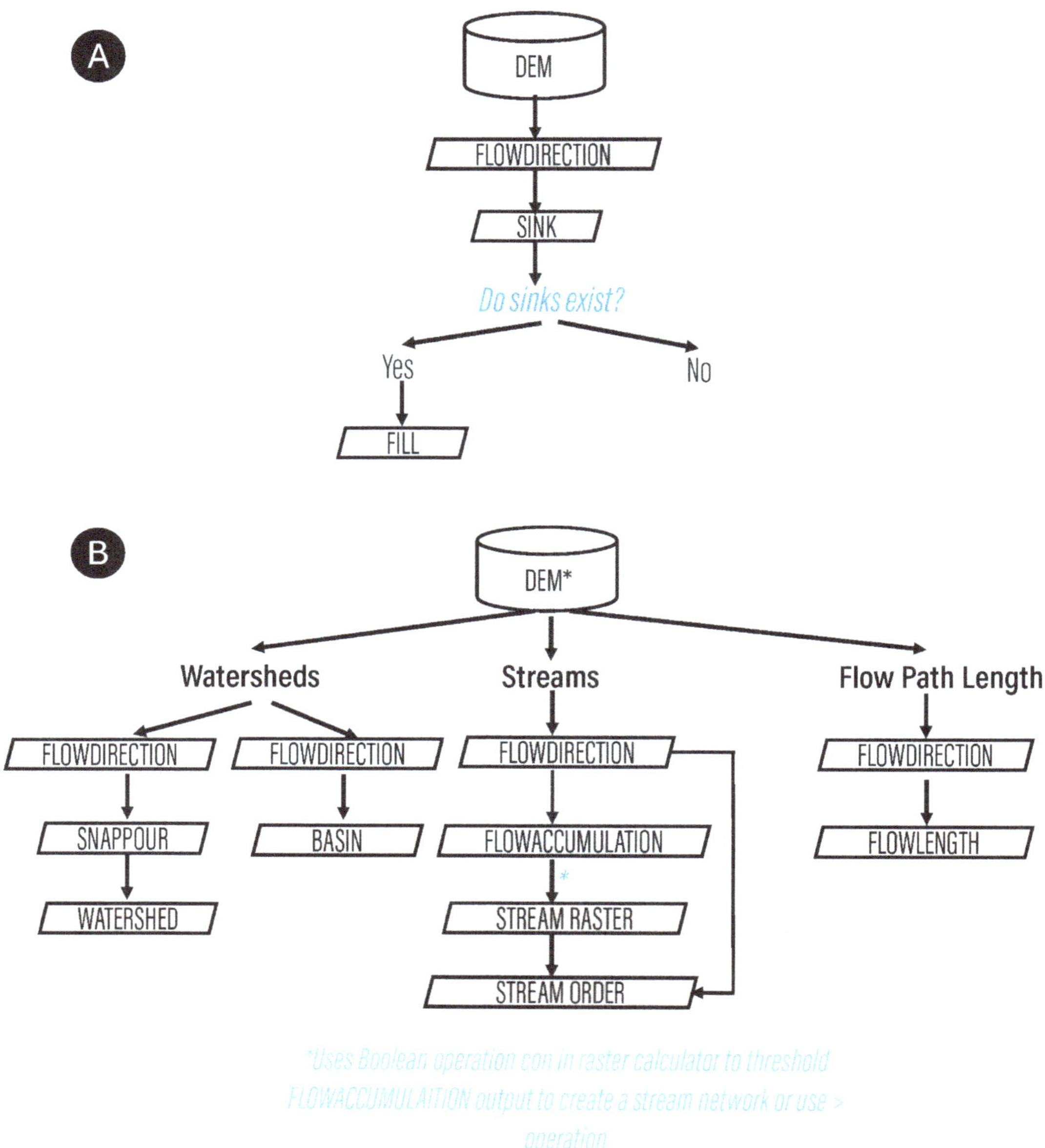

FIGURE EX6.1. The steps involved in (A) preprocessing of a DEM to fill sinks and (B) the use of hydrological tools in sequences to create watersheds, streams, and flow paths. Modified from Esri Online Help for Arc Hydro Tools (https://www.esri.com/en-us/industries/water-resources/arc-hydro; last accessed December 31, 2025).

A. FLOWDIRECTION

Flow direction is a routine in digital terrain modeling that codes the direction of water flow as it moves through the landscape. One of the most commonly used is the D8 (deterministic 8-node) flow rule (O'Callaghan & Mark, 1984). This rule assigns where water flows from a cell to its surrounding eight cells (i.e., it uses the 8-cell neighborhood rule). Figure Ex6.2A shows a hypothetical elevation model for a 5 x 5 grid. The flow direction is coded using a coding system such that 1 = directly east, 2 = southeast, 4 = directly south, and so on, as illustrated in Figure Ex6.2B. Using the D8 rule, a `flowdirection` map can be created, as shown in Figure Ex6.2C. There are several other flow direction models; the D8 model oversimplifies water flow as it directs all of the water from one cell to the lowest neighboring cell. Other algorithms use more than one cell for their flow direction rule (see López-Vicente et al., 2014).

B. FLOWACCUMULATION

The flow accumulation algorithm requires, as input, flow direction maps. Flow accumulation is the addition of all cells from anywhere in the map that flow into that cell. Let us consider the flow direction map using the coding scheme for cell location that is shown in Figure Ex6.3A and use `flowdirection` values from Figure Ex6.2C that now has directional arrows added (Figure Ex6.3B). Note that the upper left cell (Figure Ex6.3B, position [1,1]) has no cells flowing into it, so it gets a value of 0 in the flow accumulation map (Figure Ex6.3C). As we move through the grid (Figure Ex6.3A) to the right (i.e., from position [1,1] to position [1,5]), it is true that no other cells flow into each of them, so all of these positions get a 0 as well in Figure Ex6.3C. As we move down to the next array, we see that as position [2,1] has two cells that flow into it, from [1,1] and [1,2], and so this cell in the flow accumulation gets a value of 2. The `flowaccumulation` algorithm continues along the grid adding up all of the cells that flow into each cell. Note that position [5,1] in Figure Ex6.3C has the greatest number of cells flowing to it. These are from the following cells in Figure Ex6.3A: [4,1], [4,2], [3,1], [3,2], [3,3],[2,1], [2,2],[2,3], [1,1], [1,2], [1,3].

C. STREAM EXTRACTION AND STREAM ORDER ASSIGNMENTS

Creating a stream network using digital terrain modeling approaches is an easy next step although one that is not entirely established with any set rules. Let us consider that to create a stream, there needs to be a minimum number of cells that eventually flow to it. In other words, the overland flow of water that results from a precipitation event such as a rain shower needs to generate enough volume to create the stream. If, say, it is determined that it requires four cells flowing to any cell to create a stream, then we would set this threshold value across the entire grid using a Boolean expression. Setting a threshold (Figure Ex6.3D) using the con command in raster calculator would create a binary map with 1 = presence of a stream cell and 0 = no presence of a stream cell. There would be six cells that are part of a stream network using four flowaccumulation cells as the threshold from our original DEM. This command produces a binary **stream raster**.

The hydrologic tools in ArcGIS Pro can also be used to assign **stream orders** (see Chapter 5, Section 5.5). Stream orders are assigned to the output raster using one of two stream order naming systems, Strahler or Shreve.

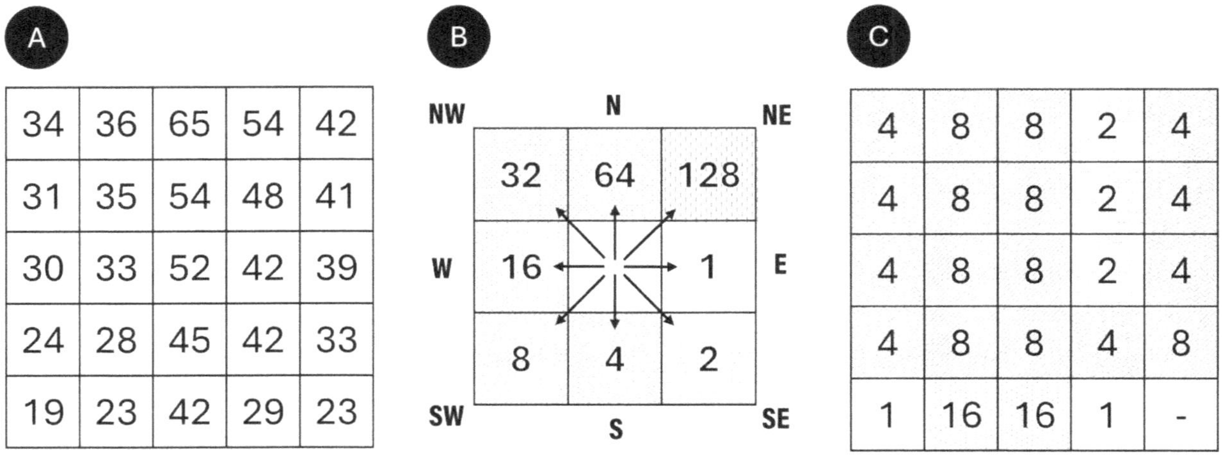

FIGURE EX6.2. Illustration of how the ***flowdirection*** algorithm works from (A) a digital elevation model, (B) the direction coding scheme that is called D8, and (C) a resultant ***flowdirection*** map with D8 codes assigned. Modified from Esri ArcGIS Pro Arc Hydro Online Help for the ***flowdirection*** command at (https://www.esri.com/en-us/industries/water-resources/arc-hydro; last accessed December 31, 2025).

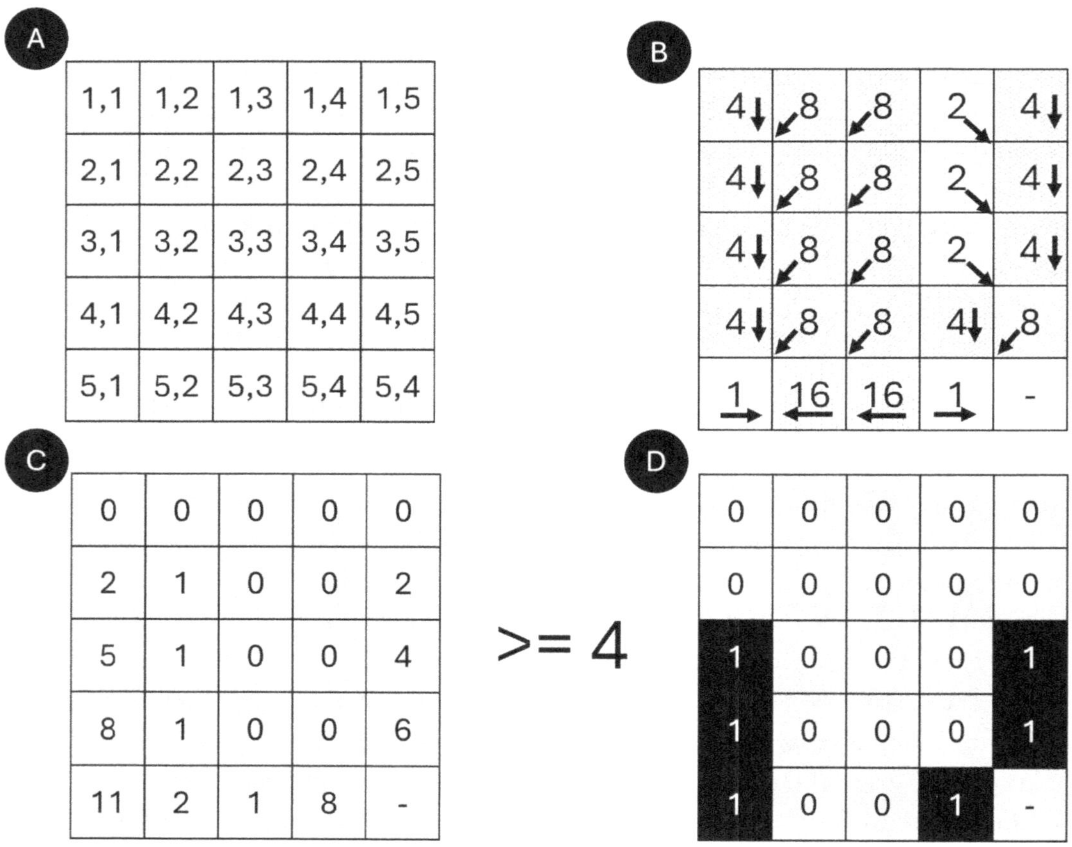

FIGURE EX6.3. Illustration of how the ***flowaccumulation*** algorithm works with (A) coding of position of cells using x and y locations, (B) the direction coding scheme that is called D8 with arrows added, (C) a resultant ***flowaccumulation*** map, and (D) a stream raster grid created using a Boolean threshold of 4 (i.e., there must be at least 4 cells draining to that cell to create enough water to form a stream). Modified from Esri ArcGIS Pro Online Help for the ***flowaccumulation*** command.

D. WATERSHEDS

There are three useful operations in the hydrologic tools of ArcGIS Pro that can be grouped into the broad category of watersheds: `basin, snappour`, and `watershed`. All of these require `flowdirection` maps as input.

The `basins` command uses the entire raster `flowdirection` map to create individual basins or catchments. The operation determines where there are ridges (see Chapter 5, Figure 5.2) and then delineates these lines creating the boundaries of the basin. The watershed command creates "watersheds" based on the location of pour points. Pour points are places along a stream network where a researcher might perform a "grab bag water sample." The upland area that flows to the point is referred to as a pour point. There are two basic kinds of pour points. The first type specify locations along streams in the watershed that are from water quality sampling projects, say, from an EPA Section 319 Project (see Chapter 5, section 5.8). A second type of pour point is to have ArcGIS Pro find all of the stream network nodes and then use these to create `subwatersheds` within the larger watershed. A potential third use of pour points is to use structural locations such as dams and bridges as these points to delineate the overland area contribution to these locations along the stream. What is important for an analyst to consider is that all pour points need to be registered to the stream network that is associated with the DEM used to create the `flowdirection` map. To do this, the operation snappour is used to "snap" any field-generated GPS locations of water quality sampling points or dam locations to stream nodes in the `flowdirection` map.

6.3 OTHER HYDROLOGIC DATABASES FOR LAB

There are many other databases that are generated and distributed by natural resource management agencies and departments of health that can be used for watershed management and/or natural resource management. These include:

1. Well-logs. These are records of the location, date, lithography (geological profile underground), depth, use (home, industry), hydrologic head (depth of where groundwater starts), depth of the well, and sometimes water chemistry. Well-logs are often collected by private well-drilling companies and are sent to state and federal agencies. Agencies include departments of public health and also state and federal geological surveys.
2. Inland lake levels. Quite a few state departments of natural resources, especially in the Upper Midwest of the United States, will often distribute lake-level data, some of which are collected by automated instruments, by their staff, or by citizen scientists.
3. Fish stocking data. Many state departments of natural resources will often distribute fish stocking data, which includes number stocked, species, and date.
4. Public facilities. Many state departments of natural resources will often distribute locations of camping facilities and boat launches.
5. Dam safety data. There are many state DNRs that store and manage locations of dams in their state. These databases often include location, dam type, owner, potential hazards, and other information that can be useful for watershed management.

6.4 NAVIGATING ARCGIS PRO TO ACCESS HYDROLOGIC TOOLS

There are several ways to access the Hydrology tools in ArcGIS Pro's Spatial Analyst Extension. One of the more useful is to view all of the tools that are available through the hierarchical tool arrangement through the long list of **Geoprocessing Tools**. For this exercise, you will use the Michigan DEM for the entire state (Figure Ex6.4A, box 1), which when loaded is rendered as a map (Figure Ex6.4A, box 2). Selecting the **Tools** in the **Analysis** ribbon brings up the **Geoprocessing** Tools as in Figure Ex6.4A, box 3. Scroll down to the **Spatial Analyst** Tools (Figure Ex6.4A, box 4) and then the **Hydrology** Tools (Figure Ex6.4A, box 5). One of the first tools that will be used is the **Flowdirection** Tool (Figure Ex6.4A, box 6) that is described in detail above.

Note that in the **Flowdirection** Tool (Figure Ex6.4B, box 1) there are several options to select from. Hovering over the command's information icon (Figure Ex6.4B, box 2) will bring up notes that are associated with that command. Note also the options that you can select from including how to define the flow direction schema (as described above) and as shown in Figure Ex6.B, label 3.

6.5 WORKFLOW FOR THE EXERCISE

The workflow for this lab (Figure Ex6.5) should follow closely the diagram provided in Figure Ex6.1. The objectives of this lab are to use the Arc Hydro tools to create a stream network that can be compared to a vector stream network created by the Michigan Department of Natural Resources (MDNR) and to create watershed and basin boundaries that can also be compared to those distributed by the MDNR.

6.6 ASSIGNMENTS

A. INTERMEDIATE STUDENTS

Complete the following tasks and include any necessary visuals and short descriptions:

- Generate a map for an area in Michigan (any area) that shows streams containing orders 1–4 that are generated using the Michigan DEM. Show in a map how these are different from streams generated from aerial photography (1:24,000 scale), which is contained in the Michigan Rivers and Streams shape file. Also show watershed boundaries that you generated using the `basins` command.
- Using `hillshade`, generate another map that shows land use/cover for 1998 for the Muskegon River Watershed and elevation/slope using `hillshade` options in the **Symbology** menu.
- Create a map with other hydrologic databases that are typically distributed by state and federal agencies. These include well-logs, base flow estimates, and groundwater recharge areas.

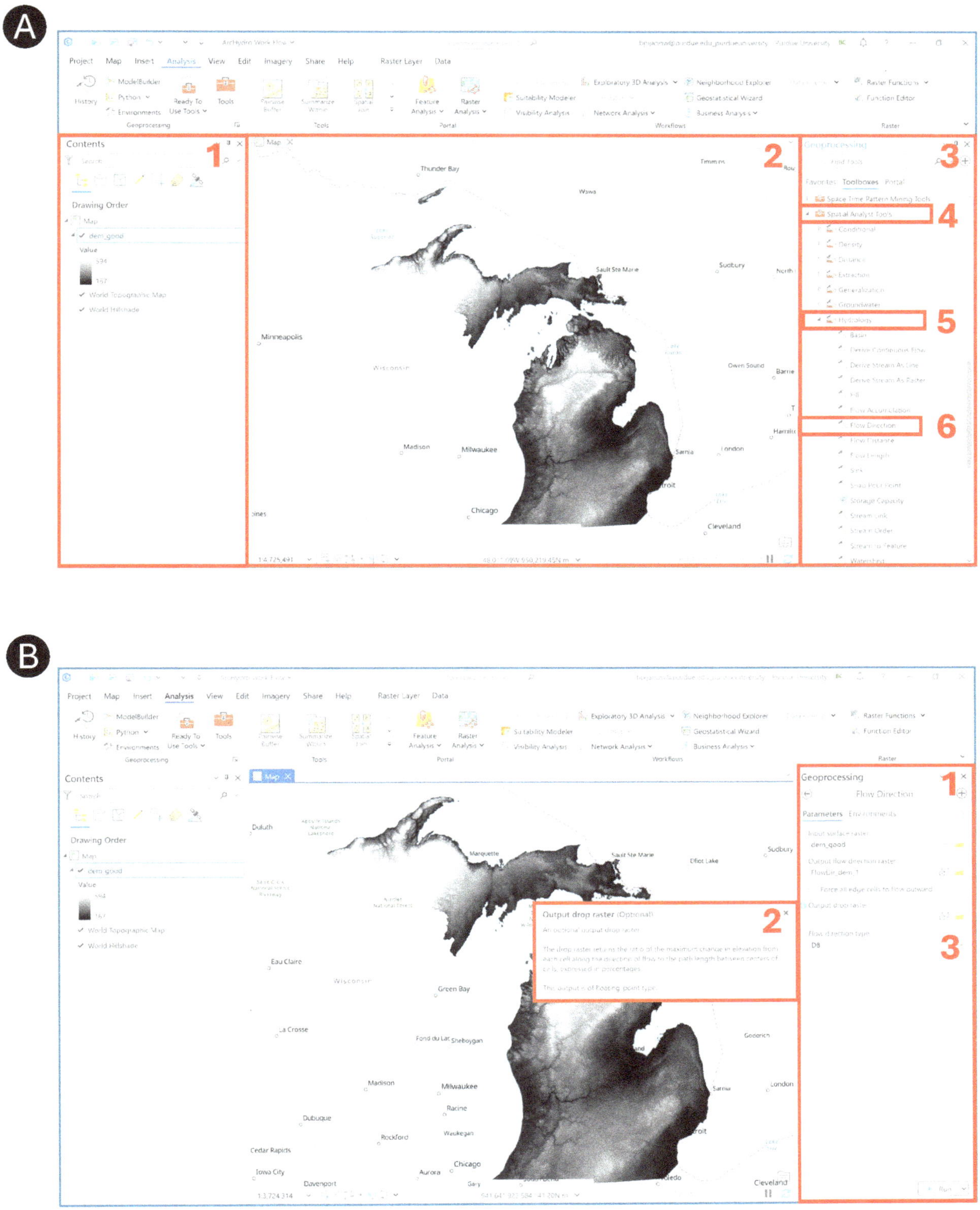

FIGURE EX6.4. The interfaces (A) for the Geoprocessing Hydrology Tools that are designed to operate with DEM data and (B) for the `flowdirection` command and the options for processing.

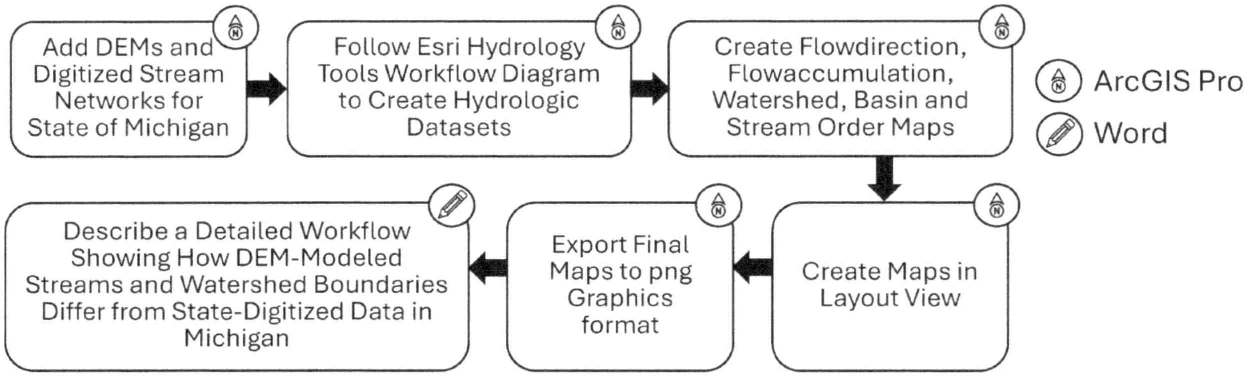

FIGURE EX6.5. Workflow for the Michigan Arc Hydro lab.

B. ADVANCED STUDENTS

1–3. Create the assignment for undergraduate students.

4. Create a watershed map from the Michigan DEMs and compare these with those from the USGS Michigan HUCs at any level (e.g., 8-digital HUC). In no less than 150 words, describe how you created this map and what it tells you about the shape of watershed boundaries in Michigan. Compare the stream networks, illustrating in a few maps where the streams line up and where they do not, and give possible reasons why they do not match well.
5. Write this exercise up using the IMRaD format.

EXERCISE 7: INTERPOLATING POPULATION POINT DATA: THE SPONGY MOTH PEST CASE STUDY

Highlights

Aims. This exercise is designed to introduce you to the spatial statistical tools in ArcGIS Pro as well as the biology and management of an important forest pest, the spongy moth.

Learning objectives. After completing this exercise in about a 2-hour time period, students should:

- Know how to create an interpolated raster layer
- Create maps that indicate threshold risks
- Combine these insect pest maps with those of Michigan forests
- Describe the history of the spread of the spongy moth
- Describe the principles of integrated pest management that are illustrative of the Slow the Spread Project started by the U.S. Forest Service

Project Data Resources. Point trap data for the spongy moth, 1985 to 1994, for the state of Michigan and a map of the forests. These data lack coordinate information.

7.1 INTRODUCTION TO POINT SURVEY DATA

Organismal data are often collected as point information. For example, field biologists collect information about animal and plant populations from traps arranged in grids or transects, and then this information is used to map distributions over space and time. There are at least four different kinds of point location data collected by spatial ecologists. The first are from live small mammal trap data where traps are set, usually baited, by the researcher and mark-recapture studies are conducted to estimate the population abundances of one or more species (Skalski & Robson, 2012). Pitfall traps (Boetzl et al., 2018) are also used for insects and these are typically made as dug-out holes in the group that are constructed so that individuals, like ground-dwelling beetles, fall into the trap and cannot get out. The second kind of point data are noninvasive (i.e., nonhandling of individuals) from acoustic and camera sensors (the latter as camera traps), which are set out in a network and species counts are made and associated with a specific point location. The third kind can be broadly classified as lure traps, often used in insect abundance studies, where a chemical or color lure is set out in the environment and the number of individuals for each species is counted. Finally, the fourth kind of point location data are created by human observers who record their location with a GPS and make standard observations. Call stations have been used for decades by herpetologists to estimate population abundances of frogs and toads (Fellers, 1995).

This lab will focus on analyzing lure trap data for an invasive insect, the spongy moth. These lure traps have been historically placed in grids, and we will use historical data from a statewide program in Michigan to create interpolated maps of population distributions and some "risk" maps, as well as presence/absence maps.

7.2 *LYMANTRIA DISPAR*—A FOREST INSECT PEST

One of the largest studies on animal distribution dynamics is that of the *Lymantria dispar* (its common name is spongy moth, but it was also known as the gypsy moth until 2022), a forest pest that was accidentally introduced to the northeast in the 1860s (Liebhold et al., 1992). The larva form of the moth (Figure Ex7.1A), the caterpillar (Figure Ex7.1B), feeds on leafy material of over 300 species of trees and plants. This moth has spread slowly westward and has caused considerable damage to forests in private and public lands over the last several decades. In the 1950s, an isolated population of spongy moths was detected in the central part of the Lower Peninsula of Michigan near the city of Lansing. In response to this occurrence, the Michigan Department of Agriculture began a program aimed to eradicate this pest using insecticides, likely DDT, in large amounts and dosages; in fact, this specific Michigan example of aggressive government action was highlighted by Rachel Carson in 1962 in her book *Silent Spring*, where she described it as "reckless" chemical use applied "indiscriminately from the skies" "at large scales" with likely negative impacts to the environment, especially to wildlife. As spongy moth populations continued to grow and spread, the Michigan Department of Agriculture established a statewide monitoring program using lure traps, starting first for the Lower Peninsula (in 1984) and then extending it a year later into the Upper Peninsula (Yang et al., 1998). Indeed, *L. dispar* males are drawn to pheromone-baited lure traps very effectively, and so this has become one of the most effective monitoring approaches of this pest; counting the number of egg cases in an area is also used. Spongy moth females are large and cannot fly, and so males with their large antenna are attracted to the females by the pheromone disparlure. A synthetic version of disparlure has been created, and this is placed in "milk carton"–style traps (Figure Ex7.1C) that have a small opening on the side. Once the moths enter, they are killed by an insecticide.

The moth has dispersed across landscapes in the United States in two ways. The first is through its natural dispersion mechanism of larval ballooning. This is the process of creating a long silk thread and, when the larva is at the top of a tree, the silk is let out allowing the wind to carry it through the air. This ballooning is likely only for a few hundred meters but there have been some reports that they could travel much farther. A second mechanism of dispersal is caused by humans, who inadvertently transport egg cases that have been laid by females on logs (i.e., firewood). The transportation of these by vehicle can be quite far. Indeed, the occurrence of the very isolated population in lower Michigan has been hypothesized to have occurred by transporting firewood from the northeast of the United States where this forest pest had already been firmly established in the early 1900s.

FIGURE EX7.1. *Lymantria dispar* (A) adult (a male moth with sensitive antenna to detect female pheromones); (B) the caterpillar stage, which is responsible for massive defoliation, and which eventually pupates into the moth during the middle of the summer; and (C) a *Lymantria dispar* pheromone trap in the style of a "milk carton."

Counts of the number of males in traps have been used as proxies for population abundances. Large-scale monitoring programs have been implemented by placing hundreds of traps in a grid arrangement and entering their locations into a GIS to map distributions. We know from research in West Virginia, Virginia, and Michigan that exceeding 300 moths is typically correlated with defoliated forest stands (Tobin & Whitmire, 2005), although in some forests, such as those in Wisconsin, defoliation can occur when there are between 100 and 300 moths per trap; this is a threshold used to trigger natural resource management response. If defoliation occurs two years in a row, tree mortality is possible.

7.3 SLOW THE SPREAD MANAGEMENT PROGRAM

A. EARLY PROGRAM YEARS

In the early 1990s, forest entomologists at the United States Department of Agriculture's Forest Service, along with a few university researchers, proposed a new management paradigm to address the invasive spongy moth; they labeled this program "Slow the Spread" (StS). The core idea of this management approach was to track population levels across a large region and to eradicate new populations along what they called the transition zone. Doing so would slow the spread and its natural coalescing of small populations that often resulted in producing contiguous populations (Tobin & Blackburn, 2007). An economic analysis during this time (Leuschner et al., 1996; Mayo et al., 2003; Tobin & Blackburn, 2007) also suggested potential economic benefits as the reduction in spread over decades would be more than the cost of monitoring with hundreds of thousands of traps combined with local eradication programs. For these early program years, participating states in this pilot program included those that contained the pest along its North American leading-edge population, which included Virginia, West Virginia, North Carolina, and Michigan. Wisconsin joined toward the end of this early StS field test site phase. In the 1990s, trapping protocols and dispersion modeling studies were conducted that resulted in the launch of a fully operational national USDA Slow the Spread program in the year 2000. The early 2000s and into the 2020s saw the states of North Carolina, Virginia, West Virginia, Kentucky, Ohio, Indiana, Illinois, Michigan, and Wisconsin participating.

B. IMPLEMENTATION YEARS AND PROGRAM DESIGN

The Slow the Spread program has two different kinds of trapping densities depending on their purpose. The first are called action area detection grids. These are trapping grids that have a 2–3 km intertrap spacing set 10 km ahead of the previous year's 10-moth line (although some states will go as far as 50 km ahead of 10-moth delineations) and are used to be the early warning detection tools that assist in determining if any leapfrog populations have been initiated. The second set of grids is called delimitation trapping grids. These are established once the early warning monitoring sites detect any new populations. The purpose of the delimitation grids is to define the spatial extent of this new population so that eradication or suppression projects can be conducted. These delimiting grids are often set up with 500 m intertrap spacing (1 km spacing for large delimitation areas). This results in about 9 traps per square mile.

Eradication efforts focus on several approaches (USDA APHIS, 2024). One of the most common is the aerial spraying of a microbial insecticide known as *Btk* (*Bacillus thuringiensis* var. *kurstaki*). This biological insecticide is applied generally from boom sprays mounted on a helicopter or plane. The substance lands on the surface of the leaf, and the caterpillar, which consumes leaves for food, ingests this microbe, which then kills the instar. Another approach is to apply diflubenzuron, also aerially applied, which, after ingestion, interferes with the molting process. Finally, mating disruption approaches have also been used by either dispersing disparlure flakes or by mass trapping. In some cases, areas where populations have existed for many years are also targeted for suppression efforts. Here, high population levels are targeted if they become a nuisance or are damaging large numbers of trees.

The Slow the Spread program is also designed as a cooperative, mult-agency program (Sharov et al., 2002; Tobin & Blackburn, 2007) where funding is provided for coordination, public outreach, and engagement efforts. One of the key events historically has been the support for local meetings where community members can come to an evening meeting at a school or a state/federal office to learn about the biology of the moth and how it is being managed. The StS program also helps to support education programs, especially messaging that focuses on not transporting firewood, as this is one of the main ways that the spongy moth is moved ahead of its natural dispersion zones. The Slow the Spread program also helps local communities with identifying funds for eradication and education efforts.

7.4 NAVIGATING ARCGIS PRO TOOLS FOR INTERPOLATION

A. OVERVIEW OF ARCGIS PRO CAPABILITIES

Geostatistical and spatial analysis functionality are distributed across four sets of tools packaged mostly as add-on extensions in ArcGIS Pro: **Spatial Analyst** (raster-based analysis and modeling), **Geostatistical Analyst** (surface modeling and interpolation), **Spatial Statistics Toolbox** (clustering and spatial pattern analysis), and **3D Analyst** (surface and 3D analysis). Each of these has different outcomes, although many of the sets of tools within each of them are similar. Spatial Analyst, the most "tool rich" of the three, is most often used to interpolate point data and to create output that are represented as raster data. Deterministic and geostatistical approaches described above are included. Geostatistical Analyst is the preferred set of tools that allow users to conduct spatial statistical analysis and to explore mean and variance across space. Output is often a statistical raster layer. These tools are best to understand the nature of spatial data and where uncertainty measures or error measures are important. Deterministic and geostatistical approaches are included along with a few interpolation model cross-validation tools that are typical of geostatistical modeling (e.g., error terms for fitting the nugget). Finally, 3D Analyst contains a set of interpolation tools where the output includes TIN spatial data models for visualizing output in three dimensions.

Within ArcGIS Pro, a lot of the workflow for interpolation and spatial statistical modeling is integrated into an interactive, step-by-step workflow set of user interfaces called the **Geostatistical Wizard**. For many of the built-in steps, it is very interactive, in that the user can change spatial statistical model parameters and have instant visualizations and calculations presented for key model output statistics (e.g., nugget value). This lab will introduce you to the Geostatistical Wizard as well as some of the ArcGIS Pro **Catalogue Tools** for **Data Management**.

B. CATALOG TOOLS FOR DATA MANAGEMENT

Management of spatial data is often best accomplished using ArcGIS Pro's Catalogue set of tools that are part of the set of data browsing and management utilities. For this exercise, we will explore some of the main features of this set of tools and use it to manipulate the data in the historical spongy moth tables.

Note that when this interface is launched (Figure Ex7.2A), there are also a set of ribbons that allow users to work with spatial data (Figure Ex7.2A, label 1). The directory information is also provided in the **Contents** pane (Figure Ex7.2A, box 2) much like Windows Explorer. The **Catalog** view (Figure Ex7.2A, box 3) shows the contents of the folder, and if a file is selected (e.g., Figure Ex7.2A, box 4), three separate views (Figure Ex7.2A, box 5) can be provided: (1) metadata; (2) geography (as a map); and (3) table (i.e., the attribute data). Since these are very old files, there are none of the traditional metadata files associated with the shape file. To enter information into the metadata file, the **Metadata** part of the ribbon (Figure Ex7.2A, box 6) allows users to Edit the metadata file. In some cases (as we will see in the next paragraph), other options for working and processing data are made available through the **Geoprocessing** pane (Figure Ex7.2A, box 7).

As you will be asked to create maps of where the spongy moth exists based on trap data, let us process the data to code the locations where there are no moths captured (e.g., moths = 0) and then for the case where moths have been detected (e.g., moths > 0). To accomplish this, right click the data layer for one of the years of trap data in the **Catalog** pane (Figure Ex7.2B, box 1) to activate viewing the Attribute Table (Figure Ex7.2B, box 2). To add a field for moth detection, select **Add** (Figure Ex7.2B, box 3), and then use the interface to Tables

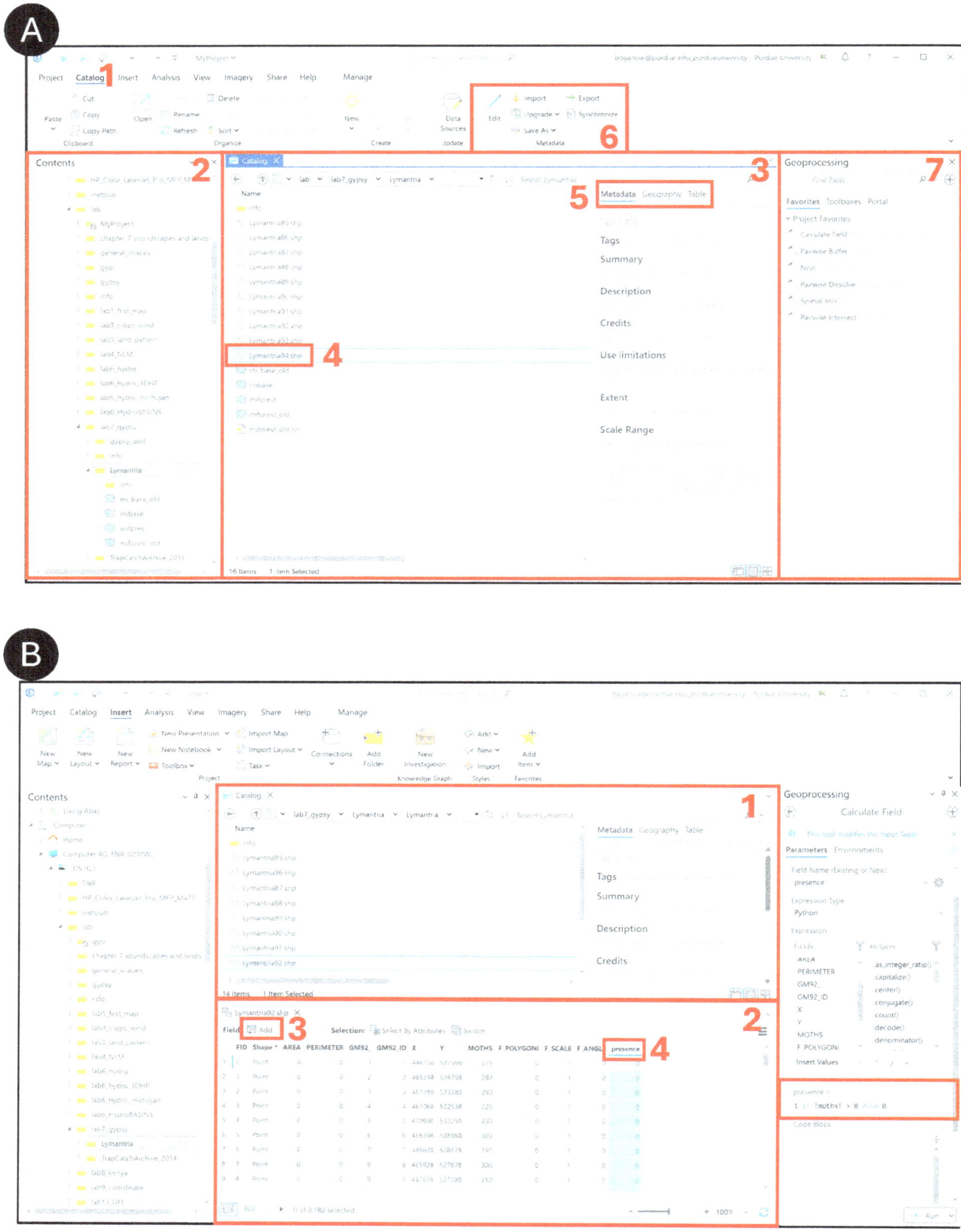

FIGURE EX7.2. The Catalog View of ArcGIS Pro for (A) exploring and viewing spatial data and (B) interfaces to manipulate tables of the spongy moth trap database.

Field view to add the new field presence (Figure Ex7.2B, box 4). Select, in the **Geoprocessing** pane, **Calculate Field** (Figure Ex7.2A, box 7) to bring up the **Calculate Field** pane. Enter the field calculation statement as shown in Figure Ex7.2B, box 5 and then select **Run**. You can double-check your work by scrolling through the table and making sure that where values of moths are greater than 0, the presence field is a 1, and when there are no moths present, the value in presence is 0. You will need to repeat this for each of the years, 1984 to 1995. You should also note that the coordinate system for these data has been lost, so the data will not be able to be viewed with any other data.

C. GEOSTATISTICAL WIZARD

Return to Map data view for the rest of this exercise. Note that what follows is only a general guide as the workflow contains many steps and a lot of interfaces. Navigating to the **Analysis** (Figure Ex7.3A, label 1) set of tools, the user is presented with analysis tools including a hotlink button to the **Geostatistical Wizard** (Figure Ex7.3A, label 2). Selecting that button will launch the interactive, step-by-step geostatistical model builder (Figure Ex7.3A, box 3). On the left of the first menu of the wizard is a list of the most common geostatistical models (note those that include kriging as well as those based on kernelling and deterministic mathematical equations e.g., **Inverse Distance Weighting**). Select **Kriging/Cokriging** (Figure Ex7.3A, label 4) and then the correct Field is Moths (Figure Ex7.3A, label 5). Once you have done this, hit Next (Figure Ex7.3A, label 6). The next menu will allow you to select from about six varieties of kriging models (use **Ordinary** and then **Prediction**, applying the default parameters) and then select **Next**. The subsequent menu provides a host of interactive features that are controlled through the **General Properties** menu (Figure Ex7.3B, label 1). Make sure you select the **Function Type** of **Semivariogram** (Figure Ex7.3B, box 2), which will then display (Figure Ex7.3B, box 3) the model as a line with the data scattered as it plots lag distances (x-axis) and semivariance (y-axis). Note that below the graph is the semivariogram equation with the parameters (Figure Ex7.3B, label 4). In the lower left is a map (Figure Ex7.3B, label 5) of semivariance that can help determine if there is a need to use an anisotropic (i.e., directional) model. If the patterns are circular, moving out from the center, an isotropic (no directional effects) model is fine. If not, then adding directional effects could improve the model fit (it might here as there is a clear east-west stretch to the semivariance). Once this interactive portion of the modeling is complete, the user can move to the next step by selecting **Next** (Figure Ex7.3B, label 6).

The resultant visual, which produces a "map" representing the model (Figure Ex7.4A, label 1) might seem like a mistake, but it is not. The visual product of the Geostatistical Wizard is a model, not a data product per se. Selecting this Model Layer (Figure Ex7.4A, label 2) will bring up the Options Menu (Figure Ex7.4A, box 3). Selecting the Export Layer and then To Raster (Figure Ex7.4A, label 4) will allow you to save the model as a raster data layer. You can use the **Parameters** tab on the **GA Export to Raster** pane to "register" the new raster to existing data dimensions (which is always a good idea). Select the 1 km x 1 km miforest raster layer to register this new model data layer.

D. COMBINING RASTER DATA LAYERS

You may encounter a challenge working with the new raster data layer in that it might have created estimates of the spongy moths that exist outside of the land boundaries. To "clean up" this model output, you can use any simple raster calculator steps. For example, you might consider using a mibase map that contains the presence

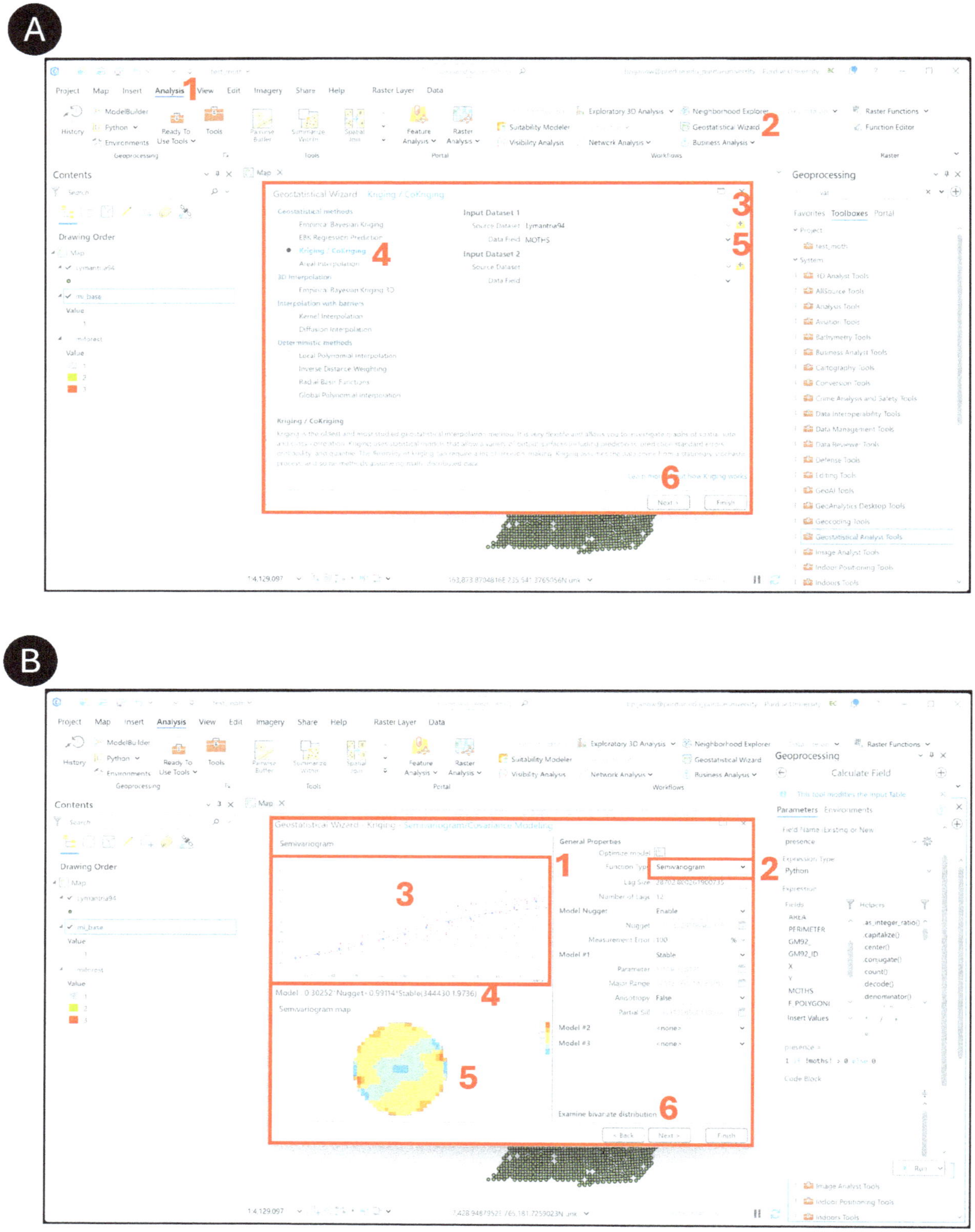

FIGURE EX7.3. Some of the menus of the Geostatistical Wizard showing (A) the initial settings and (B) the interactive model.

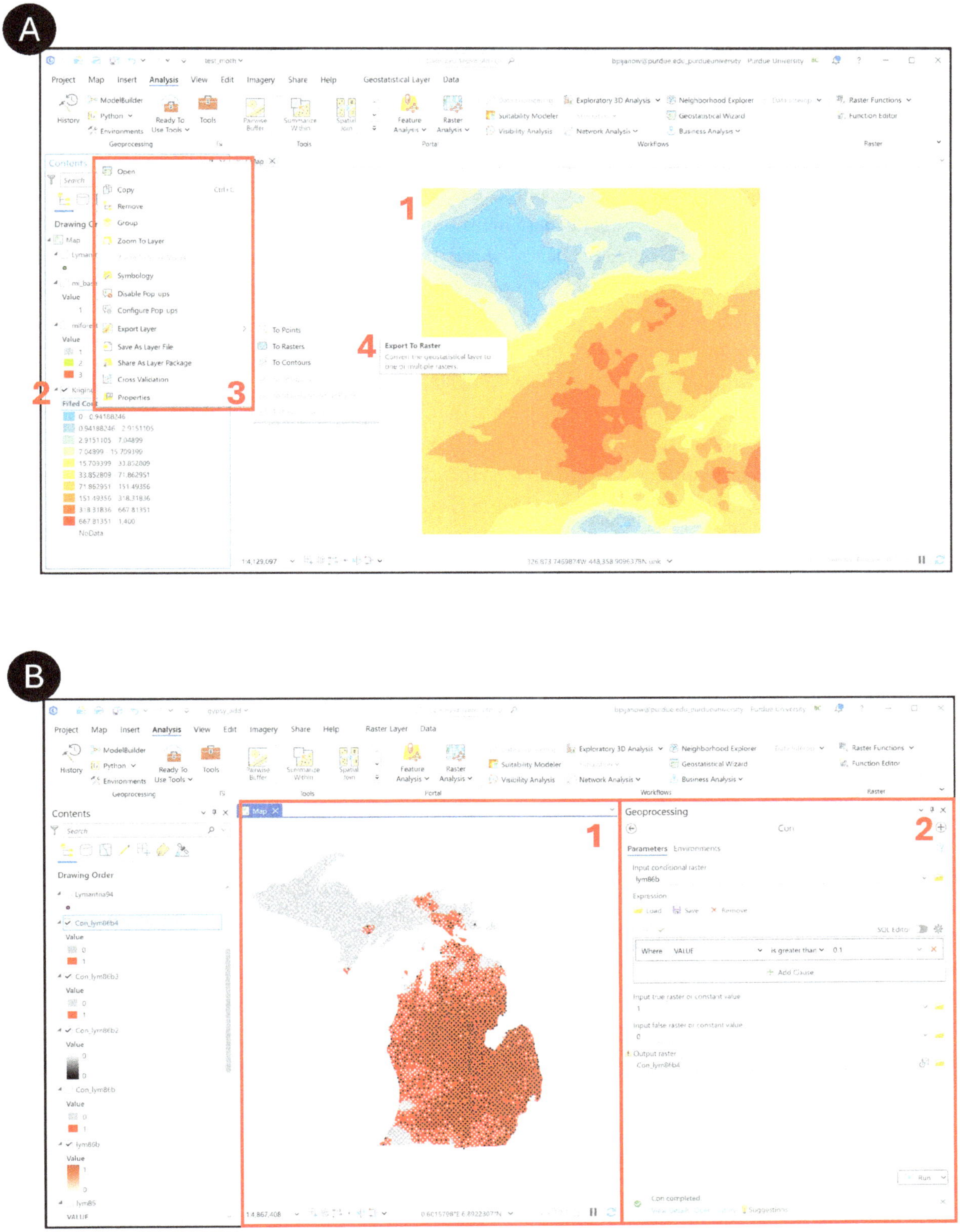

FIGURE EX7.4. Conducting analysis of the spongy moth data will involve (A) saving the results of the interpolated model and (B) visualizing the presence and absence of the trap catch data as points and as an interpolated model that was saved as a raster data layer.

of land coded as 1 and everything else as NoData. Multiplying the estimates of spongy moth population distributions by a map of 1 = land and 0 = no land will result in a map with the boundaries for the state of Michigan very clearly saved and represented in the output modeled map.

As you might expect from the large number of options that are available to model the distribution of a species given point location data, this lab only begins to allow you to explore all of the tools and possibly some of the major issues related to modeling species distributions using point data. In the case of the spongy moth, modeling distributions and their abundances would take a different set of models than modeling the presence and absence probabilities across space. Indeed, these maps could look a lot different as they each represent different kinds of risks that need to be associated across space. Take note of how raw point data maps compared to maps of interpolated data can provide a different perspective of the distribution of this forest pest and the risk across space. Figure 7.3B, box 1 shows a map of interpolated locations where maps were produced using ordinary kriging of presence and absence data that were then threshold at 0.1 (Figure Ex7.4B, box 1). This distribution is shown in red (presence = red, nonpresence and land = gray) for the year 1986. Overlaid on this map are the points coded as having moths in the trap (= black) and no moths (= white). The point data are distributed in uneven spatial distributions, whereas the red and gray map provides a less "splotchy" pattern. Indeed, one map might be very useful for local pattern understanding while another provides a larger spatial extent perspective. Note that the Geoprocessing menu (Figure Ex7.4, box 2) provides for one of the interfaces that can be used to threshold maps into binary categories from continuous maps.

7.5 WORKFLOW FOR THE EXERCISE

For this lab, we will take a database from a statewide monitoring program from the 1980s and 1990s in Michigan, an isolated introduction that spread to a nearly statewide distribution by the mid-1990s, and map that distribution relative to the primary tree species hosts. It should be noted that the coordinate system for these data is lost, so you will need to use the trap data, Michigan forest data, and the Michigan "basemap" solely for the analysis and for viewing. In other words, it will be confusing to turn on any of the Esri ArcGIS Pro base layers as the data from this lab has no coordinate assigned.

We will use ArcGIS Pro Spatial Analyst to interpolate spongy moth trap data from the Michigan Gypsy Moth Monitoring Program. Many of the interpolation tools in ArcGIS Pro Spatial Analyst (and Geostatistical Analyst or 3D Analyst) allow users to set Environment settings such as a mask. A mask in our case will be the land surface of Michigan (it makes no sense to interpolate spongy moth populations into Lake Michigan, right?); this is the mi_base raster layer that was provided for you in this exercise.

A proposed workflow is presented (Figure Ex7.5) for processing the spongy moth point data and then associating these maps with forest maps to create risk to trees in Michigan. Steps in creating time series maps and charts are also included that could form the basis of the report.

The online presentation will review the basics of interpolation through a series of videos in this folder. We will create maps that show the distribution of *L. dispar* populations and potential for defoliation through time using the Michigan Department of Agriculture's *L. dispar* database.

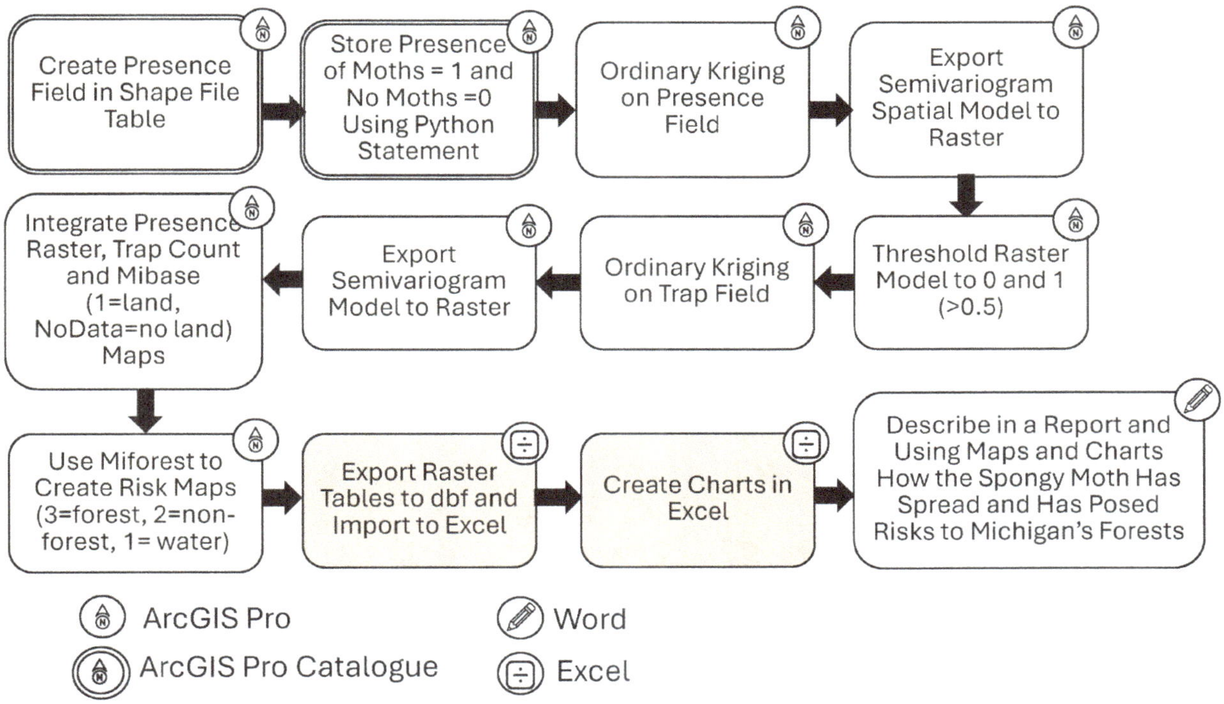

FIGURE EX7.5. Workflow for the spongy moth set of analyses.

7.6 ASSIGNMENTS

A. INTERMEDIATE STUDENTS

Complete the following tasks and include any necessary visuals and short descriptions:

- Generate a time series map (for all years) for *L. dispar* distribution using raster data of distributions that clearly show important thresholds of 0, 300, and 600 moths per trap. Use IDW to create interpolated maps.
- Also create a map of defoliation potential for the last three years. Use the Michigan forest map to conduct this analysis.
- Create a bar chart of the area for four thresholds 0, >0, >300, and >600, for all years of the monitoring study.

B. ADVANCED STUDENTS

Complete what the intermediate students are required to do. Write a report using the IMRAD format that summarizes the spread of the spongy moth across the state of Michigan. Describe the spread in relation to the risk to Michigan's forests.

Add one analysis to that of the introductory assessment. Use the kriging set of tools in ArcGIS Spatial Analyst to create spongy moth distribution maps for the final year of the dataset. Fit the data to a spherical variogram model. Compare the output of at least one map to that of an IDW-generated distribution map.

EXERCISE 8: CARTOGRAPHIC PRINCIPLES AND MAKING A SCIENTIFIC POSTER

Highlights

Aims. This lab is designed to introduce you to some of the best practices of making a scientific poster that would contain at least one map. A review of some of the principles of cartography are also put into practice.

Learning objectives. After completing this exercise in about a 3-hour time period, students should:

- Know and be able to apply some of the rules for making an effective poster
- Work with a variety of tools to synthesize the poster in Microsoft PowerPoint
- Apply some of the accessibility tools to visualizations before they are finalized

8.1 INTRODUCTION TO SCIENTIFIC POSTERS

As a spatial ecologist or GISci analyst, you will find that the science and art of making posters are skill sets that take a lot of time to develop. Posters are one of several means of conveying information, particularly at conferences and meetings of professional societies. An attractive, informative, and even creative poster takes time to make, and a successful poster will help convey your work to many people. Posters presented at conferences are often saved and then displayed in the hallways of university buildings, so taking care to make high-quality posters is vitally important as a young scientist since colleagues "down the hall" may stop and view your poster. Many poster sessions also include a large number of posters, so individuals walking through these posters will be drawn to yours not only on the basis of the topic of your poster but also whether the poster seems appealing when viewed quickly.

At government agencies or not-for-profit organizations, posters are often used to showcase the summary of a study or project during community events like local planning meetings. These posters need to convey important information to local citizens, so the information and how you present it become a vital link between your organization, its work, and potentially how that work impacts understanding of the topic or project by community members.

The purpose of this exercise is to introduce you to several features in ArcGIS Pro that can be used in addition to making maps, bring the information about how to present scientific information from Chapter 15 on report writing using the IMRAD format, and also introduce a few other tools and approaches that should be considered in making posters. Many of the latter points will leverage the information presented in Chapter 14 on cartographic principles.

8.2 SELECTING THE FINAL MAP PRODUCT FOR PRINTING

Perhaps some of the first decisions to be made are what you will use to compose and then print your poster. Many professional societies announce the sizes of their poster boards (many support poster boards that are 8 feet wide by 4 feet high; 96 inches by 48 inches). Many large-format printers can print posters that cover most of this area, but some may only print posters that are 42 inches wide (and they cannot print fully to edges). Standard sizes of paper formats include Architectural E (paper size is 36 inches by 48 inches) and ANSI E (34 inches by 44 inches). Large-format printers found at many GIS offices and in university computer labs, and those used by local and national companies that will print posters for you (and in some cases ship them to you or to a hotel you are staying at), are 36 inches, 44 inches, and 64 inches wide, with 44 inches being one of the most common sizes. In Europe, the A0 format, which is 33.1 inches by 46.8 inches, is the most common format. It is probably true that most large conference poster sessions (e.g., Ecological Society of America, American Geophysical Union, and the American Association of Geographers) have display boards that are 48 inches (4 feet) by 72 inches (6 feet).

There are several final production software platforms that can be used to develop a scientific poster. The most common for GISci analysts and spatial ecologists are (1) Microsoft PowerPoint; (2) ArcGIS Pro layout mode; (3) Canva; and (4) Adobe Illustrator or InDesign.

8.3 MAJOR DESIGN ELEMENTS OF A POSTER

There are several excellent sources for how to make a good scientific poster. One of the most highly referenced is https://colinpurrington.com/tips/poster-design/. This site contains useful design tips along with some creative ways to "lure" viewers to your poster for you to provide an overview of your work. In general, a typical poster will contain the same elements as the IMRAD contents (see Chapter 15) and perhaps a few other elements (e.g., background image). It is best to lay out a poster so that viewers follow the IMRAD storyline from the upper left to the lower right (Figure Ex8.1, upper panel). Note these several common poster layout best practices:

1. Be Readable at a Distance: Viewers at a distance, say 10–20 feet away, should be able to read the title and get a sense of the poster topic.
2. Be Easy to Follow: Use a standard sequence of presenting information (top left to center and then to bottom right).
3. Do Not Clutter with Text: Avoid a "copy from a paper" approach to building the poster. One rule of thumb is to not have more than 1,000 words on the poster (Purrington, 2025). It is also best to use left-justified text, as opposed to fully justified where text is aligned on the left and right.

4. Strive for Balance and Harmony: Make sure all elements are either centered or aligned at the top, bottom, and along the sides to create a visually "balanced" visual of the main elements. If the poster is laid out in PowerPoint, turn on the gridlines and possibly the rules. Use the Align tools in Microsoft PowerPoint or ArcGIS Pro Layout to make sure all elements are properly spaced.
5. Use Simple Fonts: Use simple fonts (e.g., Arial, Calibri).
6. Make Colors Accessible: Use the ArcGIS Pro Color Vision Simulator to determine if the map is readable by individuals with color vision deficiencies (see Chapter 14).
7. Avoid Drop Shadows and Complex Transparencies: For many printers, printer software drivers, or graphics software packages, drop shadows and some transparencies cannot properly print or print at all. So, what you see on a screen is not necessarily what you will get printed. It is best to test a sample graphic if you desire to have these graphic features.

The same poster element graphic is presented below it (Figure Ex8.1, lower panel) with added labels as the following points for each label (A–M) should be considered:

A. Title is best if it fits on one line and is at a minimum of 72-point size (i.e., 1 inch high) to around 90-point size.

B. Subtitle is best if positioned below title, in smaller font, viz, 60-point size, and in italics (i.e., in *italics*).

C. Authors and affiliations are listed below the subtitle, often in the font size for most of the text in the poster, which is recommended to be no smaller than 24-point size. Affiliations could be listed below author list with affiliations denoted with number or letter as superscript.

D. Logos are often located either on the top (short titles) or bottom (long titles). Note that the colors from an institutional logo can provide the color palette for the rest of the poster.

E. Sections should be presented in boxes with titles easily readable at a distance and centered.

F. Note how nearly all of the boxes are rectangular in shape. Squares are considered to be "not interesting" shapes visually.

G. Note how the main elements of the poster have equally spaced gaps. This includes gaps between boxes arranged vertically and horizontally. Also note how this helps to support that "balanced" look and feel of the poster.

H. Interior elements could have smaller spacings between elements.

I. Note how the main box is fully centered on the poster horizontally and vertically. Side boxes are also the exact same width. These can be controlled using the Shape size tools in PowerPoint and ArcGIS Pro.

J. The results section is likely to contain many different kinds of elements. Make sure that this section captures the most important part of your study but arranges all elements in a logical way. The results could include maps, figures, tables, text, and even infographics (see below). Consider bulleted text if that can help organize ideas and reduce the amount of text.

K. Discussion should be precise and contain potential answers to research questions and further research. Literature cited could go here or in a separate box.

L. The acknowledgment sections should include acknowledgment of funding agencies, grant number(s), and potentially the lead PI. Acknowledge sources of data, other organizations, and human subjects (in general, never by name).

M. The person who made the map or a contact name and email address could be added, along with the poster session date and the venue, often in small (< 24-point size).

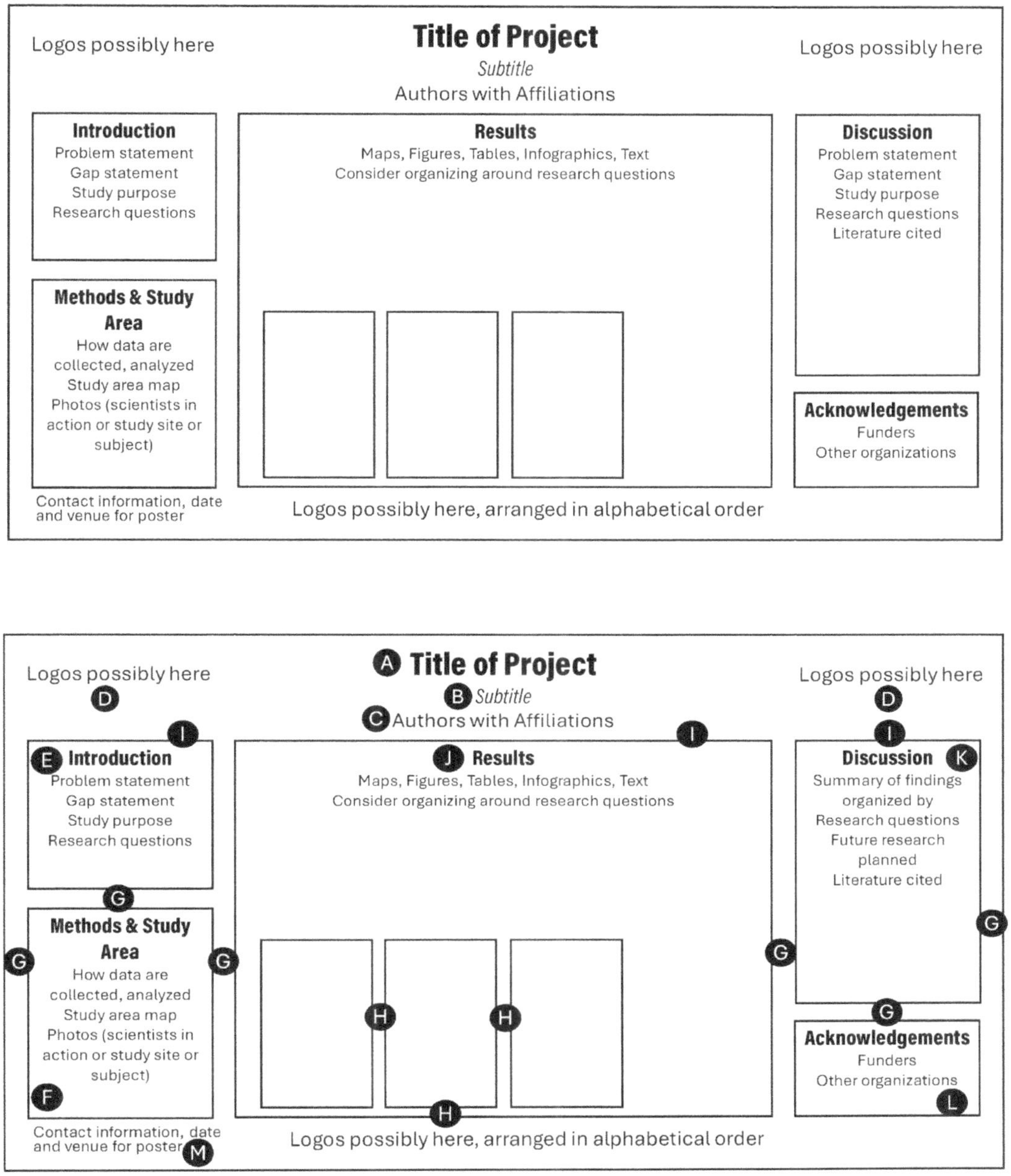

FIGURE EX8.1. Layout for a typical scientific poster presented at a conference where some of the poster elements include a map. The upper and lower panels are the same as the lower panel including letters that are referenced in the text above.

8.4 NAVIGATING ARCGIS PRO FOR CHARTS AND ACCESSIBILITY

With some careful work, the quality of the charting output can be quite good. Access to the charting features can be gained by right clicking the data layer and then selecting **Create Chart** (Figure Ex8.2A, box 1). There are several options to Create Charts (Figure Ex8.2A, box 2). Selecting the Histogram feature for a categorical land use cover map like the one that was part of Exercises 1 and 2 would give the chart shown in Figure Ex8.2A, box 3. Changing a variety of **Chart Properties** (Figure Ex8.2A, box 4) gives the chart as shown here. The chart can be saved as a graphic file (e.g., png, jpg, pdf) and then imported to Microsoft PowerPoint.

There are several color vision deficiency simulators in ArcGIS Pro. Figure Ex8.2B shows this as it appears in the ribbon for the **View** tab. The land use cover map for 1998 was brought into ArcGIS Pro and the default colors used here for this example (Figure Ex8.2B, label 1). The **Color Vision Simulator** (Figure Ex8.2B, box 2) is toggled as a two-button feature (the top button toggles the tool on and off; the bottom tool toggles the drop-down menu to allow the user to select one of the four built-in visualizing simulators (Figure Ex8.2B, box 3).

8.5 WORKFLOW FOR THE EXERCISE

This workflow (Figure Ex8.3) involves creating the necessary maps in Map View in ArcGIS Pro, exporting them to a png graphics file, composing parts of the poster components such as introduction and acknowledgments in Microsoft Word, creating charts or figures in Microsoft Excel, and then working in Microsoft PowerPoint to assemble the poster for final printing to a large-format printer. Make sure you follow as many principles of cartography as possible and check maps for visual accessibility.

8.6 ASSIGNMENTS

A. INTERMEDIATE STUDENTS

Using one of the projects from a previous lab and the report you submitted, bring the report elements together into a scientific poster and compose the poster in Microsoft PowerPoint on a 42-inch by 56-inch poster. The poster should have the following:

- All of the elements of a poster (A–M)
- All components of IMRAD with associated text, per rules mentioned above
- One map of a study area
- At least one map, one figure (using Excel or ArcGIS Pro Chart Tool), and one table for the results

Save the poster as a pdf.

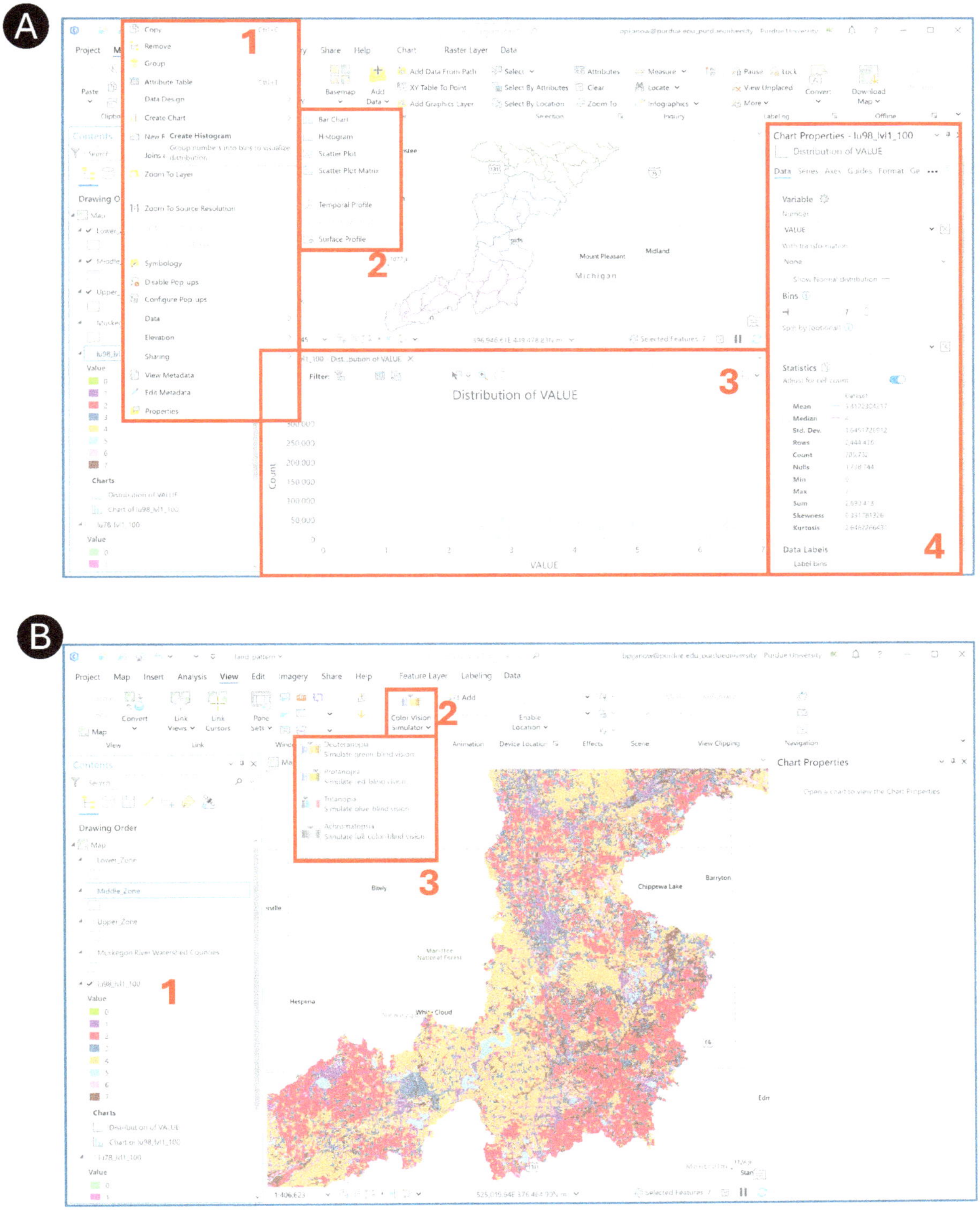

FIGURE EX8.2. Components of the ArcGIS Pro that can (A) control how a chart of spatial data appears and (B) view the accessibility features of the colors that are in the map.

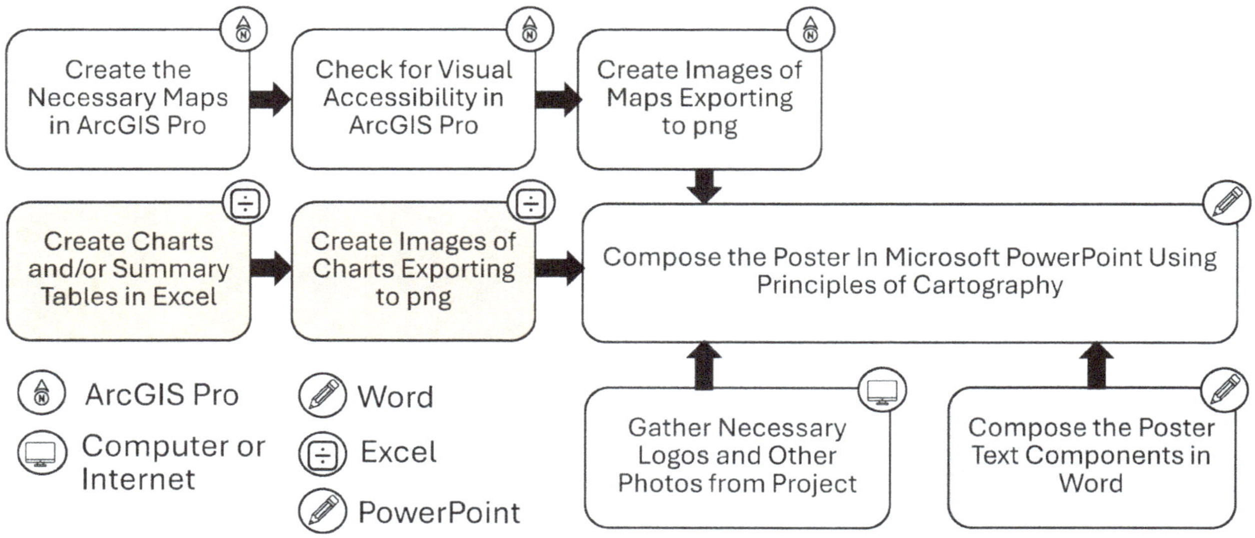

FIGURE EX8.3. Workflow for making a poster for a scientific conference.

B. ADVANCED STUDENTS

Using one of the projects from a previous lab and the report you submitted, bring the report elements together into a scientific poster and compose the poster in Microsoft PowerPoint on a 42-inch by 56-inch poster. The poster should have the following:

- All elements for the intermediate students
- An infographic that can be inserted into the poster
- References, at least in the introduction and discussion sections

Save the poster as a pdf.

EXERCISE 9: COORDINATE SYSTEMS AND PROJECTIONS: PRINCIPLES OF GEOMATICS

Highlights

Aims. This exercise is designed to introduce you to projection systems and how global and regional maps can be displayed. The background section reviews some of the fundamental concepts in geographic and projected coordinate systems.

Learning objectives. After completing this exercise in about a 1.5-hour time period, students should:

- Be able to compare and contrast geographic coordinate systems and projected coordinate systems
- Know how important coordinate system parameters such as central meridian affects how the map is projected onto paper or a computer screen
- Understand how continental and world projections affect how a map is projected onto paper or a computer screen
- Know how to convert lat-long in degrees-minutes-seconds to decimal degrees

Project Data Resources. A shape file of the world's countries and a detailed shape file of the 50 United States.

9.1 INTRODUCTION TO COORDINATE SYSTEMS

One of the fundamental sets of concepts that are needed for analysis and mapping centers on understanding coordinate systems. The study of the shape of the Earth and how a 3D globe can be projected into 2D—for printing on paper and visualizing spatial information on a computer screen—is referred to as **geometric geodesy**. **Geodesy** (Torge, 1980), in general, is the study of measuring and modeling the Earth and its gravity fields, which is heavily influenced by mathematics, the physics of gravity, and the movements of the parts of the Earth (e.g., continental drift). Geometric geodesy focuses on the study of how a 3D-shaped Earth can be represented in 2D. The closely related field of **geomatics** (Blais, 1995) differentiates itself from the field of geodesy in that geomatics considers how spatial data and remote sensing of the Earth interface with coordinate systems and the theories and body of mathematics that relate to coordinate systems. The study of both geodesy and geomatics contributes to the array of knowledge needed to produce accurate coordinate systems and thus a variety of applications in GISci and spatial ecology. Courses in geomatics are often found in departments of civil engineering and astronautical engineering, and geodesy courses would be offered in earth sciences departments. Students in geography departments could be provided with detailed content in either geodesy or geomatics, and some of this content could be delivered in a cartography course as well. A few important points of note here are that to have a firm grasp of spatial ecology and GISci (1) you need to know the basics of coordinate systems and thus some fundamentals of the fields of geomatics and geodesy; (2) these two areas of study tend to be very technical and contain a lot of mathematics; (3) all GIS software has extensive information about coordinate systems; and (4) knowing some of the fundamental concepts is necessary to produce effective, and often scientifically valid, maps. This lab is designed to allow you to explore how various aspects of coordinate systems affect the presentation of spatial information on a map.

To appreciate how technical the fields of geodesy and geomatics are, consider how one needs to take information that is contained in a 3D sphere and represent it on a 2D surface. We know that the Earth is an imperfect sphere (i.e., not completely round along all axes) that has a specific radius. Geographers centuries ago (see Torge & Müller, 2012) developed a system based on latitude and longitude (Figure Ex9.1). **Latitude** is the measure of the **angular measure** between the equatorial plane and the point on the surface; it ranges from 0 to 90. Note that locations above (to the north) of the equator are assigned positive values and those south of the equator are assigned negative values. Lines that run east and west and parallel to the equator are called **parallels**. **Longitude** is the **circular measure** of the Earth, measuring the circular distance between a location and the prime meridian (i.e., the line that goes from the north pole to the south pole crossing Greenwich, England, which is 0° longitude. As a sphere has 360 degrees, we have this to create two hemispheres, an east (those with negative longitude values) and a west (those with positive longitude values). All lines running north and south are called **meridians**.

The way that the latitude and longitude positions are laid out ends up creating a sphere divided into hemispheres. The equator divides the Earth into the Northern and Southern hemispheres, whereas the prime meridian divides the Earth as a sphere into a Western and Eastern **hemisphere**. The prime meridian is also the relative location of Greenwich mean time (the global reference clock time). The antimeridian is also the relative location of the international date line; both the global reference time and the international date line "zig-zag" around countries to make dates and time consistent for residents of the same place.

Fractions of the values for the meridian and parallel are reported in two formats: **degrees-minutes-seconds** (DMS) and **decimal degrees** (DD). Degrees-minutes-seconds divides the degrees into smaller units, called minutes; there are 60 minutes in a degree. Minutes are further divided into seconds; there are 60 seconds in a minute and 3,600 seconds in a degree. We use the following symbols for the DMS system: Degrees (°) → Minutes (′) → Seconds (″). The decimal degree system of units adds the fractional measures for minutes and seconds as follows:

$$DD = \text{degrees} + \text{minutes}/60 + \text{seconds}/3{,}600$$

Thus, for an example DMS measure of 45 degrees, 30 minutes, and 100 seconds, the same DD value for this can be calculated as follows:

$$DD = 45 + 30/60 + 100/3{,}600$$
$$DD = 45 + 0.50 + 0.02777 = 45.52778^\circ$$

Handheld global position system (GPS) units can report locations in either DMS or DD values.

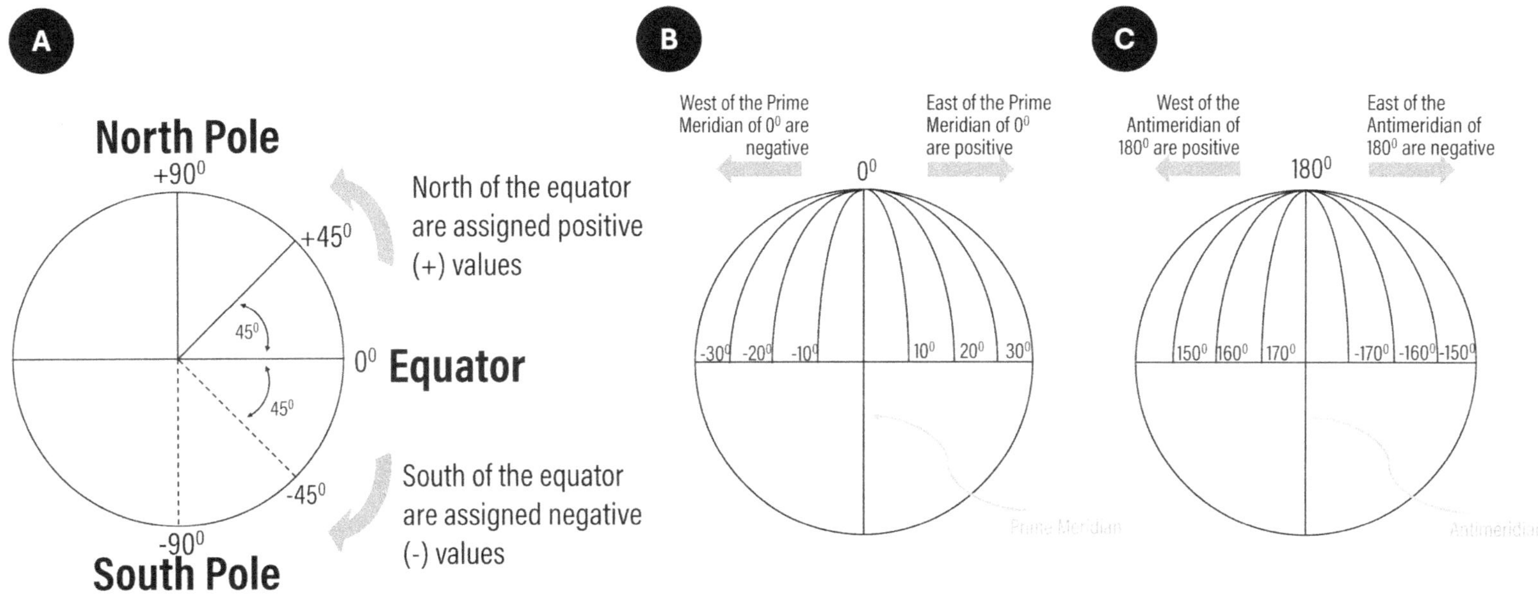

FIGURE EX9.1. Rationale of assigning values to latitude and longitude in the geographic coordinate system. Note that for latitude (A), values are assigned based on the angle from the equator and that for longitude (B and C), values are assigned to whether they are east or west of the prime meridian.

TABLE EX9.1. *Major differences between geographic coordinate systems and the projected coordinate systems*

DIMENSION OF COORDINATE SYSTEM	GEOGRAPHIC COORDINATE SYSTEM (GCS)	PROJECTED COORDINATE SYSTEM (PCS)
Shape of surface	Sphere	Flat
Measures for coordinate system	Angular units	Linear units based on northing (X) and easting (Y) measures, typically in meters
Basics of measure	Datum (measures of an ellipsoid)	Map projection references superimposed on the GCS
Examples	WGS 84, NAD 83	UTM, state plane systems
Other notes	Pure location anywhere on Earth, major distortions if shown in 2D and distance and area change in areal and length measures as one moves north and south	Due to inability to perfectly convert 3D to 2D, there is generally distortion in one or more of the following: area, distance, shape, and/or direction

TABLE EX9.2. *Comparative cell dimensions and areas for locations along different latitudes on Earth*

LATITUDE	APPROX. CELL WIDTH (M)	CELL HEIGHT (M)	APPROX. AREA (M^2)
0° (Equator)	30.92	30.92	~957 m^2
45°	21.9	30.9	~678 m^2
60° (Pole)	15.5	30.9	~479 m^2

For the purposes of mapping to all locations on Earth, there are two basic coordinate systems: the **geographic coordinate system (GCS)** and the **projected coordinate system (PCS).** The major differences between them are summarized in Table Ex9.1.

Some spatial data that are released by organizations are in GCS with angular units. These are typically specified in units such as arc-seconds or 1/3,600 of a degree. A 1 arc-second unit, which is common, has x-y lengths and areal measures that vary depending on latitude but is approximately around 30 x 30 m at the equator. One of the important patterns is that the meridians become closer together as they approach the poles, and this affects width (but not height). Table Ex9.2 summarizes these for several different latitudes; note that the largest values are at the equator.

9.2 GEOGRAPHIC AND PROJECTED COORDINATE SYSTEMS

The geographic coordinate system (Figure Ex9.2A,B) needs to be converted to a 2D planar system that is referred to as a projected coordinate system. All projected coordinate systems are developed from dimensions specified by the geographic coordinate system such as semi-major axis (largest radius of the Earth at the equator), semi-minor axis (largest radius of the Earth at the poles), and several shape parameters. There are three

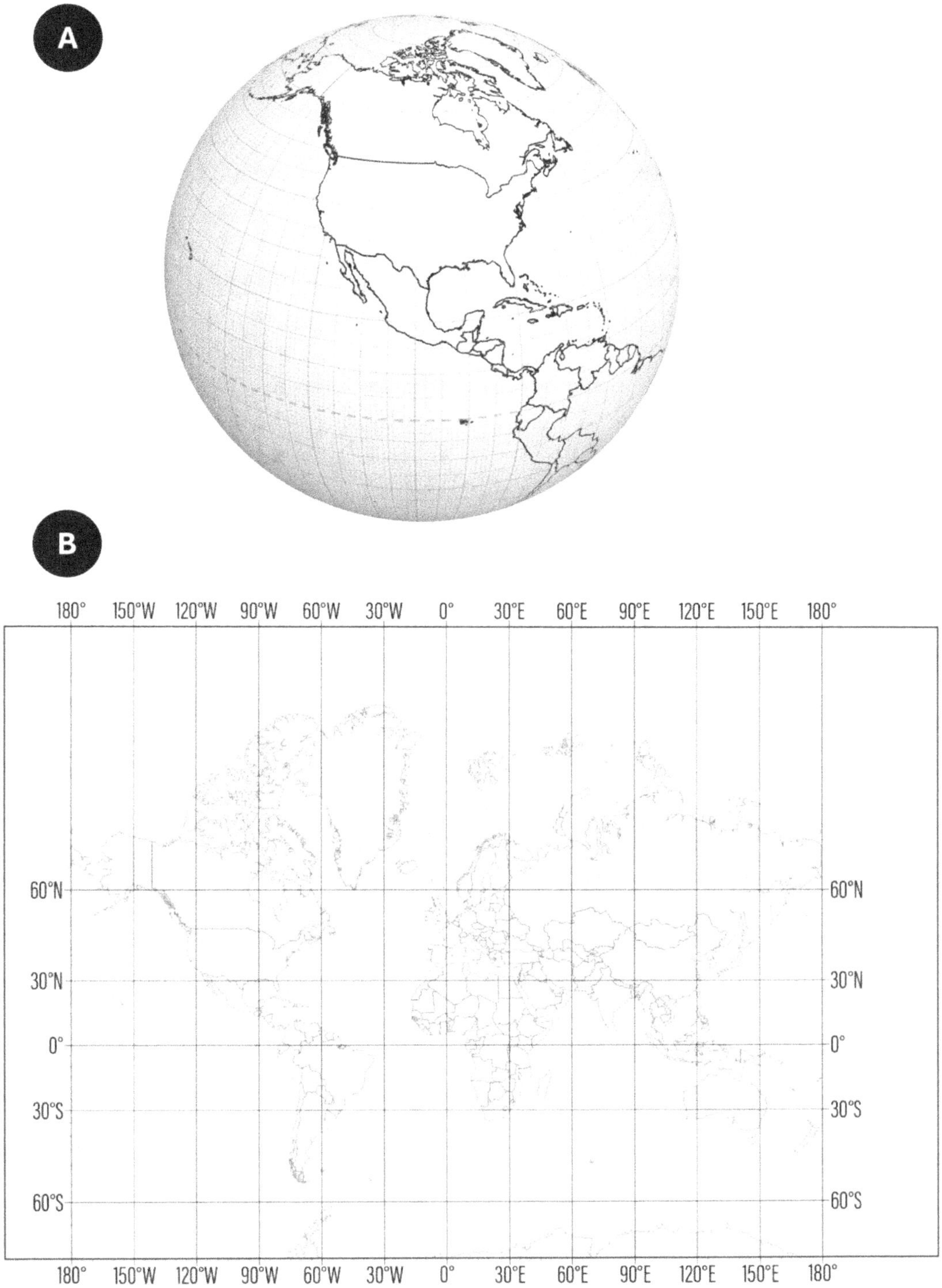

FIGURE EX9.2. Visual differences between the (A) 3D representation of the geographic coordinate system (GCS) and a (B) 2D representation of GCS.

standard PCS referencing systems, which are called **datums**. The datums most commonly used by the U.S. federal agencies are NAD27, NAD83, and WGS84. NAD stands for North American Datum, and the two numbers refer to the year that the datum was adopted. NAD27 was adopted in 1927 and NAD83 was adopted in 1983. WGS84 refers to the World Geodetic System, which was adopted by the U.S. Department of Defense in 1984 and is used by satellite technologies.

It should be noted that, although most uses of the GCSs are based on global measures, there are many versions of GCSs that have been optimized for local and regional applications. Thus, when you examine GCSs in ArcGIS Pro's list of possible projections, you will find several dozen GCSs that have specific names, mostly associated with continents and large regions. Finally, there is another set of published GCS standards that frame the basic NAD and WGS system by year. This coordinate referencing system, the **ITRF** (**International Terrestrial Referencing Frame**), considers Earth's dynamics such as very small shifts in the continents. For most spatial ecological applications, it is appropriate to use NAD83 or WGS84.

Projected coordinate systems are based on several standard measures. One measure is the **central meridian** (usually the meridian in the center of the area being mapped) and one or two **standard parallels** (upper and lower limits of the map for two), a **GCS reference**, a **scale factor** (often a value less than but very close to 1.0), a shape used to convert from a sphere to a surface (either a cylinder or cone), and, in some cases, northing and easting. This latter case shifts standard map locations so that the area and lengths are optimized. Because there are places on Earth that have unusual shapes to map (e.g., the Aleutian Islands off the coast of Alaska), the cylinders for reference can be at an angle; these are called oblique. Oblique refers to the tilt of a coordinate system. Secant refers to whether the cylinder or cone bisects the sphere once or twice. There is yet another measure that is often used, and that is the **latitude of origin**. For some PCSs, the latitude of origin is not the equator, but another value that helps to center the projection. Finally, given that there are so many PCSs and that regional and local specifications are common to help improve accuracy of spatial data at small extents, all of the coordinate systems are catalogued as **WKIDs** or Well-Known ID (also as Well-Known Identifier). Figure Ex9.3 shows all of the projection information for a commonly used projected coordination system (Figure Ex9.3A) for the lower 48 states (CONUS) and its GCS (Figure Ex9.3B).

Many projections end up displaying maps using a slight curvature. One of those used frequently by the U.S. federal agencies is the NAD83 Albers Equal Area Conic Projection. The moderate curvature of this projection creates a slight "smiley face." Note that in the Albers Projected Coordinate System, there is a -96.0° central meridian, which is used to map the lower 48 states (hence called the NAD83 USA Contiguous Albers Equal Area Conic USGS projection). This projection has the map with the central meridian perfectly vertical around Omaha, Nebraska (Figure Ex9.4A); all other meridians slant to the left or right. Using a -83.3° central meridian (Figure Ex9.4B) will center the map around Cleveland, Ohio; this projection is called the NAD83 Great Lakes and St. Lawerence Albers (WKID 3175) since it is most often used to display all of the Great Lakes as well as the St. Lawrence seaway as it connects with the Atlantic Ocean to the east.

It is generally agreed that maps should not be displayed using the geographic coordinate system. The NAD83 Albers Equal Area Conic Projection with a -96.0° longitude coordinate system preserves land area representation equally in the south compared to that in the north. Figure Ex9.5 shows how GCS differs greatly from a PCS like Albers.

A

Projected Coordinate System	NAD 1983 Contiguous USA Albers
Projection	Albers
WKID	5070
Authority	EPSG
Linear Unit	Meters (1.0)
False Easting	0.0
False Northing	0.0
Central Meridian	-96.0
Standard Parallel 1	29.5
Standard Parallel 2	45.5
Latitude Of Origin	23.0
Area of Use	USA - CONUS - onshore
Top	49.38°
Left	-124.79°
Right	-66.91°
Bottom	24.41°

B

Geographic Coordinate System	NAD 1983
WKID	4269
Authority	EPSG
Angular Unit	Degree (0.0174532925199433)
Prime Meridian	Greenwich (0.0)
Datum	D North American 1983
Spheroid	GRS 1980
Semimajor Axis	6378137.0
Semiminor Axis	6356752.314140356
Inverse Flattening	298.257222101
Area of Use	North America - NAD83
Top	86.45°
Left	167.65°
Right	-40.73°
Bottom	14.92°

FIGURE EX9.3. Parameters and names for the Albers Equal Area projection (A) based on the NAD83 GCS (B). Image based on screen grabs of Esri's ArcGIS Pro 3.5.

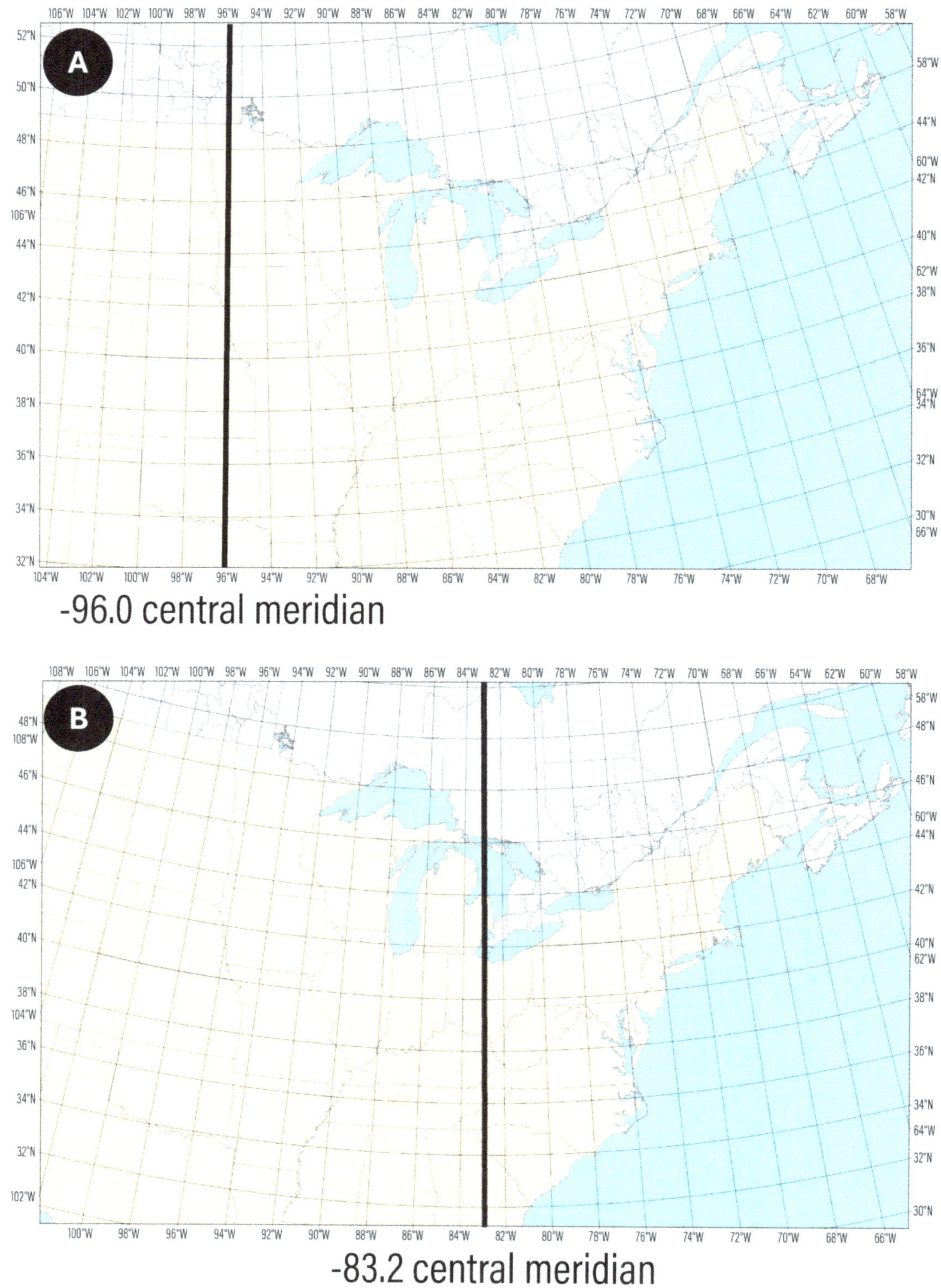

FIGURE EX9.4. Comparison of maps for North America that differ in their central meridians. The standard one used for the lower 48 states of the United States has a central meridian of -96.0 longitude (A), whereas maps that would display spatial information for the northeastern part of the country with the Great Lakes would have a central meridian of -83.2 longitude (B).

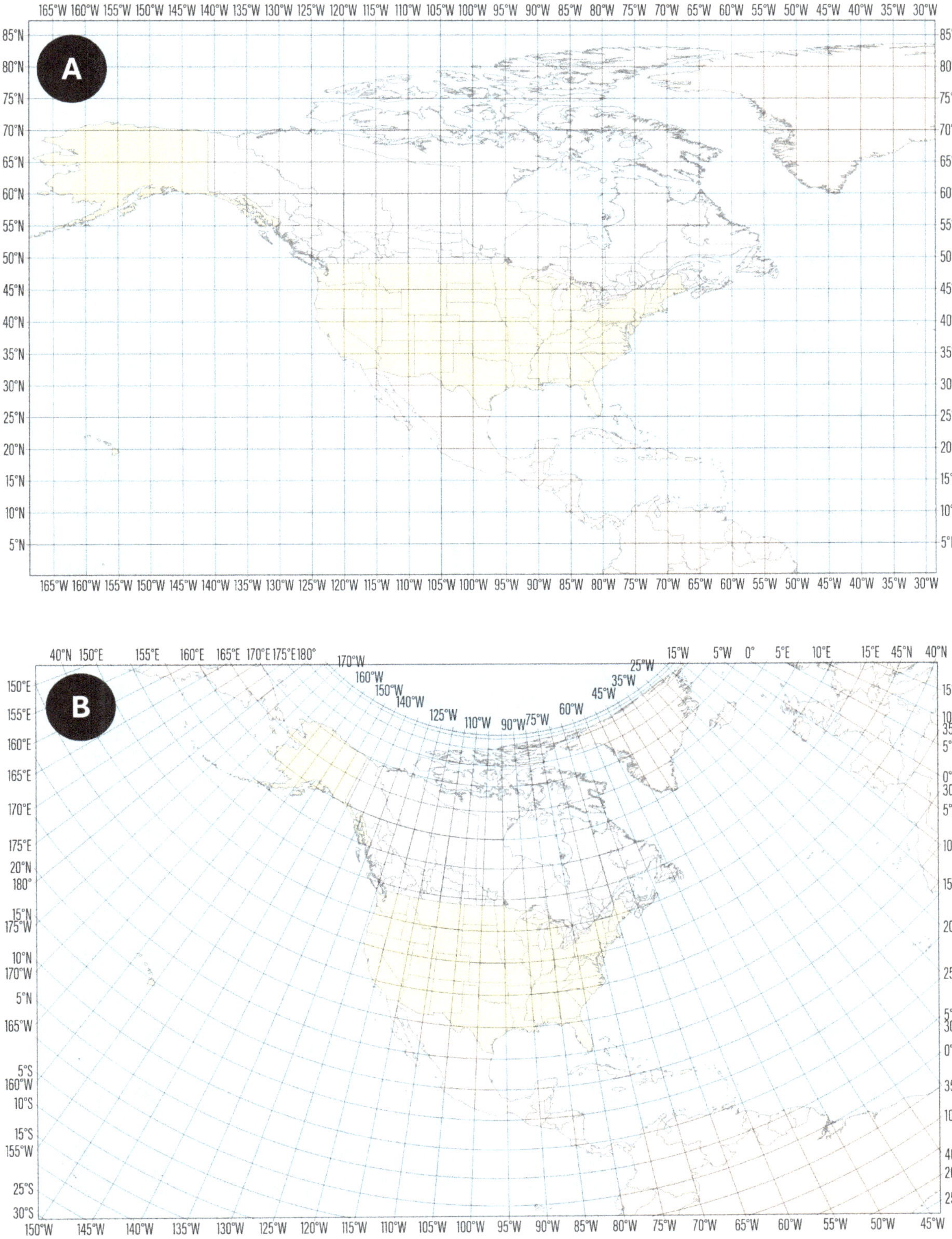

FIGURE EX9.5. Map of North America for a GCS (A) and a PCS (B) using NAD83 Albers Equal Area Conic Projection.

9.3 NAVIGATING ARCGIS PRO FOR COORDINATE SYSTEM MANAGEMENT

There are many options to change projections of maps and some of the key map elements. Changing the coordinate system is done only in **Map** View. **Layout** View has no spatial references so in **Layout** View, you can combine maps of different coordinate systems.

Figure Ex9.6A shows the important menus and menu options for viewing and then changing the coordinate systems. In the **Contents** pane (Figure 9.6A, box 1), note that right-clicking the **Map** item (Figure Ex9.6A, box 2) will bring up the **Map Properties** Menu (Figure 9.6A, box 3) and from there selecting **Coordinate Systems** (Figure 9.6A, box 4) activates the **XY Coordinate System Menu**. For this shape file, the **WGS 1984** (Figure 9.6A, label 5) geographic coordinate system is associated with this data layer. Note that two sets of coordinate systems are listed: **Geographic Coordinate System** and **Projected Coordinate System**. With the projected coordinate Systems, many are **Continental**, and from within this, there are projected coordinate systems that are based on the name of the continent. Note that there are a host of projected coordinate systems for **North America** (Figure 9.6A, label 6).

Paging through **Projected Coordinate Systems** for the **World** shows that there are dozens of possible 2D representations of the globe (Figure Ex9.6B, box 1). Right-clicking on any one of these, such as **Berghaus Star AAG**, and then selecting **Details**, brings up the parameters for this world projection system. Indeed, selecting this coordinate system projects global data in the shape of a star.

Creating a **Layout** View gives the user access to additional visualization capabilities in ArcGIS Pro (Figure Ex9.7A, box 1). Using the **Insert** tab collection of tools in this ribbon (Figure Ex9.7A, box 2) gives access to an array of options to insert the map element called **Grid** or **Graticulates** in ArcGIS Pro (Figure Ex9.5A, box 3). Note that there are several types of grids (Figure Ex9.7A, label 4). The term **Grid** in ArcGIS Pro refers to evenly spaced lines placed on a map. **Graticulates,** on the other hand, are grids that can be drawn using irregular spacing as the distance between latitude and longitude can be irregular. Once the grid has been created, it is then added to the **Contents** pane (Figure Ex9.7B, box 1), and right-clicking on the **Grid** data layer name (Figure Ex9.7B, box 2) in the **Contents** pane (Figure Ex 9.7B, box 3) for the map (Figure Ex9.7B, box 4) gives rise to the **Elements** pane (Figure Ex9.7B, box 5). The **Elements** pane contains the options for the display of the grid. At the top of this pane are grid display option tabs (Figure Ex9.7B, label 6) that can be toggled. The right option tab activates more detailed controls for the grid such as for its **Components** (Figure Ex9.7B, label 7). You should explore all of the options, which include the those that turn labels on or off, ticks, and so on to understand what can be displayed in the **Layout** view. Font type and size can also be changed.

9.4 WORKFLOW FOR THE EXERCISE

Using the countries of the world shape file, let us first explore some of the amazingly cool world projections by just changing the map's projection system, viewing the base Topographic data layer. Adjust the projections and explore each of the following projections under the Projected/World projections menu system:

- Azimuthal Equidistant
- Berghaus Star AAG
- Bonne

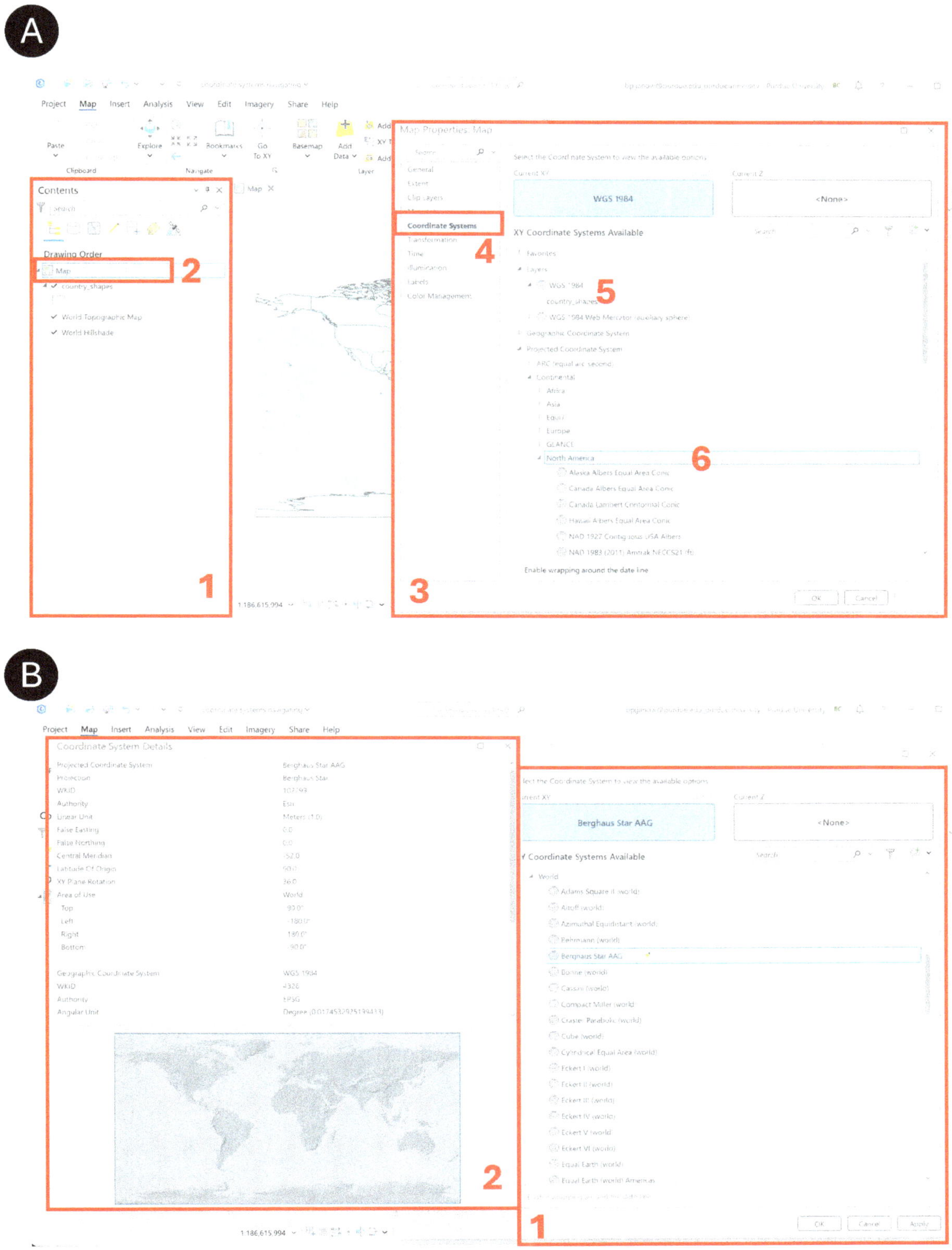

FIGURE EX9.6. Exploring (A) coordinate systems for map views and (B) with examination of details of coordinate systems.

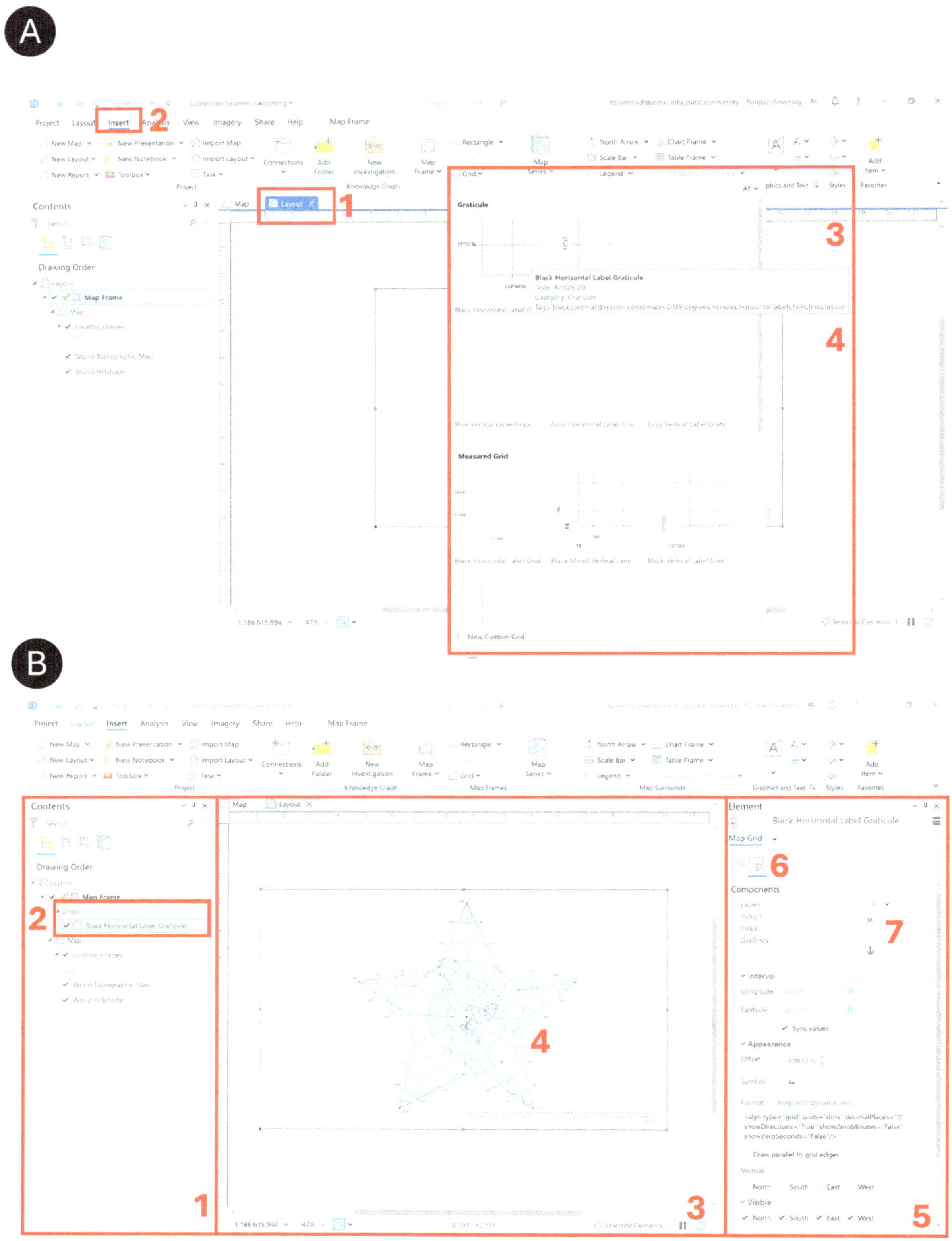

FIGURE EX9.7. Selecting (A) grids and graticulates for a map in Layout View and (B) configuring the grid map elements using the Elements pane.

- Craster Parabolic
- Eckert I & IV
- Gall Stereographic
- Polyconic
- Robinson
- World from Space
- Vertical Perspective and then Geographic WGS 1984 (in Geographic)

For at least three of the projections listed above, go to the projection parameters menu that lists the PCS and WCS parameters. Select all of these parameters (you might consider pulling down the small view finder at the bottom to allow you to select them all) and paste these into a Word document for THREE different projections. Note that you can do this by going to the Map View, then selecting the map layer at the top, right clicking the top map layer, and the Properties. In Map Properties, select Coordinate Systems and then the coordinate system you want. Right click the Coordinate System and then Details. Note that the parameters for the coordinate systems will look like Figure Ex9.3 above.

Next, select the Azimuthal Equidistant Projection above to view the countries of the world (without Antarctica). Create a layout and then insert a grid (graticulate) to illustrate the locations of meridians and parallels and latitude and longitude values around the world. Create some layouts that display at least three different world projections illustrating how diverse these can be.

The workflow is easy (Figure Ex9.8). This lab is designed to allow you to explore how different 2D mapping of a globe can be but also allows you to explore how the grids can be set up to help visualize the meridian and parallels of the coordinate system.

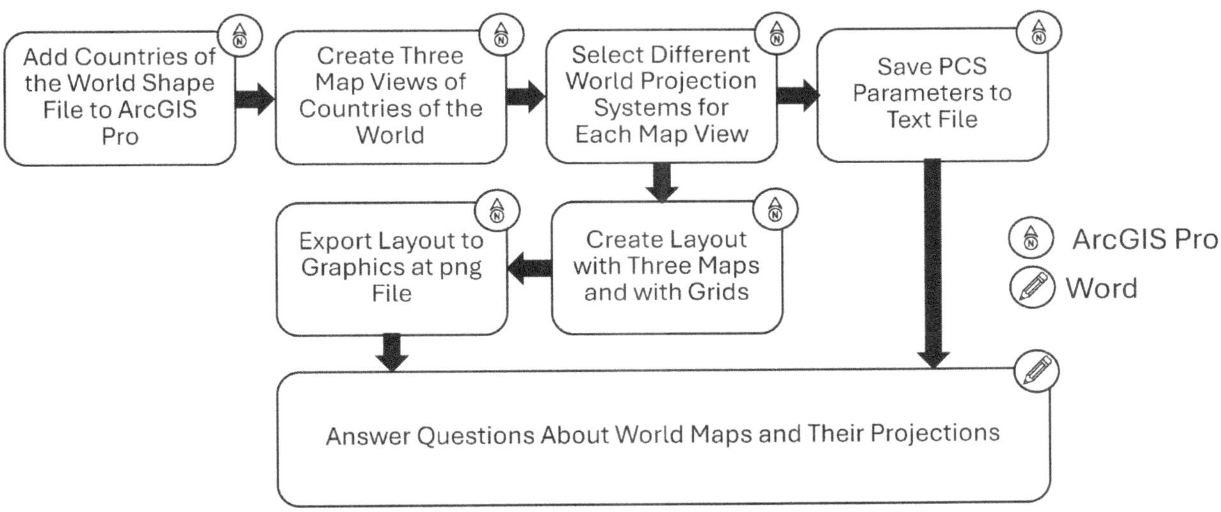

FIGURE EX9.8. Workflow for the coordinate system mapping lab.

9.5 ASSIGNMENTS

A. INTERMEDIATE STUDENTS

Complete the following tasks and include any necessary visuals and short descriptions:

- Produce three maps that vary in the way that they arrange countries around the world. In no less than 100 words, describe in your own words how these three maps differ. What are the positive and negative aspects of each of these projections, in your opinion? Which one is your favorite and why?
- Create a figure that shows how each of the three projections vary in their projection parameters. In no less than 50 words, compare and contrast these three.

B. ADVANCED STUDENTS

Write a short (~3 pages of text) IMRAD report with the following as a guide. Using the maps and tables that are produced for questions #1 and #2 for intermediate students, write a report that provides a general overview of map projections (in your own words) for global datasets. You should mention the importance of WCS datums as well as the challenges of mapping spatial information in 2D given that we start with a globe that is a 3D sphere. In the methods describe your steps for producing the maps and tables (you might consider a workflow diagram that is more detailed than what is provided above), with a brief description of the results (how these maps differ to you and how the parameters create the visuals that are the map shapes). Your discussion should focus on what you learned about map projections (select at least three to present).

EXERCISE 10: MT. KENYA LAND USE 3D MAPPING CASE STUDY

Highlights

Aims. This lab is designed to introduce you to ArcGIS Pro's 3D Mapping Tools and analysis of land use cover change using a variety of spatial data and mixed methods approaches in the context of land adjudication policies and coupled human and natural systems (CHANS).

Learning objectives. After completing this exercise in about a 3-hour time period, students should:

- Know how to visualize data using 3D visualization tools in ArcGIS Pro's Global Scene set of tools
- Understand and be able to apply mixed methods approaches to land change analysis
- Describe some of the common land use/cover change patterns in several of the Mt. Kenya communities (named for the towns of the surrounding agricultural landscapes)

Project Data Resources. Data resources from two previously funded projects, LUCID and CLIP, that have shape files and imagery of land use/cover.

10.1 INTRODUCTION TO MT. KENYA

Mt. Kenya is Africa's second tallest mountain behind Mt. Kilimanjaro. It is slightly taller than 17,000 feet (5,199 meters) and is a UNESCO World Heritage site because of its cultural significance (the country of Kenya is named for this structure). The mountain also harbors large numbers of endangered species, such as the African savanna elephant, the blue monkey, the mountain bongo (O'Donoghue et al., 2017), and the Mt. Kenya sentinel chameleon (Malonza, 2015). It is the home of the indigenous community's god; and, at its highest elevations, areas are protected as a national park. It is one of the few equatorial mountains of the world to support glaciers. Along its slopes are many small towns, natural forests, agroforest plantations, and numerous rainfed and stream-fed crops, such as tea and coffee. In the 1980s and 1990s, land adjudication policies changed land ownership that eventually led to landowner fragmentation, with some landowners having large farms while others own small, barely profitable farms.

10.2 LUCID AND CLIP PROJECTS

Two large, Michigan State University–led projects, LUCID (Land Use Change Impacts and Dynamics), and CLIP (Climate-Land Impact Project), were funded to study long-term land use/cover change along the slopes of Mt. Kenya. The LUCID project was funded by the Ford Foundation, and the CLIP study was funded by the U.S. National Science Foundation. These studies combined remote sensing, spatial analysis of environmental and socioeconomic data, interviews, and forecast modeling to understand the coupled natural human system dynamics, especially in the context of large-scale land use policies, high human population growth rates that increased food demand, and climate change (see Olson et al., 2004; Olson et al., 2008). Some of the unique datasets include digitized land use/cover maps that were made from aerial photographs of the 1950s through 1990s, providing a times series of landscapes that spanned important periods; the country of Kenya transitioned from being a British colony to existing as an independent, sovereign county in 1964.

Shifts in policies that affected land ownership occurred multiple times between 1964 to about 2010, which eventually affected the ways that land was managed for crops and whether land ownership was fragmented or consolidated. At specific times, local and national government agencies proceed through a process referred to as **land adjudication**, which is "the process of final and authoritative determination of the existing rights and claims of people to land" (FAO, 2002b). Land adjudication is typically a process that describes the rules for assigning and documenting (e.g., owner, boundaries, use, etc.) land rights based on general rules for claims of ownership.

Both the LUCID and CLIP studies were designed to be **mixed methods**; that is, they combined quantitative approaches with those that were qualitative (see Olson et al., 2008 for more information about how these are used in combination and how each provides useful perspectives of landscape dynamics where human behavior, demographic change, and economics are key elements of drivers of change). Water resources are also in short supply in lower elevational areas, and uses of ground and surface water are creating a variety of environmental problems (Notter et al., 2007).

Qualitative methods attempt to describe patterns in data given the general context for how the data represent the real world. Qualitative methods examine several data sources along with information that is difficult to quantify, such as policies and behaviors. Quantitative methods, on the other hand, attempt to create numerical outcomes that are driven by specific questions, typically research questions or scientific hypotheses. Quantitative methods often include the use of empirical or deterministic models; statistical tests are performed in the context of scientific hypotheses.

10.3 GLOBAL AND LOCAL SCENES IN ARCGIS PRO

As you have learned in other parts of this textbook, ArcGIS Pro has four distinct kinds of projects: Maps, Catalogues, Global Scenes, and Local Scenes. The latter two are projects that display data in 3D. Scenes in ArcGIS Pro require there to be at least one layer that is an elevation surface. The Global Scene project, which is most often used for large-scale visualizations such as those at regional (e.g., national) continental extent, use the WorldElevation3D/Terrain3D elevation database that is linked to a server managed by Esri. We will use that DEM to visualize shape files and satellite images in 3D for an area that includes Mt. Kenya. To properly do this, the DEM must be added as an elevation layer.

10.4 NAVIGATING IN ARCGIS PRO GLOBAL SCENE FOR MT. KENYA VISUALIZATION

After adding several of the data layers from the East Africa CLIP Project, you will see a set of maps organized in the **Contents** pane (Figure Ex10.1A, box 1) of an ArcGIS Pro Global Scene (Figure Ex10.1A, box 2). Three sets of maps can be displayed in this 3D visualization and analysis set of tools (Figure Ex10.1A, box 3). These include those that are organized as **3D layers** (Figure Ex10.1A, box 4), which are special data layers that are created for 3D visualization such as buildings); **2D layers** (Figure Ex10.1A, box 5), which are the shape files most spatial ecologists will be working with; and then **Elevation Surfaces** (Figure Ex10.1A, box 6). The latter can be any DEM that you might have, but you can also use the global digital elevation model (**WorldElevation3D/Terrain3D**). When one of the **Elevation Surfaces** is selected, the special tab for **Elevation Surface Layer** toggles on. Having this toggle on gives options for viewing. One of the most important is **Vertical Exaggeration**. For you to properly view these data, it is necessary to exaggerate the height of the mountain and surrounding places. Select **Ground from the Elevation Surfaces** list of **Scene Layers**; this will activate the **Elevation Surface Layer** tab at the top next to **Help**. When that occurs, you will be presented with several options. Select **Vertical Exaggeration** and experiment with values between 3 and 10. You should also explore what happens when **Shade Relative to Light Position** is selected on and off (Figure Ex10.1A, box 9).

A special Global Scene feature is the 3D navigation controller in white (Figure Ex10.1A, box 10). It allows users to pan, rotate, and tile. Three small buttons change the controller to one of two versions of the controllers.

For this lab, we will use a three-band satellite image from the year 2000 to visualize the land surface around Mt. Kenya. When you bring in an image, you will see ArcGIS Pro show this data layer with its three bands (Figure Ex10.1B, box 1). Selecting it will turn on the tab for **Raster Layer** (Figure Ex10.1B, box 2), which then shows the options in the ribbon. Note that you can change how the intensity of the image is displayed by changing the **Stretch Type** (Figure Ex10.1B, box 3). ArcGIS Pro will provide over a half dozen **Stretch Types** (Figure Ex10.1B, box 4). Exploring how each of these changes to the visualization is easy. Note that for this scene, the 3D navigation controller (Figure Ex10.1B, label 5) is toggled to the tilt version of tools. The central tool, which looks like Saturn with rings, can be used to tilt the view (by moving the rings).

10.5 WORKFLOW FOR THE EXERCISE

Figure Ex10.2 shows a high-level overview of a possible workflow for this lab. Note that given the rich databases from the NSF CLIP and Ford Foundation's LUCID Projects, analysis of the changes in land use/cover change are potentially full of rich research questions about the impact of policies and economics on long-term changes in cropping systems. The number of life zones along Mt. Kenya is also extensive, and so there are many questions that can be posed relating to environmental drivers of land use/cover change.

Begin by exploring the various data layers that are in these folders: Africa Continental Data, Case Studies, East Africa Data Layers, and Kenya. Case studies are organized by the name of the local community; these include Embu, Ivondo, and Upper Embu. There is also a folder called Mt Kenya, which includes a variety of land use cover, topos (scanned paper maps), and satellite images. We will use the sharpen_2001.img satellite image to create a 3D visualization of his mountain. The richest dataset is for Upper Embu with land use/

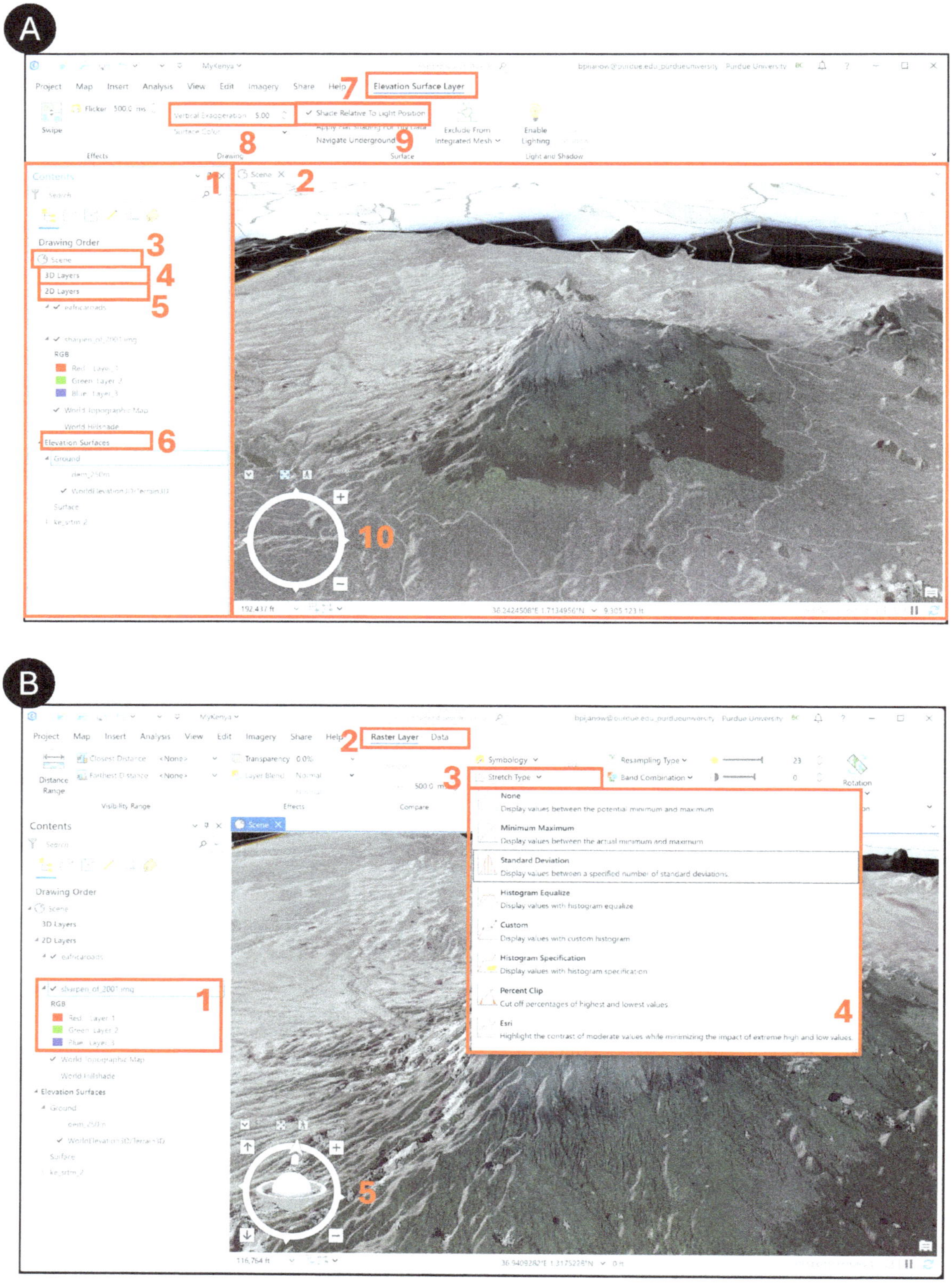

FIGURE EX10.1. Interfaces to (A) global 3D scenes for Mt. Kenya and (B) how to change visualization of a satellite image that is composed of three bands: red, green, and blue.

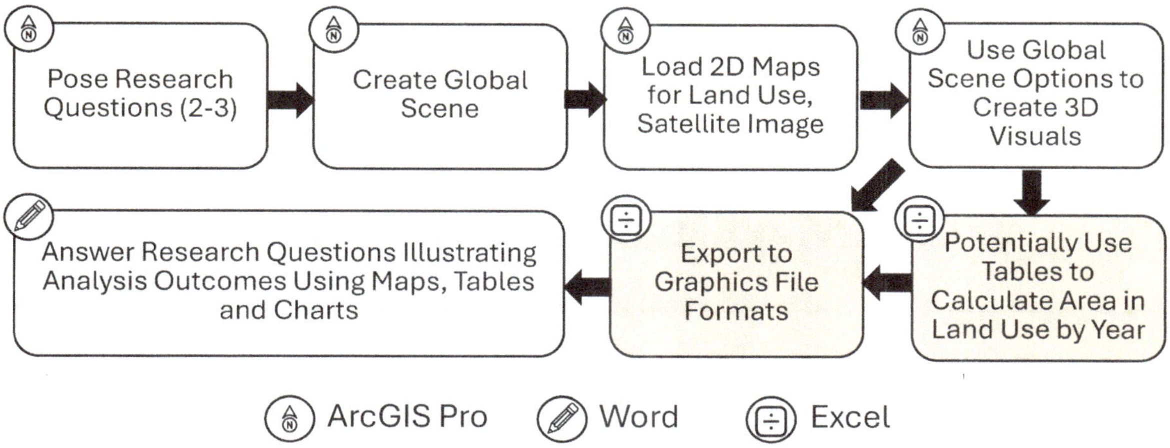

FIGURE EX10.2. Workflow for the Mt. Kenya 3D visualization lab.

cover maps created for each major decade from 1950 to about 2000. Land use/cover classes are standardized and embedded in the shape of files.

Also note that there are a variety of other useful files. These include pdfs that describe the data, legacy *.lyr files that contain legends for raster files, and sample maps made by the team in the early 2000s.

10.6 ASSIGNMENTS

Using data packaged as MtKenya3Dlab.zip, complete the following.

A. INTERMEDIATE STUDENTS

Produce these maps and writeups:

- 3D map of Mt. Kenya with the proper map elements. Provide a short writeup of the ecological and cultural significance of Mt Kenya in your own words that should be supplanted with information from Olson et. al. (2008) and what you can find online. Minimum number of words is 100.
- A map of time series of land use/cover change for one of the communities along the slopes of Mt. Kenya. Describe the changes that you see in no less than 100 words.

B. ADVANCED STUDENTS

Pose a set of research questions that can be addressed using the Mt. Kenya dataset. Create a series of maps and charts that describe the changes in land use with visuals that include at least one 3D map. Submit this as an IMRAD-style report. Make sure you clearly state your research questions (best practice is to use 2–3) at the end of your introduction.

EXERCISE 11: MAPPING AND ANALYZING GLOBAL CLIMATE ZONE SHIFTS

Highlights

Aims. This exercise is designed to introduce you to analyzing global models of climate and bioclimate variables.

Learning objectives. After completing this exercise in about a 1.5-hour time period, students should:

- Be able to compare and contrast two global bioclimate models
- Be able to describe how the Leemans (1990) and Beck et al. (2023) models were developed (in general)
- Know how to take text information and layer file information and use this in ArcGIS Pro's Symbology to create informative maps

Project Data Resources. Leemans (1990), Holdridge (1967) Life Zone maps of current and future conditions, and Beck et al.'s (2023) online distributed projections for 30-year periods at five different spatial resolutions and for six socioeconomic future scenarios.

11.1 EARLY WORK ON CLIMATE AND VEGETATION ZONE MAPPING AND FORECASTING

One of the first attempts to model current and future climate integrated with associated vegetation patterns was carried out in the late 1980s by Rik Leemans (1990) of IIASA (International Institute for Applied Systems Analysis in Laxenburg, Austria). Leemans's approach involved synthesizing global meteorological records and summarizing these as maps using spatial interpolation methods into 0.5 global grid cells (about 55 km x 55 km at the equator). These climate data, principally annual temperature and precipitation, were then used for the three bioclimatic variables proposed by Holdridge (see Chapter 9). Using these three bioclimate variables as proxies, a map was created to predict current vegetation life zones (i.e., circa 1940–1990). Then, using the U.K. Meteorological Office's global climate model (GCM) that assumed a doubling of CO_2 concentrations from the current time, Leemans recalculated where the life zones would be given the future changes (circa 2100) in the three key bioclimatic variables. The models had the same 38 life zones that were proposed by Holdridge in his classic *Science* paper (Holdridge, 1947). Although advanced for its time and fairly simple based on current technical capabilities, the work was heralded as a breakthrough since (1) it was one of the first global models that linked to vegetation models; (2) it resulted in a series of maps that could be useful for policy analysts interested in understanding the impacts of climate change on global ecosystems; and (3) it set the stage for many more sophisticated forecast models that were global in scale and that involved the integration of global datasets and bioclimatic models.

11.2 RECENT MODELING OF THE KÖPPEN-GEIGER CLIMATE ZONE

There have been several global climate classification schemes proposed (see Chapter 9). One of the most widely known and used today is the **Köppen-Geiger Climate Zone** system (Beck et al., 2018). Most climate classification systems use annual measures of air temperature and precipitation that are bracketed into groups that yield a set of climate zone classes. The Köppen-Geiger Climate Zone system was first proposed by the German climate scientist Wladimir Köppen in 1884. It was designed to follow some of the suggested biogeography global vegetation classification system of Alexander von Humboldt in the early 1800s. Köppen later updated these maps, after acquiring additional meteorological data from around the world, producing the final map in 1936 along with his student (Köppen & Geiger, 1936). As additional meteorological data become available, there are attempts to update these maps and distribute them for use by the spatial ecology community.

The Köppen-Geiger Climate Zone system is based on a hierarchical classification system, identifying five (A–E) major climate zones as listed in Table Ex11.1.

The second and third levels of the Köppen-Geiger Climate Zone classification system focus on specifics for precipitation and temperature. By combing the highest level and the two sublevels, most versions of the Köppen-Geiger Climate Zone system yield about 30 classes globally. Figure Ex11.1 shows the system of climate classes that exist around the world currently. Note that some climate scientists include a sixth higher level class, Highland, that is given the letter H.

The Köppen-Geiger Zone Classification System is considered especially useful for conservation biology for a variety of reasons. First, these breakpoints for annual temperate and precipitation are widely regarded as breakpoints for broad vegetation communities. For example, Zone B is arid and in hot, arid regions, which would be hot, dry deserts, there are certain kinds of plants and animals that have adapted to live in these extreme environments. Second, understanding these distributions, especially across continents and nations, helps to force a more ecologically relevant focus to management of all species on Earth within the context of climate parameters.

TABLE EX11.1. *The major climate zones (sometimes called groups) for the Köppen-Geiger Climate Zone classification system. This is the first level of three.*

ZONE	NAME	DESCRIPTION
A	Tropical	High rainfall and average annual temps > 18°C
B	Dry	Can be hot or cold, evaporation exceeds precipitation
C	Temperate	Moderate temps and rainfall, coldest month > 0°C
D	Continental	Moderate temps and rainfall, coldest month < 0°C
E	Polar	Warmest month < 10°C

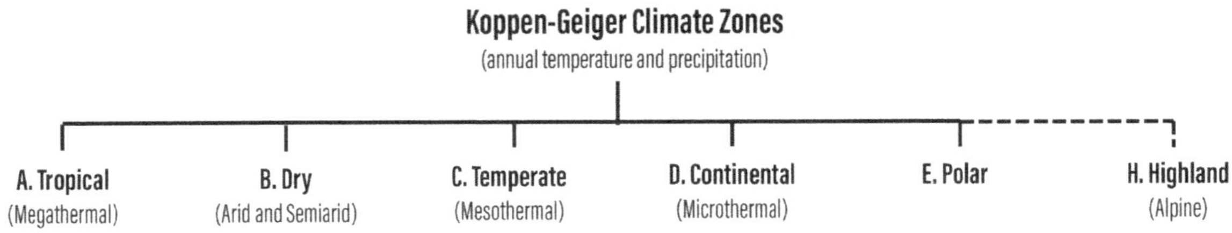

FIGURE EX11.1. The 30 Köppen-Geiger Climate Zones as organized by the six major zones (the sixth, Highland, is not always considered by some researchers).

11.3 BECK ET AL.'S CURRENT AND FORECAST CLIMATE MAP PACKAGE

Beck et al.'s (2018, 2023) study is a second focus of this lab exercise. This group of climate and ecological scientists used output from the Coupled Model Intercomparison Project (CMIP), phase 6, which produces harmonized data from a variety of global climate models (GCMs) to produce Köppen-Geiger Climate Zone maps. Output from the CMIP was used to create these maps for historical (prior to 2000), current (circa 2020), and future (circa 2100) periods for the entire world with future forecasts based on six shared socioeconomic pathway (SSP) scenarios. Their simulations are released at different resolutions, with the finest resolution at approximately 1 km x 1 km cell sizes (see Table Ex11.2), associating each terrestrial location to one of the 30 Köppen-Geiger Climate Zones. Model output are distributed as GeoTIFFs with embedded colors set to specific RGB values for each of the 30 climate zones. The Köppen-Geiger Climate Zone codes are aligned to values that range from 0 to 255 with the value of 0 being the oceans. Values 31 and larger have no climate zone associations; indeed, codes above 30 are set to black. A separate set of files was distributed that contain model confidence scores (see Beck et al., 2023, for details of how these are calculated and stored). The text file legend.txt was also distributed, which contains Köppen-Geiger integer code (1–255), the Köppen-Geiger two- or three-character codes, the climate zone names, and the values for the RGB colors.

TABLE 11.2. *Filename conventions for the Beck et al. (2023) model distribution, with the degrees and then approximate resolution in km at the equator. Note that 1 km model output can take some time on desktop computers, so for this lab it is recommended to select resolutions that are coarser than the ones distributed as 0p01.*

FILENAME SUFFIX	DEGREES	APPROX. KM AT EQUATOR
0p01	0.01°	~1 km (≈1.11 km)
0p1	0.1°	~11 km
0p5	0.5°	~56 km
1p0	1.0°	~111 km

11.4 NAVIGATING ARCGIS PRO FOR CLIMATE ZONE MODEL ASSESSMENTS

Output from Leemans's models is straightforward raster data layers with current conditions named holdnorm and future predictions as holddouble, as shown in the **Contents** pane (Figure Ex11.2A, box 1) with a **Map** of the current Holdridge Zones (projection was changed to Eckert VI World) displayed with the default gray scale (Figure Ex11.2A, box 2). A view of the **Attribute Table** indicates that there are 38 records (i.e., life zones) with 0 coded for the ocean. A layer file (*.lyr) was distributed with the data by IIASA, which links the grid codes to the names of the life zones. In the **Symbology** pane (Figure Ex11.2A, box 4), you can right-click the hamburger (Figure Ex11.2A, label 1) to bring up the list of actions that includes **Import from layer**. Navigate to the folder containing the Holdridge.lyr file and apply it.

After applying the lyr file to the two datasets, the list of life zones is now shown with the data layer in the **Contents** pane (Figure Ex11.2B, box 1). As one of the tasks is to determine what areas will be expected to undergo change, we will need to remove the ocean cells as this will, obviously, not change in life zone class in the future. To accomplish this task, the local operation **Set Null** can be used to assign original values that are changed to null or no data (Figure Ex11.2B, box 2).

Creating a map of areas that are the same (i.e., status) and those that have changed can be accomplished using a Boolean operation (Figure Ex11.3). Using **Raster Calculator** (Figure Ex11.3, box 1), a conditional test (i.e., use of the Con operator, which is technically the if/then/else) asks what locations of the two raster maps are the same. If they are, then this is a true condition and a value of 1 is assigned. If they are different, then the condition is false and a value of 0 is assigned. The Raster Calculator operator is:

Con("holdnormal" == "holddouble", 1, 0)

The resultant map is made of 1 (no change between normal and future) and 0 (normal and double maps are different). To a spatial ecologist, an analysis of the sizes and distributions of 1s and 0s would be of interest, especially as they relate to (1) continents and (2) latitudes. Other analyses might also be of interest, including how much change in any of the life zones occurs and if any are lost and where those locations might be.

11.5 WORKFLOW

A proposed workflow (Figure Ex11.4) could involve the following tasks that use some of the Navigating ArcGIS Pro workflow interface steps described above combined with others that have been presented in earlier labs.

Step 1: Follow the section *Navigating ArcGIS Pro* for the Holdridge Life Zone maps of Leemans (1990) above. Using the holdnorm and holddouble raster data layers, prepare the data and then conduct a change analysis as described above.

Step 2: Inspecting Köppen-Geiger Climate Packages. Unzip the data package for this lab and explore the files that are contained in this package at https://www.gloh2o.org/koppen/. Note that there are files for historical present and future at three resolutions. Bring each of them into an ArcGIS Pro map project. Note the largest differences in coarseness for future and present climate zones. Using Notepad, insert the legend.txt file. Note the value for each of the

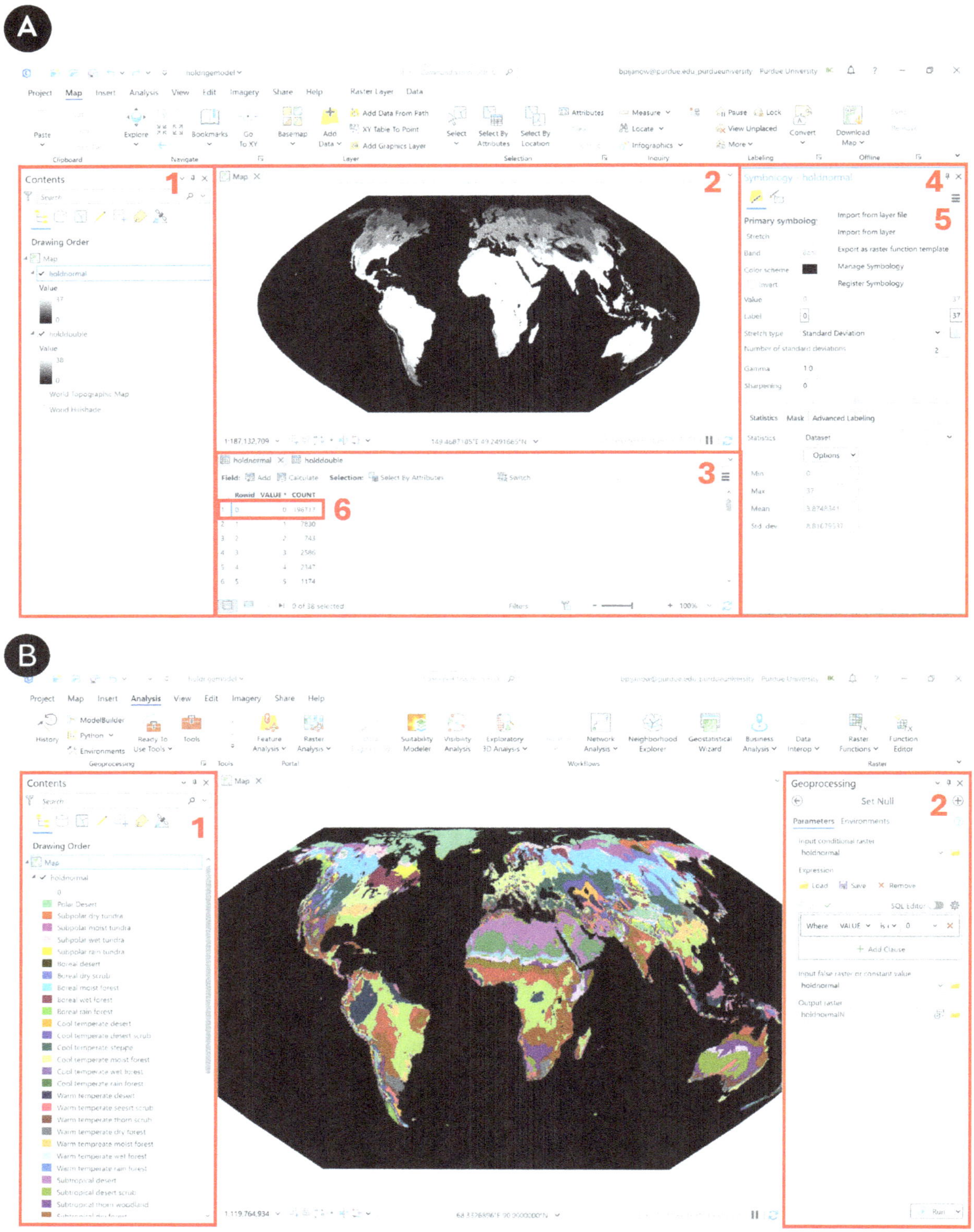

FIGURE EX11.2. Interfaces for the workflow of (A) inspecting the original model output and (B) preprocessing the Holdridge Life Zone maps from Leemans (1990).

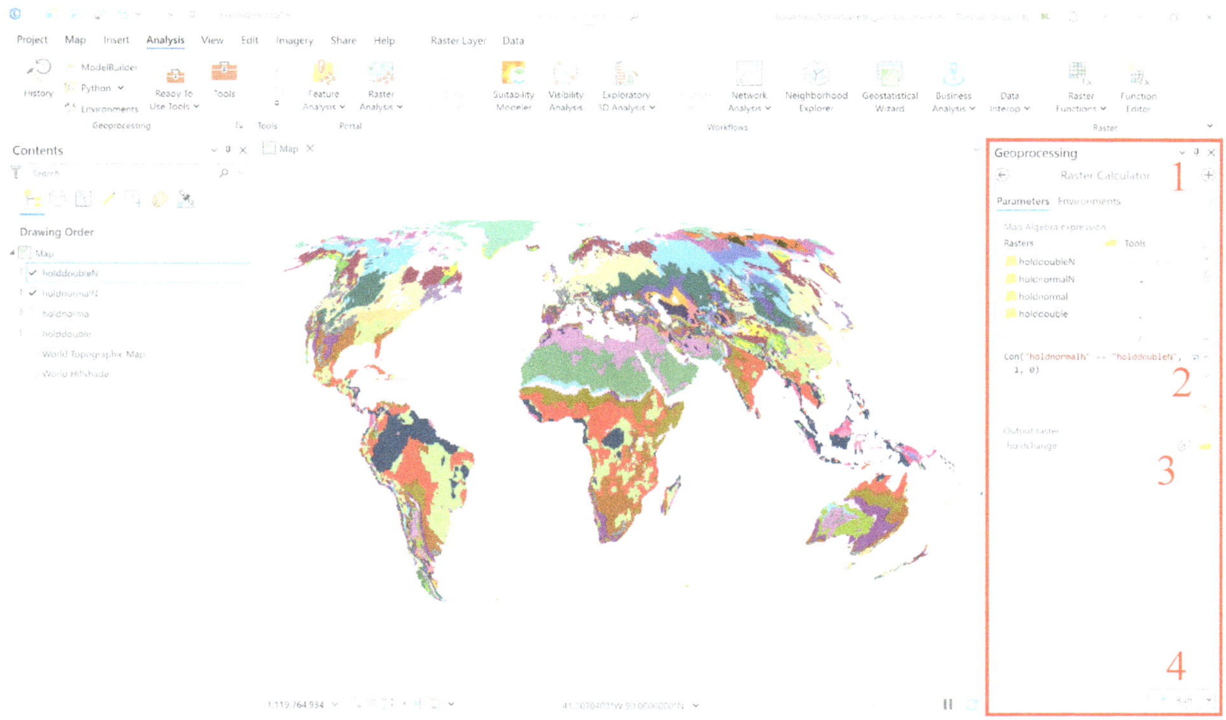

FIGURE EX11.3. The use of raster calculator operators to specify a Boolean command to create a change/no change map.

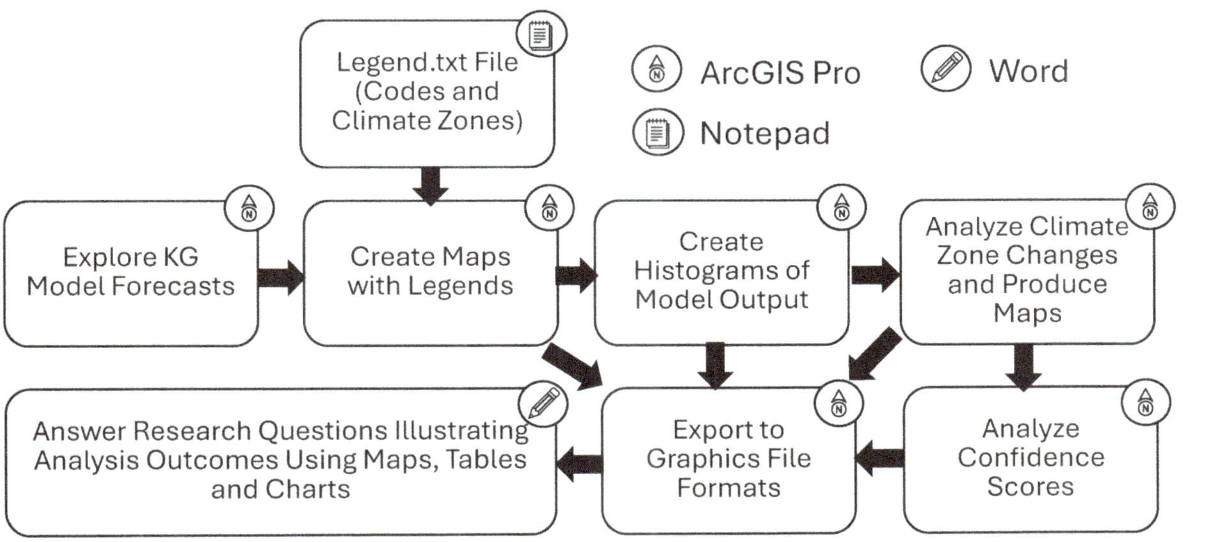

FIGURE EX11.4. Proposed workflow for the climate zone modeling lab. See also the more detailed workflows with interfaces and also the steps described below.

30 zones, their names, zone codes, and the RGB codes. Notice in your display that value is blue. You should also notice that the RGB values for 1 are [0,0,255], which is no red, no green, and all blue. Next, bring up the maps with confidence scores that range from 0 to 100; a value of 100 indicates that the model has a 100% confidence that the cell location is the climate zone it assigned and no other zone. Note the spatial patterns for the confidence scores as they relate to the borders of the climate zones.

Step 3: Histogram Features in ArcGIS Pro. Let us next explore how we can use the built-in chart features of ArcGIS Pro. There are a lot of parameters that can be modified to produce a chart. Watch the video "Using histograms and changing parameters to make this look good.mp4." Note that many adjustments are needed to make this chart one that can be exported to your lab report. Follow this video and then a second one, "Adjust histogram for KG present map.mp4."

Step 4: Create Raster Tables. Let us next convert each of these GeoTIFF files (present and future) to an Esri ArcGIS Pro grid file. To do that, right click on the GeoTIFF file, select Data, and then Export Raster. Select Grid as Output format, and select a name for the file (make it compliant with grid filenames!). Once that is created, bring up the Attribute Table (right click the grid file and then select Attribute Table from the long menu). You should see the Attribute Table for the grid placed below the map. Select (left mouse click) the hamburger symbol in the Attribute Table view. The last item in the list is Export. Export the Attribute Table to a separate file (do not place this in the geodatabase) and assign the file a name and .dbf. Do this for present and future raster data layers.

11.6 ASSIGNMENTS

A. INTERMEDIATE STUDENTS

Complete the following tasks and include any necessary visuals and short descriptions:

1. Using Leemans's Holdridge Life Zone maps (normal and double), conduct an analysis following the workflows described above that show locations of change between the current and future climate. Create a map of these changes and no changes. In not less than 50 words, describe the spatial patterns of change and stasis.
2. Using the 2018 or 2023 spatial data as GeoTIFFs posted at Beck et al.'s (2018, 2023) model output distribution sites, create maps of the Köppen-Geiger Climate Zones for the present (e.g., 1991–2020) and future (2071–2091), following the same workflow for the Leemans model output. Use the legend.txt information to generate a legend for the 30 climate zones. Compare and contrast these maps in areal amounts and distributions. Use at least 50 words.
3. Produce a map of the confidence scores for the Köppen-Geiger Climate Zones climate zone models for the present and future. Describe the patterns that you see in reference to the borders of the climate zones. Use at least 50 words.
4. Summarize the dbf tables of climate zones for the present and future using the Holdridge Life Zone maps. Describe why this straightforward comparison might not be a good one. Use at least 100 words to answer this question.

B. ADVANCED STUDENTS

Write up this exercise as a lab report that addresses this research question: What are the major changes in the Holdridge Life Zone maps and the Köppen-Geiger Climate Zones between the present and future? Use the information and outcomes from the exercises for this report. In addition, address how you can use the confidence scores from the model output of Beck et al. (2023) to consider additional information about this research question. Cite at least five peer reviewed papers that have used information or output from these models, presenting this in your discussion.

EXERCISE 12: MAPPING AND ANALYZING ECOREGIONS AND NPP

Highlights

Aims. To analyze ecoregions and their net primary production at a subcontinental scale.

Learning objectives. After completing this exercise in about a 1.5-hour time period, students should:

- Be able to understand how hierarchical data are structured within a spatial data model and then use this to create new maps
- Know how to gather satellite remote sensing data at large scales and then analyze these data for long-term NPP calculations
- Know how to analyze raster and vector data for analyses at large spatial extents

Project Data Resources. From the internet, Bailey's and Omernik's Ecoregion maps and from Google Earth Engine or the NASA EROS Data Center, Annual NPP for the US CONUS region.

12.1 INTRODUCTION TO ECOREGION DATABASES FOR THE UNITED STATES AND NORTH AMERICA

Chapter 9 summarizes how plant and animal community assemblages are grouped at global (i.e., realms and biomes) to regional (e.g., ecoregions) scales. One feature of these assemblages is that they are distributed across space based on annual precipitation and temperature or by plant functional type (PFT). As was described in Chapter 9, these broad patterns of organismal assemblages were recognized and studied by the early 19th-century naturalists like Humboldt, Darwin, and Wallace. With the advent of satellite technologies and the ability to measure broad features of the landscape from space, approaches such as using light-based sensors to calculate net primary production (NPP) allow contemporary spatial ecologists and GISci analysts to create very precise maps of ecoregions.

In the United States, there are two ecoregion maps that have been developed by federal agencies to assist in natural resource management at large (e.g., states, multistate, to national) spatial extents. The first, **Bailey's Ecoregions**, was developed by Robert Bailey of the United States Forest Service in the early 1980s (Bailey,

1983). It is a hierarchical classification system of ecological regions that is defined at the highest level by long-term climate patterns of annual temperature and precipitation. More fine-level organization is defined by topographic features (e.g., elevation, slope, aspect), soils, and vegetation patterns. Bailey's Ecoregion maps contain polygons at levels referred to as domains, divisions, provinces, and sections, often referred to as Level I, II, III, and IV, respectively (Bailey, 1995, 2009). Bailey's Ecoregion boundaries have been used to develop natural resource management plans in forested ecosystems that span state boundaries and thus is considered an excellent resource for the management of areas at large spatial extents.

A second common ecoregion database is the one referred to as **Omernik's Ecoregions**, also developed in the 1980s (Omernik, 1987). Omernik's Ecoregions were created by the U.S. EPA, in collaboration with the U.S. Forest Service and Bailey, to provide for regions that would allow for environmental monitoring and reporting to state and federal agencies. These ecoregions more explicitly include hydrologic features such as rivers, lakes, and wetlands so that they are organized within spatial units that are similar (Omernik & Griffith, 2014). Omernik's Ecoregions are also hierarchically organized at four levels, I–IV, and are not given any other names for their levels. The most used levels are III and IV (EPA, 2013). Researchers that study spatial-temporal patterns at this spatial extent are often using these boundaries to examine important trends. Pidgeon et al. (2007), for example, used Bailey's Ecoregions to summarize bird species richness patterns with landscape patterns and housing development patterns. Stoddard et al.'s (2008) work applied Omernik's Ecoregion maps at Level III organization to demonstrate how stream monitoring data could be summarized and examined across space at regional scales. An excellent comparison of these databases and their potential uses is provided in Wright et al. (1998), who examined how the boundaries of the ecoregions lined up with vegetation maps from satellite remote sensing (e.g., Landsat-derived measures).

12.2 VEGETATION INDICES FROM SATELLITE-BASED SENSORS

As was described in Chapter 2, mapping vegetation patterns to quantify land cover patterns is now done successfully using a host of Level 1 through to Level 3 products from many of the space-based remote sensing platforms like Landsat, MODIS, or Sentinel-2. Indeed, there are likely dozens of remotely sensed measures of vegetation patterns that can be downloaded for large areas, even at global extents.

One of the commonly used products from the NASA MODIS sensor is the suite of gross primary production (GPP) and net primary productivity (NPP) products, grouped as part of the MOD17 suite of products. **Gross primary production** is the total amount of carbon that is fixed through the process of photosynthesis. **Net primary productivity**, on the other hand, is the "net" or difference of gross primary production and the amount of energy used for plant metabolism. Thus,

$$GPP = NPP - R_a$$

where R_a is autotrophic respiration.

For this lab, we will use a valuable MODIS product called MOD17A3, which are modeled annual NPP values for each pixel (500 m x 500 m) derived from the application of a variety of light and meteorological models that use land cover maps and meteorological data as inputs. The models estimate net primary productivity at each pixel summarized across a year with specific rules for replacing missing data (e.g., locations where cloud cover creates a missing value). Several data portals (e.g., the NASA EROS DAAC, Google Earth Engine) store these data products, and they can be extracted from near global coverage maps using shape file boundaries (e.g., US CONUS boundaries). The value for each NPP pixel is the amount of carbon stored per year as expressed in units of grams of carbon per square meter per year with a scale factor of 0.1 (e.g., pixel value of 4,500 means 450 g C/(m^2 yr). The value is not a direct measure of biomass but rather an estimate of the amount of photosynthesis occurring (a type of measure called flux by physical scientists).

12.3 SUMMARY SPATIAL STATISTICS USING ZONAL, FOCAL, LOCAL, AND GLOBAL OPERATIONS

ArcGIS Pro, as well as nearly all GIS software platforms, can calculate and then summarize spatial statistics using one of several levels of operations. For this lab exercise, we will use the Zonal statistics set of operations to calculate averages of annual NPP values and then summarize these using zonal statistics. These operations are important to know for this lab and for the next few, so a summary of them is included here. These operations have the following characteristics and are performed either entirely on raster data layers or using vectors (e.g., polygons) to summarize spatial statistics.

A. ZONAL

Zonal operations perform a summary of spatial statistics within **zones**. Zones are defined using either categorical raster maps or polygons. For this exercise, we will calculate summary spatial statistics for zones defined as the area within each ecoregion polygon. Summary spatial statistics include the calculation (within each zone) of the mean, median, range, maximum, minimum, majority, minority, and standard deviation of all pixel values. Another way to conduct a zonal summary statistical operation is to calculate these same statistics based on a raster data layer that contains land use/cover classes. In short, zonal statistical operations are a powerful set of tools that spatial ecologists and GISci analysts can use. Output is typically in two formats, such as (1) a separate raster data layer that stores summary statistics to a grid, or (2) in tables, where the summary statistics are compiled as columns in a table with each row as a separate zone (e.g., ecoregion).

B. FOCAL

Focal operations perform summary spatial statistics within specific neighborhoods. These operations perform **moving window** statistics, and these can be useful if you are interested in, say, the variability of elevation in a defined window (e.g., a 10-cell x 10-cell square window or a 5-cell x 5-cell circle, also called an **annulus**). Such a visualization can help identify areas of high variability within specific areas that might be important to consider for local natural resource management. Focal statistics can allow users to specify weights (e.g., for distances) within a focal area as well; in such cases this is referred to as **kernelling** (i.e., passing a moving window over a grid with values weighted across the window, such as specifying a distance decay algorithm).

C. LOCAL

Local operations perform summary statistics across multiple raster data layers, storing the output as a new raster layer with that summary statistic. The local operations used the most include the command **Cell,** which provides a user with the ability to calculate, viz., the average value across all raster data layers. In this exercise, the 7-year annual net primary production (NPP) will be calculated across these raster data layers for the analysis of how NPP relates to Bailey's Ecoregion zones. One very useful local operation is the int function, which converts floating-point raster data layers to integer; this is sometimes necessary if the floating-point raster data layer is very large.

D. GLOBAL

This set of raster operations is described by Esri as a group of operations that fall into a more general category of statistics set up to do very specific spatial calculations. Raster dataset summary statistics can be stored in a table but most often are calculated and then stored in an associated linked file created by the simple operation **Get Raster Properties** (which can be retrieved with program scripts). Another set of global statistics is configured around specific uses, such as those that are used in ArcHydro toolsets (e.g., flowdirection) or use vector input (e.g., city centers as points, roads as vectors) to calculate the distance of each cell from the nearby vector data point. Finally, another useful set of commands are those that use slice. This command takes a raster table and places output raster tables into values based on ranges and distributions. For example, you could take a raster data layer that has distance to road and then place the cells into 10 classes of distances based on the range of values. The output could be values 1 through 10 assigned based on distance categories, area, or another spatial function. Doing so means that this global operation could create a raster map that is used for a zonal calculation.

12.4 NAVIGATING ARCGIS PRO FOR ECOREGION ANALYSIS

Both the Bailey's Ecoregion Database and Omernik's Ecoregion Database are constructed using a hierarchical data coding scheme that is common in spatial databases. This coding scheme is based on a spatial data model design of creating polygons for the most detailed level (both at Level IV) and then harmonizing these very detailed polygons at higher levels of organization by introducing fields that contain the same codes for that higher level set of polygons. Figure Ex12.1A shows the ArcGIS Pro interface to Bailey's Ecoregions and how this is accomplished. Note that in the map in Figure Ex12.1A (box 1), the Symbology is set to display, using different colors, domains (Level II) with boundaries, outlined in black, for ecoregions classified at the section (Level IV) level. We are able to create this map since we have coding for Level II and Level IV.

To create polygons at higher levels of ecoregion organization, the ArcGIS Pro tool `dissolve` is a great operation to perform on the data when "like" polygons have the same code (Figure Ex12.1A, box 2). In this case, the names of the ecoregions at each level (see Figure Ex12.1A, box 3) serve as those harmonizing codes. If we were to create a set of polygons at level II, we would dissolve polygons using the coding of that level. To fully appreciate how this works, scan the coding for **Domain,** noting that all have the same name and thus code; for **Division,** noting that most have the same code; for **Province,** noting that most have the same code; and for **Section,** where none have the same code in the list of this visual.

Once we have created these polygons, let us now pay direct attention to processing of annual NPP data for 7 years (Figure Ex12.1B). Let us calculate a long-term NPP average using the Local Cell statistics tool in the **Geoprocessing** Toolbox (Figure Ex12.1B, box 1), where the **Local** set of operations (Figure Ex12.1B, box 2) is selected and then the **Cell Statistics** (Figure Ex12.1B, box 3).

Continuing on to the **Cell Statistics** pane (Figure Ex12.2A, box 1), a user is provided with options to enter all of the data that would be performed for each cell across multiple raster data layers. Here, there are 7 years of annual NPP data that are listed (Figure Ex12.2A, box 2). The operation to perform across all 7 datasets is **Mean** (Figure Ex12.2A, box 3).

A note about using remotely sensed imagery or products like NPP in ArcGIS Pro to visualize spatial patterns. Defaults in the **Symbology** panel generally give very poor results as values in the raster data layer tend to be skewed, or in many cases have outliers that influence the scaling of the color ramp. Note what the default for the 2024 annual NPP raster data layer looks like (Figure Ex12.2B); it has no gradation across the color green, which typically occurs when data are skewed or there are outliers present. Once a stretched histogram-scaled algorithm is applied (Figure Ex12.2C) and the color is scaled with the **Invert** option on, high levels of NPP in the northwest and across the southeast are observed (as expected).

The ArcGIS Pro interface controlled through the Symbology pane (Figure Ex12.3, box 1) illustrates how one can control the visualization of data like that for annual NPP. Note that one option is to explore, as we did to produce Figure Ex12.2C, the **Stretch Type** of **Histogram** (Figure Ex12.3, box 2). This selection can then be used to activate a visual of the histogram that is shown below box 2. Note that the histogram for the raw data values is shown as gray bars. There is a distinct pattern here, common with remotely sensed data, where data are distributed unevenly across values. Note that most of the values in this histogram are found to the left side (Figure Ex12.3, label 3). There are, however, a few values of the x-axis that occur to the very right (almost as a spike). These are outliers of the processing of the data; they occur over water and over areas with a high degree of impervious surfaces (e.g., pavement). Note that the histogram "correction" that is applied gives a distribution of values that are shown with red bars (Figure Ex12.3, label 4). In this case, I have taken the bracketing arrows using the vertical slider (see Figure Ex12.3, label 5; there is one to the left and right) and shifted the visualization for the high bracket to "cut out" those high anomalous values.

12.5 WORKFLOW FOR THE EXERCISE

Figure Ex12.4 gives an overview of the steps necessary to conduct the analyses required to complete this assignment. You are going to explore how (1) Bailey's and Omernik's Ecoregion boundaries differ at Level III and Level IV for the state of Indiana; and (2) determine if Bailey's Level III boundaries indeed define areas of unique NPP, based on MODIS remotely sensed model output.

12.6 ASSIGNMENT

A. INTERMEDIATE STUDENTS

Answer the following questions and include any necessary visuals and short descriptions:

- How do Bailey's and Omernik's Ecoregions differ at Level III and Level IV for the state of Indiana? Compare and contrast the number of features and their distributions across the state.

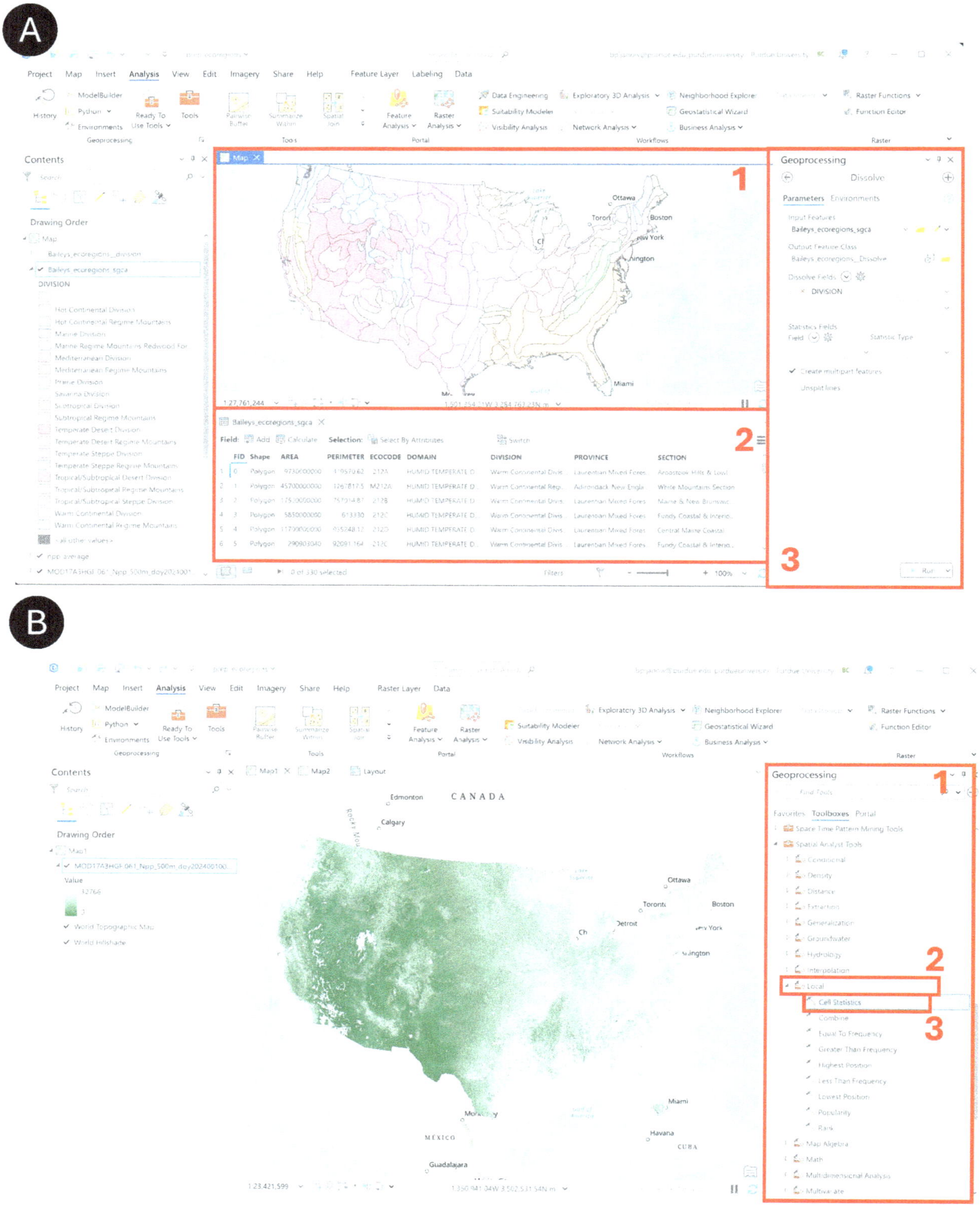

FIGURE EX12.1. Viewing (A) Bailey's Ecoregions as a map and as a table also showing the dissolve tool pane and (B) viewing the MODIS Annual NPP map and the cell statistic Geoprocessing pane.

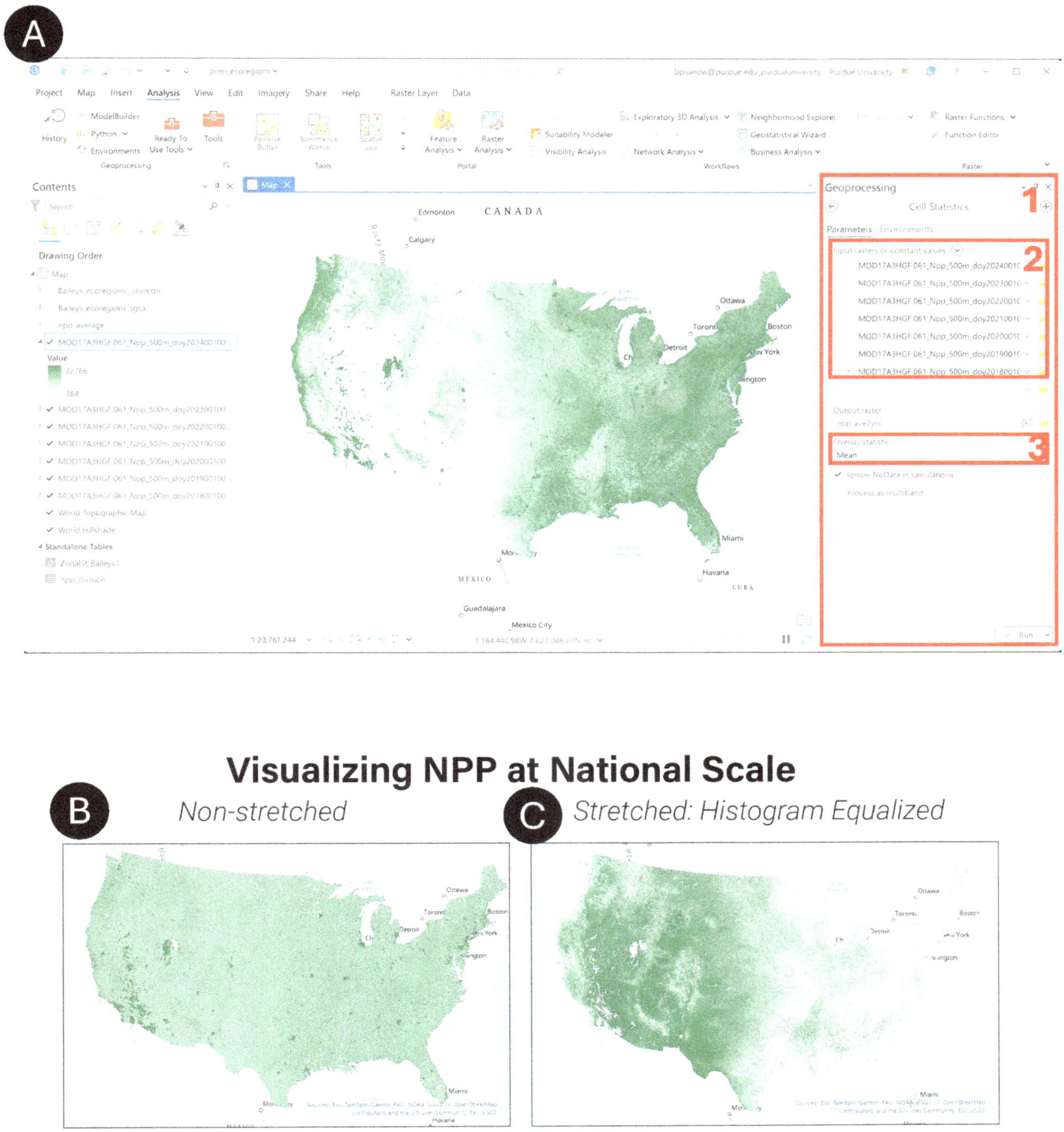

FIGURE EX12.2. How a user can view the unadjusted and then the adjusted distribution of values of average annual NPP across the U.S. CONUS region. See text for details.

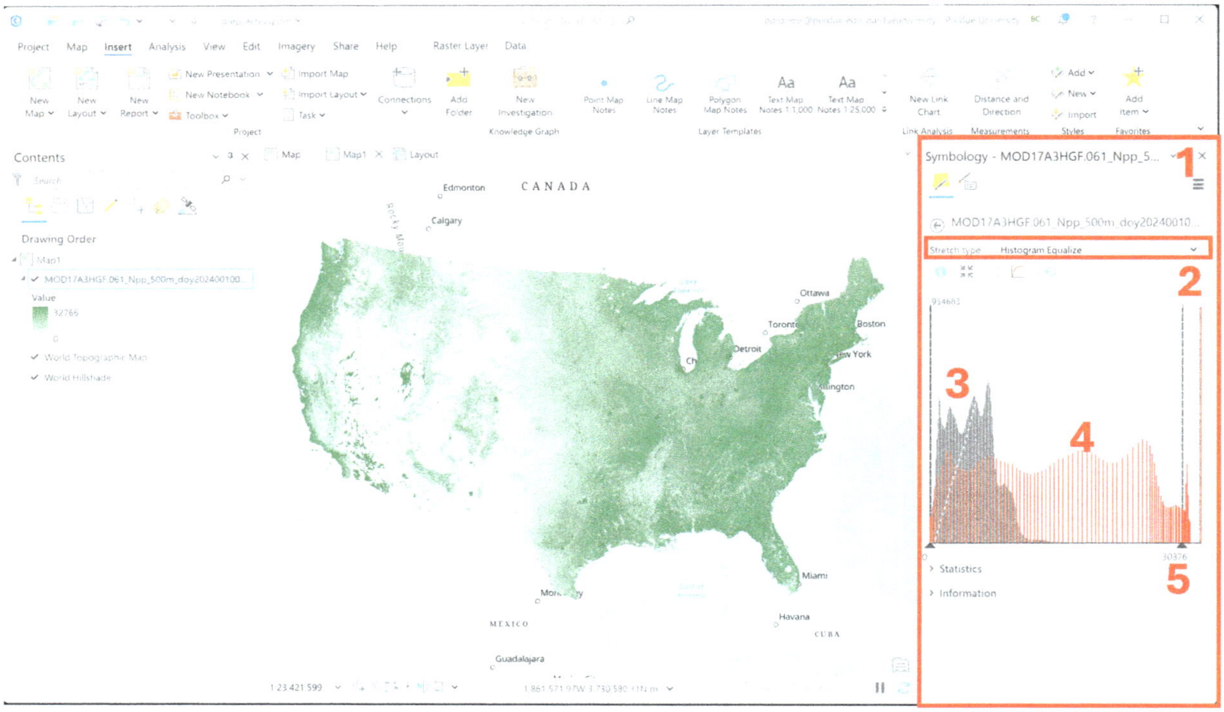

FIGURE EX12.3. Interface to (A) Cell Statistics pane that allows users to select functions to apply across multiple raster data layers (here to calculate the mean value for all 7 years of annual NPP; (B) how the default visual in Symbology produces a nonuseful visual; but (C) applying a reversed, stretched algorithm provides a better understanding of the spatial distribution of NPP.

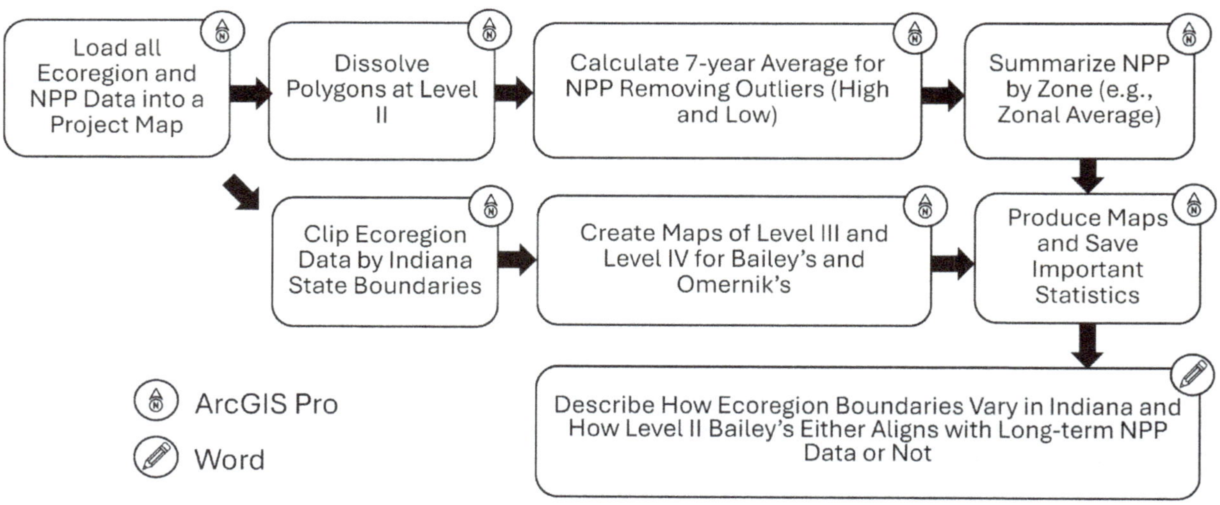

FIGURE EX12.4. Workflow for the analysis of ecoregions and MODIS land cover data.

- Does NPP vary across Bailey's division-level ecoregions in ways that you think are correct? Explain your answer in no less than 50 words.

B. ADVANCED STUDENTS

Using the IMRAD format, write up this lab as an exploration of how these two ecoregion maps are distributed across space in the lower 48 states of the United States. Use the following research questions to guide your research:

- Given that there are some similar histories to both Bailey's and Omernik's ecoregions, one expects them to match up well. Do they? Address this question by developing maps and doing some simple spatial analysis of both ecoregions' shape files at Level III and Level IV for the state of Indiana.
- Continuing along these lines, compare and contrast how Bailey's lines up with the 7-year NPP long-term averages that are then averaged as Level II zones. Consider creating maps of NPP values that have Level II boundaries. Do the boundaries line up with sharp breaks in NPP? In your answer, place your results in the context of the work by Wright et al. (1998).

In your report, describe the steps you took in your analysis, including how you went about removing any potential outliers of the NPP data.

EXERCISE 13: CALCULATING SPECIES RICHNESS AND WORKING WITH TABULAR DATA

Highlights

Aims. This exercise is designed to introduce you to tabular data and to spatial data that will need to be integrated using joins and relates.

Learning objectives. After completing this exercise in about a 2-hour time period, students should:

- Be able to join tabular data to spatial data and understand one-to-one and one-to-many relationships
- Know how to calculate species richness using standard tabular data
- Understand how landscape features can give rise to high levels of species richness detected at subcontinental scales

Project Data Resources. Tabular species presence and absence data for butterflies of the U.S. CONUS region, DEM data for U.S. CONUS, and county shape files with FIPS codes.

13.1 TRENDS IN INSECT ABUNDANCES

Insects are grouped in the class Insecta in the phylum Arthropoda. It is the most speciose animal group, with systematists recognizing over 1 million species worldwide; there are some estimates as high as 5.5 million species (Stork, 2018; Tihelka, 2021). Decreases in this group of animals have been widely reported, with reports documenting declines in the number of species, abundances, and/or total biomass over time at local to regional scales (Hallmann et al., 2017; Seibold et al., 2019). Insects are important groups as they provide critical ecosystem services to humans, including the pollination of crops and serving critical roles in decomposition, as well as providing for stable ecosystems as they are critical food sources for many species (e.g., amphibians, reptiles, birds, and mammals). Causes of the declines in insects have been attributed to the use of insecticides, changes in land use/cover, alterations to biogeochemical cycles (e.g., increased nitrogen and phosphorous in water bodies), and to increased artificial light at night, which disrupts diurnal behaviors (Forister et al., 2019; Sánchez-Bayo & Wyckhuys, 2019; Wagner, 2021).

Within the class Insecta are a large group of important volant (i.e., animals capable of flying) insects that are members of the order Lepidoptera, called the scaled-wing insects; these are often referred to as the butterflies and moths. There are about 160,000 species of lepidopterans described (Pogue, 2009), with most species occurring in the tropical regions of the world. In the continental United States, there are about 790 species of butterflies that have been identified. What is also noteworthy of this group is that their home ranges tend to be very specific to one or more species of plants; in other words, they have very narrow host ranges, many are pollinators, and a large group use pheromones to attract mates. Many species use visual cues so many tend to be "showy."

Lepidopterans are a group of insects that have a particular developmental pattern, called holometabolous, which means that individuals go through a complete metamorphosis from egg → larva (caterpillar) → pupa (cocoon) → adult. The larval phase is primarily the body form that feeds and grows, and the adult phase focuses on dispersal and reproduction. A handful of lepidopterans have gained media attention (e.g., monarch butterfly) for their long-distance migration and their peculiar overwintering behavior (e.g., converging at small, highly specific locations in the world in massive population numbers). It has been estimated that about 3% of all lepidopterans migrate. Most species have adopted the ability, in any phase, to undergo a slowed or nearly halted dormant phase, allowing them to pause development, a pattern called diapause.

In terms of the biodiversity patterns of lepidopterans, it is hypothesized that there is a greater species richness in areas where there is also a large number of species of flowering plants. This often occurs in mountainous areas where there are many types of environmental gradients that give rise to gamma diversity patterns.

This lab focuses on the use of a very unique butterfly database that was compiled by a large number of lepidopteran entomologists in the 1990s. This U.S. CONUS county database represents a significant effort coordinated by the U.S. Geological Survey's National Biological Information Infrastructure Program (this program was retired in 2012). This unique database was carefully constructed as a presence/absence table (i.e., 1 = present in the county, 0 = absent in the county) compiled from museum and citizen science efforts with a large effort to verify their presence. The U.S. CONUS area contains about 3,100 counties (not including Alaska, Hawaii, and Puerto Rico). A U.S. CONUS digital elevation model (DEM) at a moderate resolution (90 m x 90 m) will be used to summarize spatial statistics of elevation over each county. A set of simple statistical tests will be used to assess the validity of the hypothesis that increased regional elevation variation will be positively related to species richness of butterflies.

13.2 NAVIGATING TABLES AND CHARTS IN ARCGIS PRO

Data for butterfly distributions are provided as Microsoft Excel tables. These data will need to be summarized in Microsoft Excel as there are over 790 species of butterflies that have presence and absence codes for each county record. Bringing in that summary data as county names, FIPS codes, and species richness totals is done through the **Map, Add Data** (Figure Ex13.1A, box 1) button. Doing so brings up the **Add Data** menu where the Excel spreadsheet can be selected (technically, the worksheet within the spreadsheet needs to be selected).

Once that table is added (you can double-check that in the Contents Panel), a join between the county shape file can be made by selecting the data layer (Figure Ex13.1B, box 1), which launches the Actions menu (Figure Ex13.1B, box 2), and **Joins and Relates** is selected next. Doing so will then launch the **Add Join** option (Figure Ex13.1B, box 3). Note that the **Join** that will be performed will be with tabular data that is partially displayed in Figure Ex13.1B, box 4.

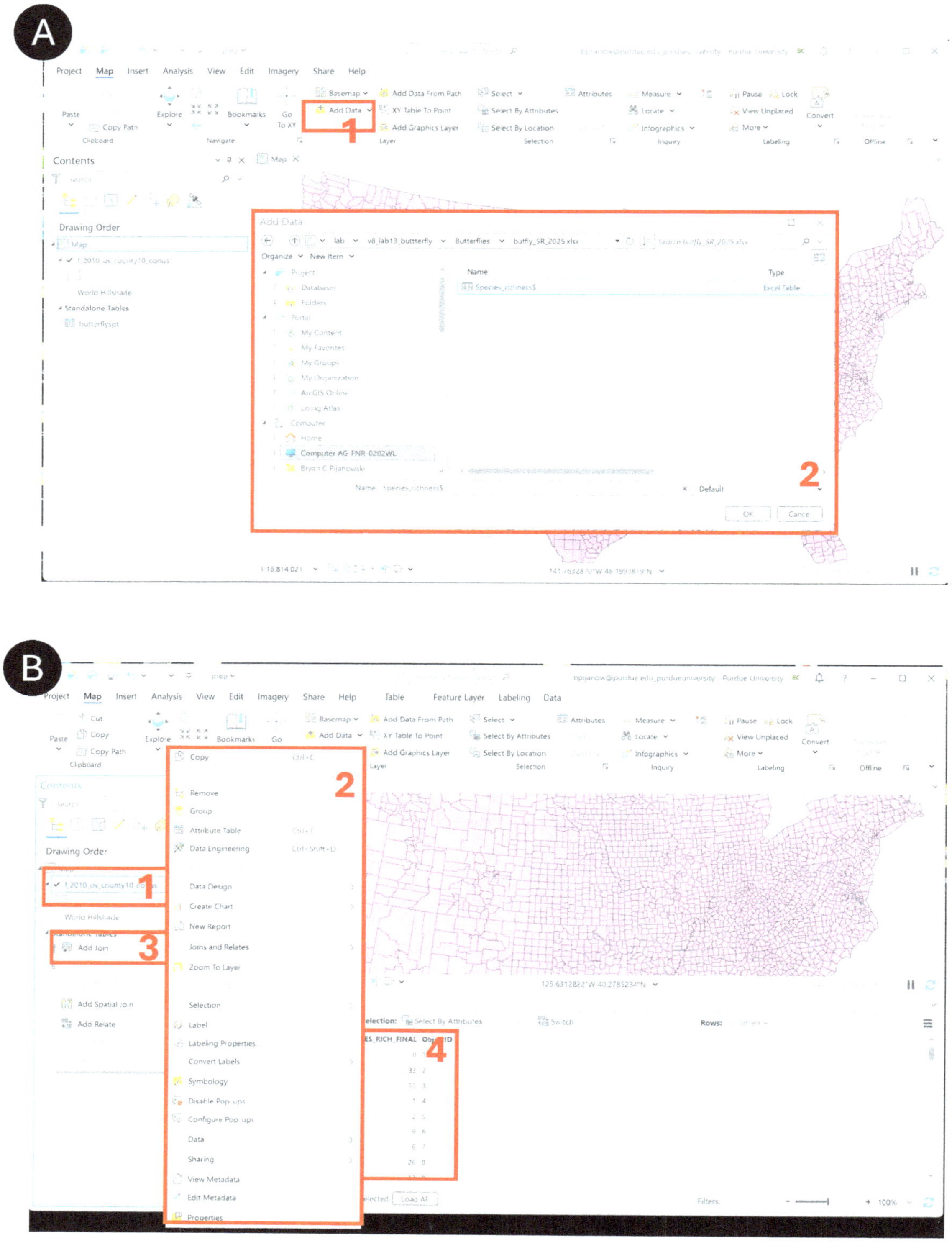

FIGURE EX13.1. The very common task of (A) adding data to a Map View and (B) selecting the Joins and Relates from the options menu.

The Add Join menu is shown in Figure Ex13.2A, box 1. This is where the shape file and the tables are joined using a common field; in this case, county FIPS code. In this display, the US_Counties_CONUS is also shown here partially in Figure Ex13.2A, label 2 and the US Counties_CONUS map is shown in the map (Figure Ex13.2A, label 3).

A map displaying species richness from the resultant join is shown in Figure Ex13.2B. In this display, the Symbology panel was used to select **Graduate Colors, Field** of Species Richness, display method of Natural Break (using the Jenks method) and a color ramp scheme of yellow (low) to red (high). Note that in this map of butterfly species richness, the greatest richness occurs in the western states.

For us to test the hypothesis that there is high variation in elevation at regional scales (e.g., at a county summary statistic), a national digital elevation model (Figure Ex13.3, box 1s) can be used and these summary statistics calculated using US_Counties_CONUS polygons as zones. Doing so creates another table, ZonalSt_UScounty (Figure Ex13.3, box 2) that can then be joined to the US_Counties_CONUS shape file to create a final shape file that contains the butterfly species richness total and the DEM summary statistics for each county (as a record).

A chart with a linear regression model can be created that examines how the independent variable of elevation standard deviation affects the dependent variable, species richness (Figure Ex13.3, box 3). The **Chart Properties** panel (Figure Ex13.3, box 4) allows the user to select the two variables to plot and the kind of model (e.g., linear) to explore. Note that the linear regression model is displayed at the bottom of the **Chart Properties** panel.

13.3 WORKFLOW FOR THE EXERCISE

Table Ex13.1 lists the tables and spatial databases to be used for this exercise. The workflow for the assignment, in very broad terms, is illustrated in Figure Ex13.4. Note that there is a lot of "up front" work in Excel to make sure the butterfly data are properly summarized and prepared for importing to ArcGIS Pro for eventual joins to spatial databases (e.g., county shape files). An important table that will not be used within the GIS is the table of species number and their species names and associated known taxonomy. This database will be consulted in the next exercise.

13.4 ASSIGNMENTS

A. INTERMEDIATE STUDENTS

Complete the following tasks producing short descriptions for accompanying visuals:

- Create maps of species richness by county and digital elevation.
- For two counties, calculate the conventional species richness for each county and then apply the Jaccard similarity index to these two counties and describe the differences.
- Create a chart that illustrates the relationship between digital elevation variation and species richness of butterflies.

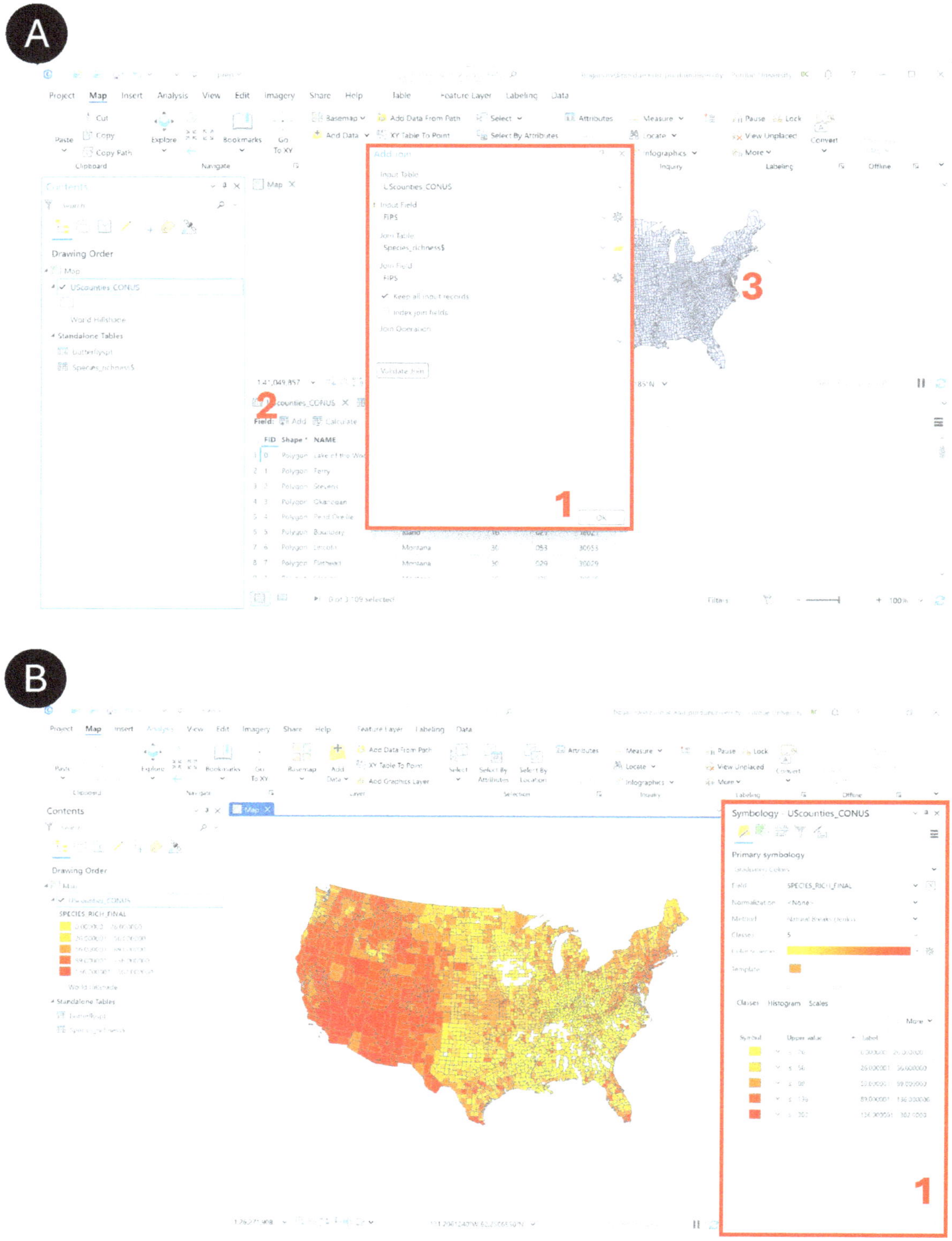

FIGURE EX13.2. The interfaces to (A) Add and Join and then to (B) view a map using selected features in the Symbology pane.

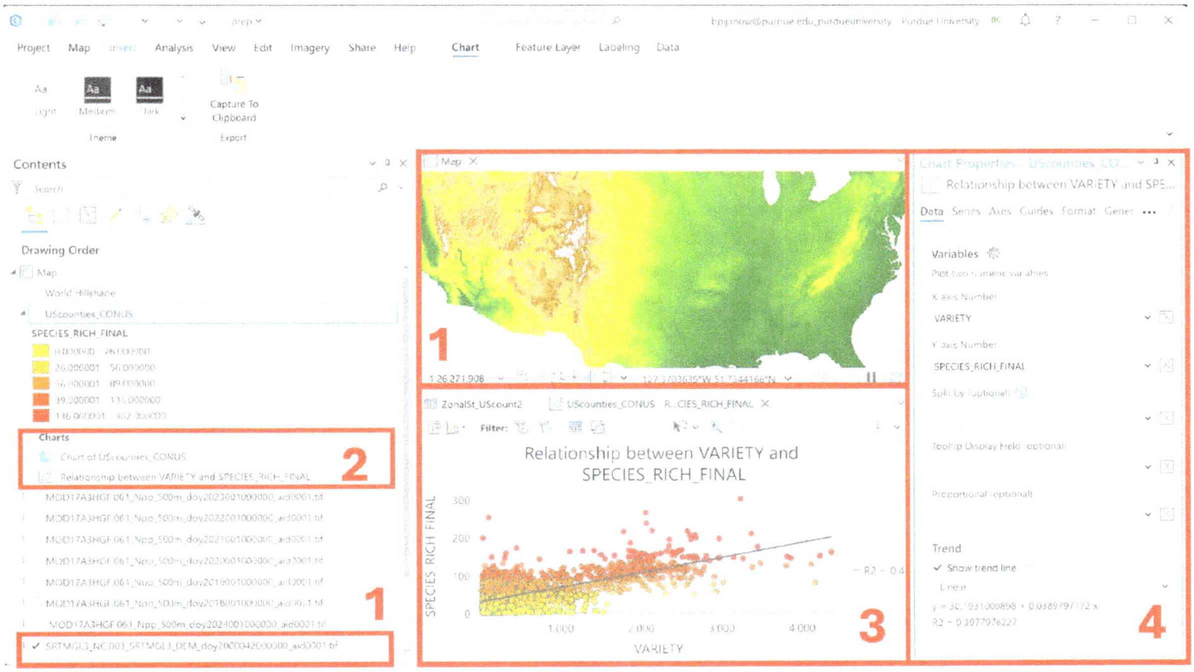

FIGURE EX13.3. Chart properties features that include the ability to create an x-y plot with a linear statistical model fit to the line.

TABLE EX13.1. ***Databases for use in the butterfly mapping and analysis lab***

TABLE	DESCRIPTION	KEY VARIABLES
butterflyspt	Butterfly species table	Species (with genus species name), family, subfamily, reference or voucher photo
Butflyt2020.dbf	County list (with FIPS code) of butterfly species presence and absence	Species lists (by column) with 1 (presence) and 0 (absence)
Butfly_SR_2025	Species richness summary table	Species richness, which is a sum of all 1s in butflyt2020.dfb
UScounties_CONUS	Shape files for the counties in the lower 48 states of the United States circa 2010 (from Esri)	FIPS code for the county, which is in field GEOID10
SRTM DEM for CONUS	A raster data layer that is from the NASA SRTM mission summarized at 93 m x 93 m pixel sizes at 1 m vertical resolutions	Elevation in masl (meters above sea level)

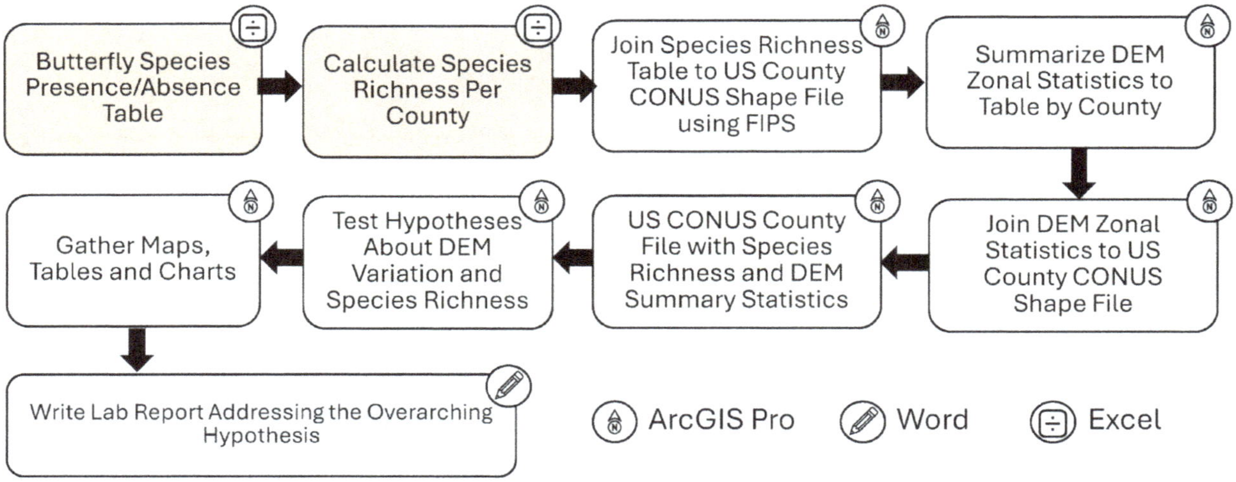

FIGURE EX13.4. Workflow for the analysis of butterfly species richness and variability of regional elevation across the continental United States.

B. ADVANCED STUDENTS

Create the same items that undergraduate students have to produce but write this up using the IMRAD format. Also add to this a map of digital elevation that displays a focal spatial statistics of standard deviation of elevation in a 10 x 10 movable window. Use the map to demonstrate where there are patterns of high variability in elevation in the lower 48 states of the United States. Add at least five citations to your report that support the need to understand declines in butterfly populations and also the factors necessary for linkage of environmental factors to species richness.

EXERCISE 14: QUANTIFYING SPECIES-AREA DISTRIBUTIONS

Highlights

Aims. This exercise focuses on different approaches one can use to calculate species-area relationships used in conservation science.

Learning objectives. After completing this exercise in about a 2-hour time period, students should:

- Know how to calculate species-area curve and estimate its parameters
- Know how to calculate the area of occupancy
- Know how to calculate the extent of occupancy
- Be able to calculate geometry when that is not stored in the attribute table

Project Data Resources. Butterfly and county shape file data from Exercise 13.

14.1 INTRODUCTION TO SPECIES AREA ANALYSES

We will use the databases we developed for the butterfly species richness analysis to explore some further species-area relationships in this lab. These will include using the data to (1) estimate c and z from a species-area curve developed from selection of the county shape files and the species occurrence spreadsheet; and (2) for two species of your choice, calculate species area of occupancy (AOO) and extent of occurrence (EOO) based on county databases of their presence and absence. For how a species-area relationship (SAR) curve is calculated, readers should refer to Chapter 6 for details.

14.2 NAVIGATING ARCGIS PRO FOR CONCEPTUALIZING SPECIES-AREA RELATIONSHIPS

This exercise will involve calculating species richness totals for several county groupings so that species richness totals can be plotted against area. The calculations are a bit nuanced, so let us work through the steps conceptually and also sequentially.

Figure 14.1A (box 1) shows a map of eight counties in north-central Indiana. County FIPS code for Tippecanoe County in Indiana (which is where Purdue University is located) is 18157 (it is coded as state=18, county=157). Seven counties surround Tippecanoe County (Figure 14.1A, box 2). Note that Tippecanoe County (Figure 14.1A, box 3 and label 3) is 1314 km^2. It should be noted that the calculated area of counties is not currently in the shape file. To make this calculation, you need to have the table view of the shape file open, then **Add** Field (Figure 14.1A, box 4), insert the new name, like Area, and then select **Calculate Geometry** (further note: what is shown here in this Table view are the fields, including the added Area, that were made after Calculate Geometry had already been applied). Area is needed for all three of the exercise components here, so this is an important step.

Let us now consider how we need to gather the data necessary to create a **species-area curve** plot. Figure 14.1B shows a hypothetical table for eight possible species of butterflies with their presence and absence scores per county (as county represents a record or row in the table). Note that for Tippecanoe County, there are three butterfly species present of the eight considered (Figure 14.1B, box 1) and that the county area is 1314 km^2. Note that the surrounding seven counties, plus Tippecanoe, have seven butterflies present (Figure 14.1B, box 2) and that these eight counties total 8972 km^2. These two steps illustrate how one needs to proceed with the species-area curve plot. Species richness (S) values for one county and its area (A) is then saved as the first *x,y* pair to into a spreadsheet ($S = 3$, $A = 1314$ km^2). The second *x,y* pair is the number of butterfly species present ($S = 7$) within the eight-county area that they occupy ($A = 8972$ km^2). Note the nuanced way to calculate species richness for the eight counties; each species only needs to be present in any county to be counted since this represents a presence and absence calculation for that entire area of eight counties! You will need to consider several other larger groupings of counties to eventually create the table of species richness and area (aim for 5).

Let us now consider another measure of species area that is commonly used in spatial ecology and conservation science (IUCN, 2019, 2024); that is the **area of occupancy (AOO)** (see Chapter 6). We can use the area for each county and then the presence/absence of each species to calculate this simple statistic; for species 7 in the table in Figure Ex14.2A (box 1), the area that this species occupies is 5673 km^2. This is the largest AOO of the eight species contained in the table. Another common way to express AOO is to calculate the proportion of the counties where the butterfly species is present; for species 7, it would be $5/8 = 0.625$. Finally, yet another way to calculate AOO is to use the measure of the proportion of area that this species occupies as a fraction of the total area, which for species 7 is $5673\ km^2/8972\ km^2 = 0.632$. Note that the last two measures, in this example, are fairly similar. This would not be the case where there are counties to be used in the calculation that vary a lot in their size.

Finaly, let us consider, conceptually, how **extent of occurrence (EOO)** (see Di Marco et al., 2017; Joppa et al., 2016) for any species could be determined using data such as county records for presence and absence of butterflies. If we were to select one species, such as species 7 from our table in Figure Ex14.2A, we would need to determine the EOO boundary using a convex polygon geometry. A convex polygon for counties is typically determined by using county centroids as the boundaries of any convex polygon that is created. Creating centroid points from polygons is accomplished using the **Feature to Point** command (Figure Ex14.2B, box 1). Note the simple set of **Parameters** that are needed here. A **Run** of this tool creates the point feature database shown in the **Contents** panel (Figure Ex14.2B, box 2). These points (i.e., county centroids) are illustrated here in Figure Ex14.2B, box 3).

Although not described here illustratively, creating a convex hull polygon can be accomplished in ArcGIS Pro relatively easily; indeed, it is a specific command. From the point files of those counties where a species of

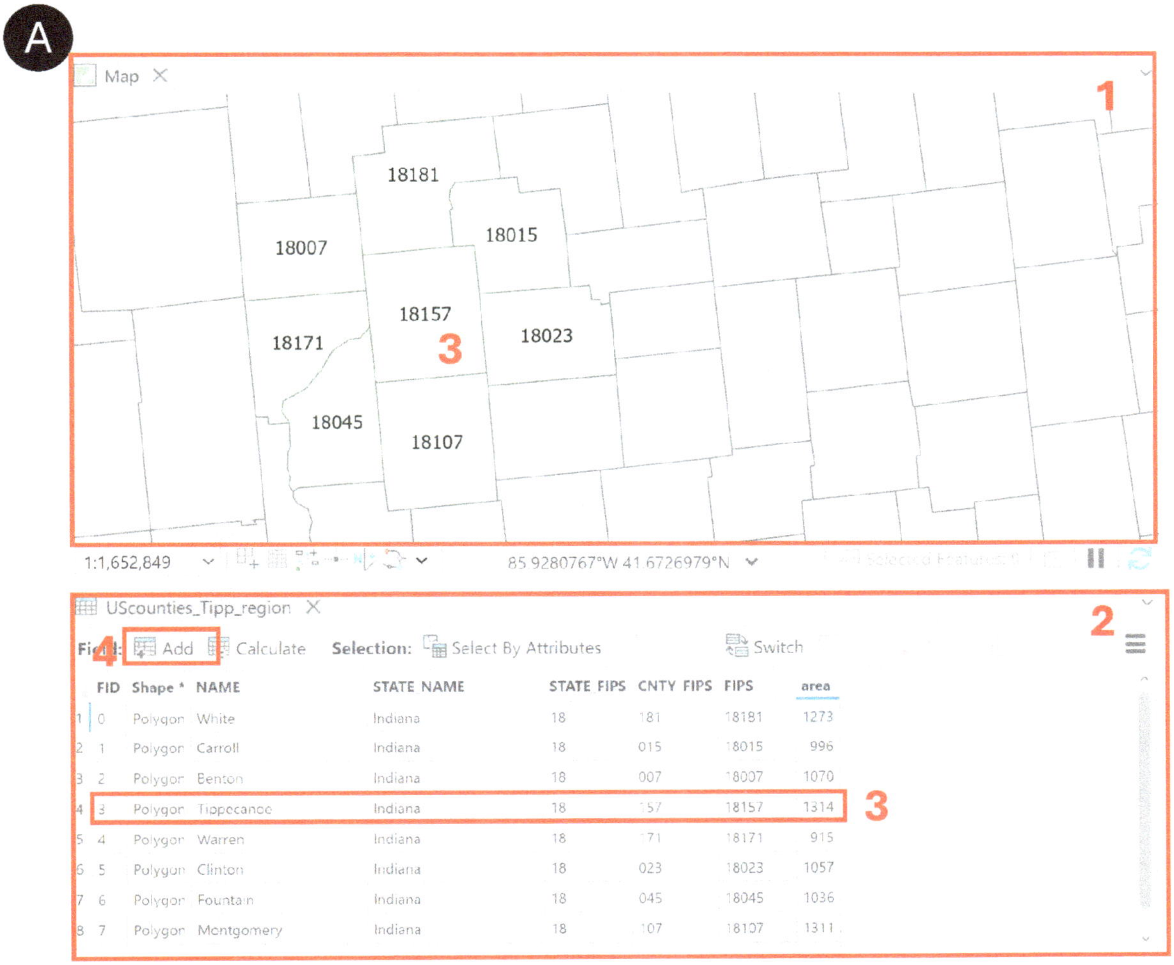

B

Presence/Absence

County FIPS	Area	Species 1	Species 2	Species 3	Species 4	Species 5	Species 6	Species 7	Species 8	Row Totals
18157	1314	1	0	1	0	0	1	0	0	3
18181	1273	0	1	0	1	0	0	1	0	3
18015	996	0	0	0	0	0	0	1	0	1
18007	1070	0	1	0	1	0	0	0	1	3
18171	915	1	0	0	0	0	1	0	1	3
18023	1057	0	0	1	0	0	0	1	0	2
18045	1036	0	0	1	0	0	1	1	0	3
18107	1311	0	0	1	0	0	0	1	0	2
Total	8972	2	2	4	2	0	3	5	2	
8 county presence		1	1	1	1	0	1	1	1	

(Callouts on spreadsheet: 1 — row 18157; 2 — 8 county presence row; 3 — total area 8972.)

FIGURE EX14.1. How one can "hand count" the number of species and its relationship to area. Shown is the polygon and table view (A) and then how one would calculate species number and area using a spreadsheet (B).

A

Area Calculations

	Area	Species 1	Species 2	Species 3	Species 4	Species 5	Species 6	Species 7	Species 8
18157	1314	1314	0	1314	0	0	1314	0	0
18181	1273	0	1273	0	1273	0	0	1273	0
18015	996	0	0	0	0	0	0	996	0
18007	1070	0	1070	0	1070	0	0	0	1070
18171	915	915	0	0	0	0	915	0	915
18023	1057	0	0	1057	0	0	0	1057	0
18045	1036	0	0	1036	0	0	1036	1036	0
18107	1311	0	0	1311	0	0	0	1311	0
total for area		2229	2343	4718	2343	0	3265	5673	1985

1

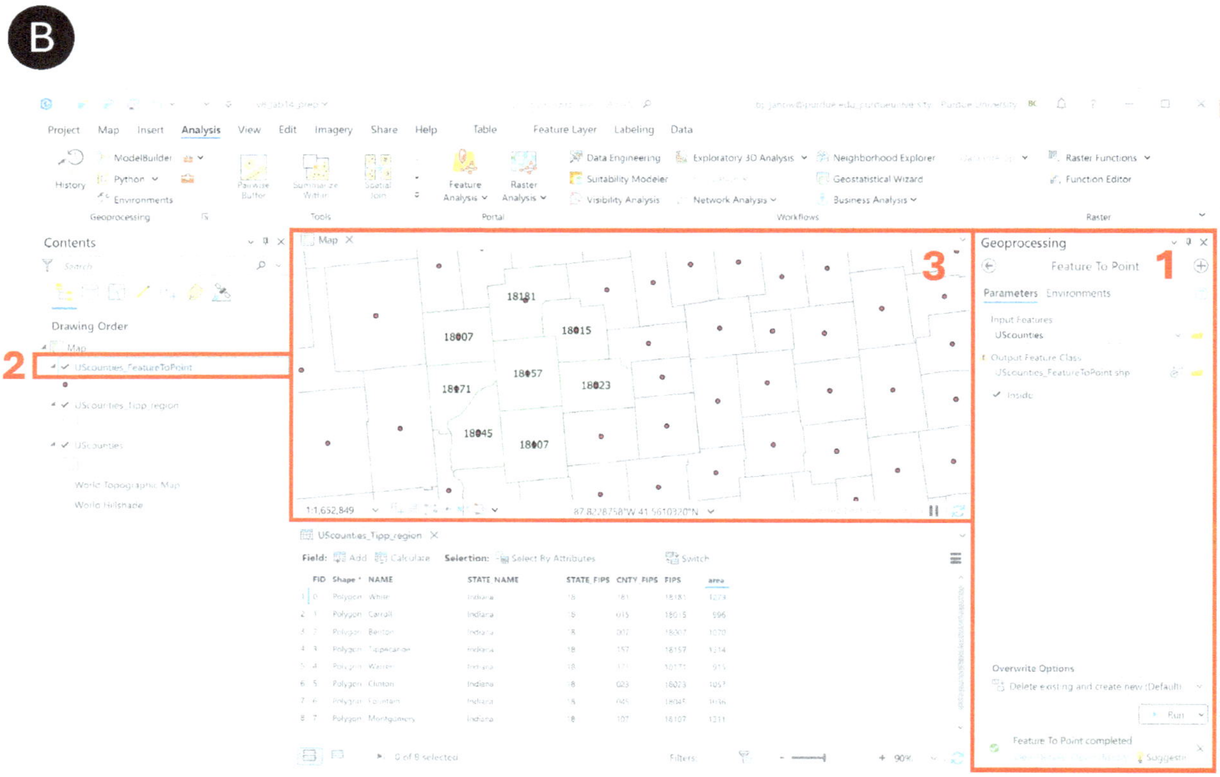

FIGURE EX14.2. How one can "hand count" the area of occupancy (AOO) and extent of occurrence (EOO) using spreadsheet and polygon areas from the geometry calculation if county data does not have that already calculated.

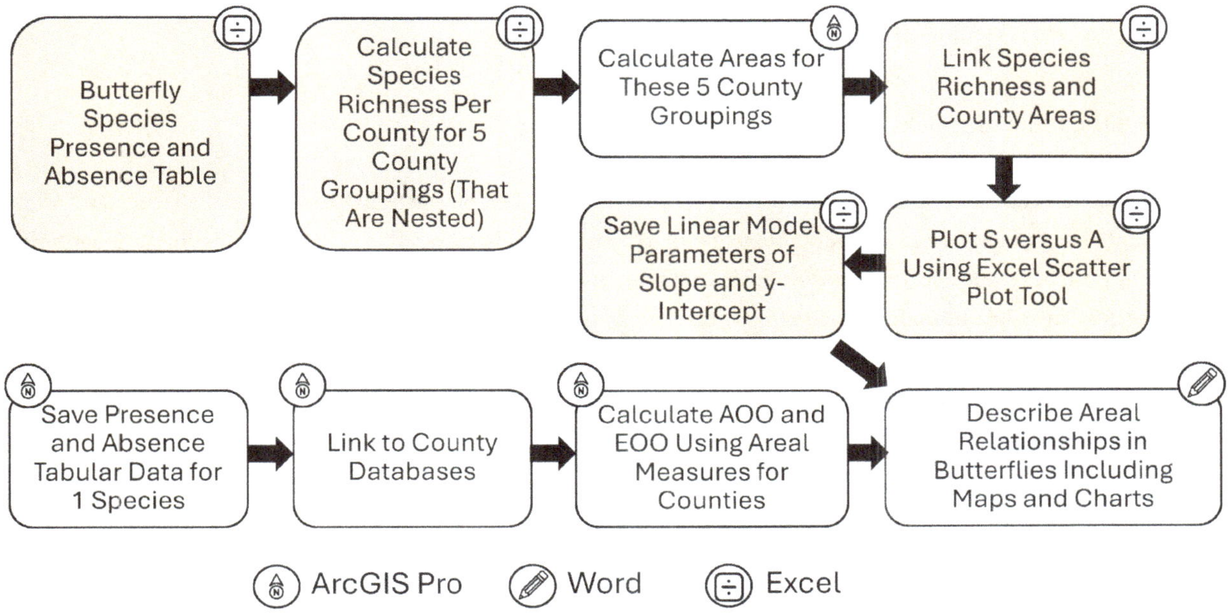

FIGURE EX14.3. Workflow for species area calculations.

interest is being assessed for its spatial distribution, those point centroid features of the counties that it occupies would be used for input to the **Minimum Bounding Geometry** command in the Toolbox; this execution would create the required polygon using convex geometry.

14.3 WORKFLOW FOR THE EXERCISE

Figure 14.3 shows a simplified workflow for these exercises. The objective of this lab is threefold. The first task is to use species presence and absence data for butterflies and areas of counties to create a species-area curve where the parameters of the curve, c and z, are estimated using log transformed data that are then presented to Microsoft Excel for estimating a linear model. It is recommended that you start with the largest set of counties (viz., around 50) and then work "backward" to select fewer groupings of counties. You should display a map of counties and then use any of the **Feature Selection** tools found in the **Map** view of ArcGIS Pro (the **Lasso** feature selection tool works particularly well for this task). For this to work, however, you will need to remove counties as you work backward to eventually have one county. There are numerous ways to approach this task; part of the exercise is for you to explore different ways. A fair warning here is that (1) this always works; and (2) it can be tedious if you are doing it correctly.

The second task is to use the butterfly county presence and absence spreadsheet to select one species that is of interest and to use the presence and absence data and the county shape files with their calculated area to produce estimates of AOO and EOO. Use the conceptual presentation outlined above to work through the AOO and EOO steps.

The third task is to synthesize these assessments into a report that describes how species richness varies by scale and also how one can use a variety of species areal occupancy measures to describe their distributions.

14.4 ASSIGNMENTS

A. INTERMEDIATE STUDENTS

Complete the following tasks producing short descriptions for the work:

- Estimate *c* and *z* for a species-area curve from the provided data on butterfly presence and absence and U.S. County shape files. Use at least 5 points to estimate *c* and *z* in Microsoft Excel. Read the section in Chapter 6 that summarizes the species-area curve. Make sure that you use the correct log transformation of the data prior to calculating a linear model to estimate *c* and *z*. Briefly describe, in no less than 50 words, how you prepared the data for this model.
- Following the conceptual description for calculating EOO and AOO, summarize your findings for the distribution of species that you selected.

B. ADVANCED STUDENTS

Complete Tasks #1 and #2 above but write this up as a technical report that summarizes the ways that spatial ecologists estimate species-area relationships for communities (i.e., as species richness) and for individual species. Include in your report at least five references total selected from either an NGO (e.g., IUCN), government organization (e.g., USGS), or from the primary literature.

EXERCISE 15: IDENTIFYING SPECIES CLUSTERS USING HOTSPOT ANALYSIS

Highlights

Aims. This exercise is designed to introduce you to the analysis of multiple species boundaries and to hotspot mapping.

Learning objectives. After completing this exercise in about a 1.5-hour time period, students should be able to:

- Use and interpret Spatial Analyst's Moran's *I* and Getis-Ord Gi* statistical modeling tools in ArcGIS Pro
- Interpret Moran's *I* and Gi* statistic for hotspot assessments

Project Data Resources. Shape files of bat distributions in the U.S. CONUS region.

15.1 INTRODUCTION TO TRENDS IN BAT ABUNDANCES

Bats are the second most diverse group of mammals—behind rodents—with over 1,400 species recognized globally (IUCN, 2025; Jebb et al., 2020). There are about 49 species in the United States and Canada and, when Mexico is added, that number reaches 150 (Adams, 2024). Most bat species occur in the tropical regions; many islands also support considerable numbers of endemic species.

A reduction in species numbers and abundances have been documented (Frick et al., 2020), especially among island species (Kingston et al., 2023), but one of the main challenges of understanding trends has been the lack of adequate data on species distributions and abundances. Threats to bats include disease (white nose syndrome or WSN), wind turbines, habitat change, and reductions of their food supplies (e.g., insects, nectar).

Bats support a variety of ecosystem services. Bats suppress insect populations, especially those that affect human health and insects that damage crops. Bats are also pollinators of many plants and crops (e.g., bats pollinate wild bananas, not the ones you purchase in a store) and in high population numbers contribute to nutrient loads in high numbers through their guano. Long-distance migration is rare in this group of animals (Fleming & Eby, 2003), with only about 3% of all species having been found to migrate long distances. Major organizations that are involved in bat preservation efforts include Bat Conservation International and the North American Bat Conservation Alliance.

15.2 HOTSPOT MAPPING

We will use the databases developed for the butterfly species richness analysis of Exercise 14 to explore spatial autocorrelation using Moran's *I*, and then, we will apply the Getis-Ord Gi* statistical model to create hotspot maps. As part of this exercise, we will also determine if there is a potential bias in the data that could result in the use of county data to determine species richness. The foundational information for this work is found in Chapter 6 ("Landscapes and Species Distributions") and Chapter 13 ("GIScience and Spatial Statistics").

The bat distribution data that we will be working with, in addition to revisiting the butterfly species database, is from the National Atlas of the United States. Data to generate these species distribution features were derived from data provided by Bat Conservation International and the IUCN Red List of Threatened Species. Records of species locations date back about 150 years.

15.3 CHARACTERIZING SPECIES RICHNESS CONCENTRATIONS

As part of any statistical analysis of spatial clustering, it is customary (following Anselin, 1995) to conduct a global analysis of spatial autocorrelation, typically using Moran's *I*, before one assesses local clustering of data (i.e., **hotspot** and **coldspot analyses**). As you might recall, Moran's *I* tests for the occurrence of autocorrelation to a random spatial distribution of values with positive values associated with values near one another being similar (i.e., clustering) and negative values with neighbors being different (called dispersion). We will do both analyses—Moran's *I* and Getis-Ord Gi*—using the U.S. County CONUS polygon database with species richness field that was joined from the Microsoft Excel spreadsheet in the previous lab. In ArcGIS Pro, there are several algorithms that define spatial neighborhoods or spatial relationships. We will use the "queen" polygon arrangement, which is also called the **Contiguity Edges Corners** option for the Conceptualization of Spatial Relationships.

We will also explore how we can use multiple species distribution polygons to create an overlay of bat species ranges in North America. These **species richness zones**, not to be considered technically as hotspots in a spatial statistical sense, give natural resource managers the ability to know what areas could contain the greatest number of species as one considers the geographic overlap of species occurrences. This analysis uses the **Count Overlapping Features** tool to count the overlapping polygons, creating a new set of polygons with the attribute of count (more specifically as COUNT_).

15.4 NAVIGATING ARCGIS PRO FOR HOTSPOT AND OVERLAP ANALYSIS

Figure 15.1A shows a screen grab of the Moran's *I* **Parameters** panel (Figure 15.1A, box 1). Note that you should select **Generate Report** (Figure 15.1A, label 2), which will produce output to a folder that has the summary statistics and some associated graphics in an html (i.e., web page) format. Select **Contiguity Edges Corners** (Figure 15.1A, box 3) for the neighborhood algorithm. The resultant report will provide easy to understand graphics for how the output should be interpreted.

Figure 15.1B is a screen grab of the Getis-Ord Gi* statistic to the **Parameters** panel (Figure 15.1B, box 1). Use the same **Conceptualizing Spatial Relationship** settings. After selecting **Run**, this statistic outputs a map with hotspots (as red counties), coldspots (blue counties), and areas that are neither (not color coded).

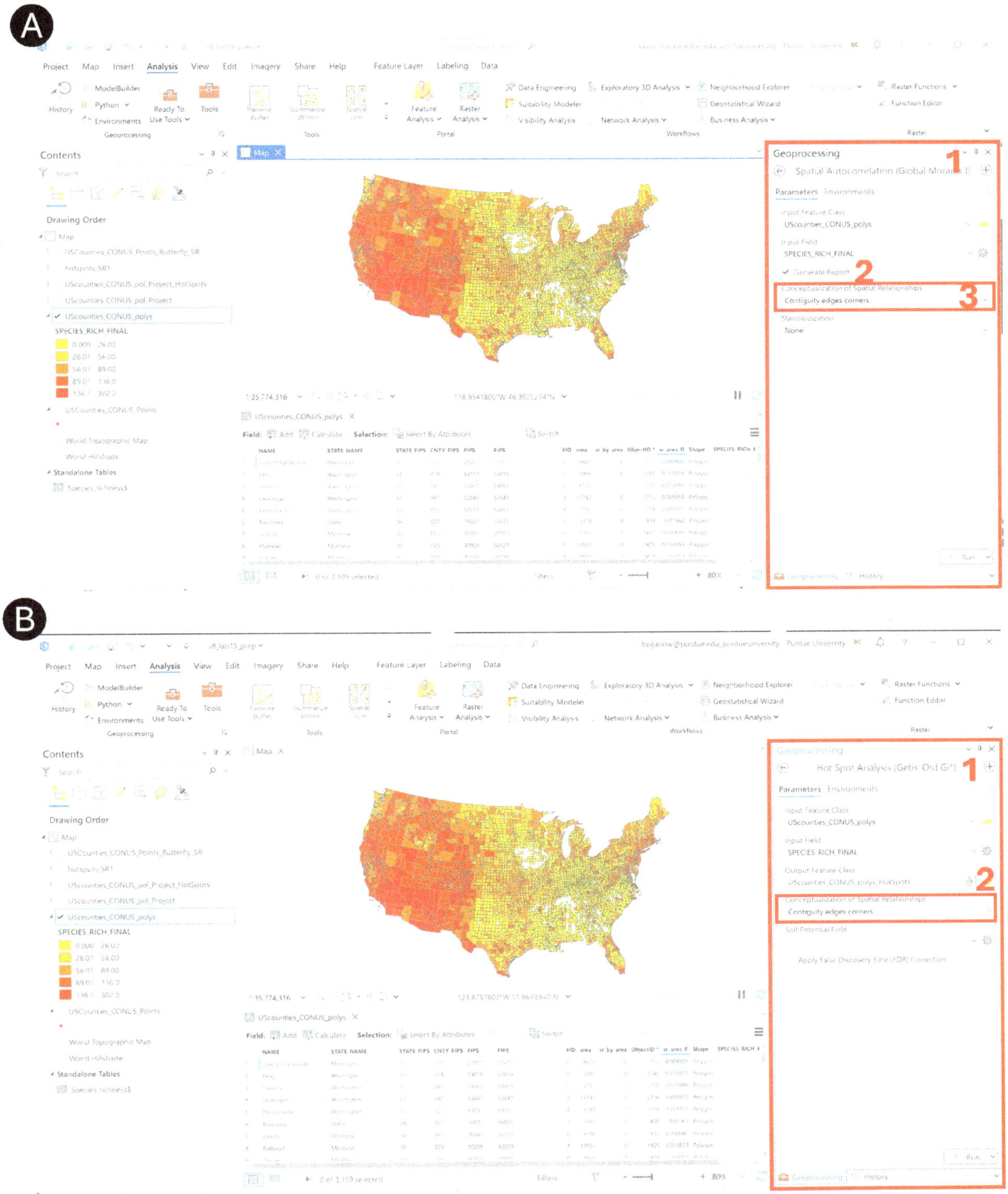

FIGURE EX15.1. Interface to (A) Moran's *I* Geoprocessing pane and (B) Getis-Ord Gi* statistics Geoprocessing pane.

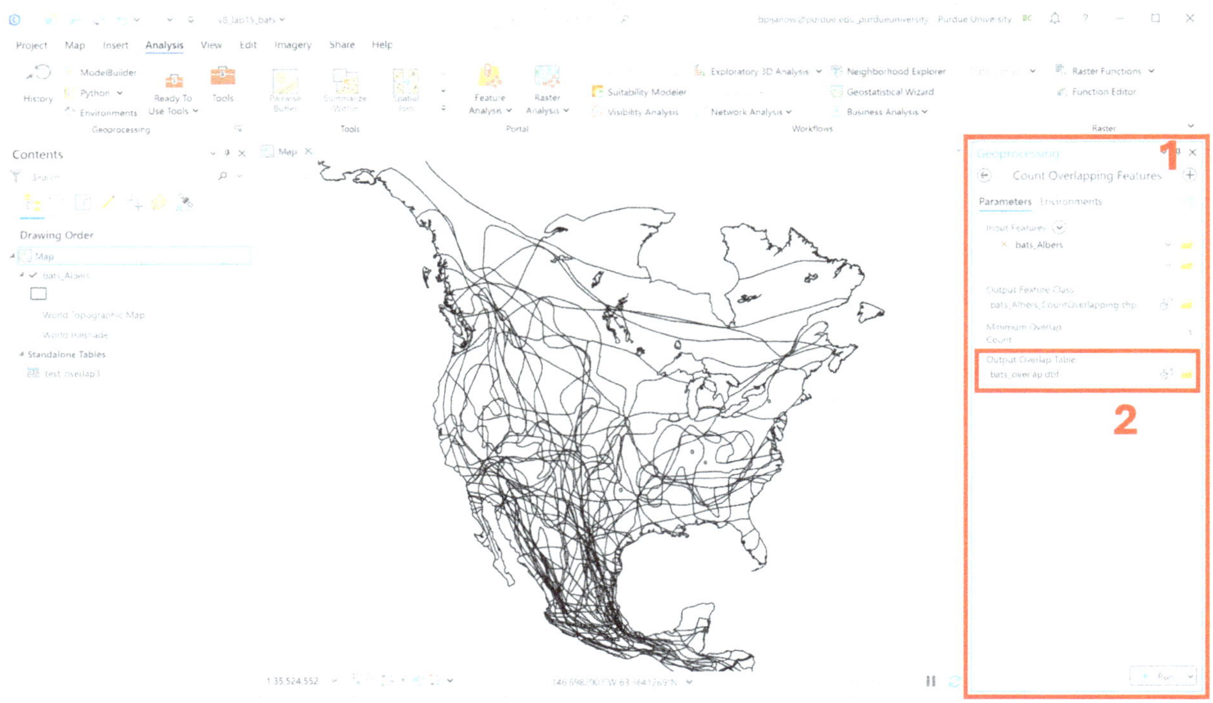

FIGURE EX15.2. Interface to the Count Overlapping Features tool that can be used on multiple species boundary files.

If you made a map of the distribution of species richness across the lower 48 states, you probably noticed a few trends: (1) Species richness for butterflies is greatest in the western states; and (2) there are some areas of high species richness of butterflies in the Smoky Mountains. What is also characteristic of western states is that counties are very large. So, one might be suspicious that the high species richness could be due to the larger area of these "reporting units." Let's test that assumption here by scaling the species richness data for counties by area. Doing so creates another variable, species density, which is calculated as

$$SD = S/area$$

where SD is the number of species per unit area (e.g., km^2) and S is the number of species in the county. To accomplish this in ArcGIS Pro, you will have to open the table containing species richness and county area. Then **Add** a new field (e.g., SR_area), making sure that the field type is Float. Then, using the **Calculate** feature in Table view, calculate this new value in that field.

Figure Ex15.2 shows the interface for the **Geoprocessing** tool, **Count Overlapping Features**. Note that this can be accomplished using overlapping polygons in one or more files, making this an ideal tool when a spatial ecologist or GISci analyst has species distribution files that are separate. The output are polygons that contain the field COUNT_ , which is a count of the number of features that overlap in space. As there are 49 polygon features (i.e., there are some records where there are more than one polygon for a species, meaning that it has a disjunct population distribution), the resultant operation will create hundreds of polygons.

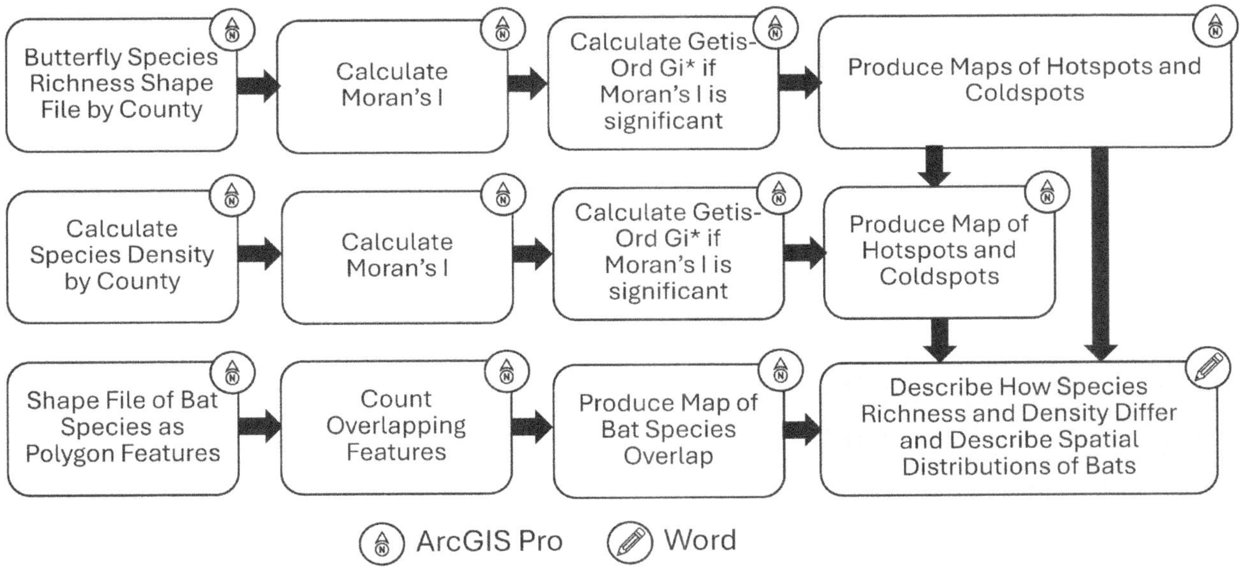

FIGURE EX15.3. Workflow for the multispecies hotspot analyses.

15.5 WORKFLOW FOR THE EXERCISE

A workflow for these exercises is shown in Figure Ex15.3. There are three major tasks, two of which involve the use of Getis-Ord Gi* to calculate hotspots on raw species richness tallies and then areal adjust values (e.g., species densities). A separate task is focused on creating a map of overlapping polygons that represent species distributions.

15.6 ASSIGNMENT

A. INTERMEDIATE STUDENTS

Prepare three maps, two of which show either hotspots or coldspots, using different representations of species richness, and a map of bat species distribution overlaps. In no less than 100 words, describe these species richness distribution patterns.

B. ADVANCED STUDENTS

Complete the assignment above but write this up as a technical report that discusses challenges and biases in each of these databases and approaches to mapping species richness. For the butterfly species richness data, you might consider how some states have either lower or higher numbers compared to their neighbors.

REFERENCES CITED

Adams, A. M., Trujillo, L. A., Campbell, C. J., Akre, K .L., Arroyo–Cabrales, J., Burns, L., Coleman, J. T., Dixon, R. D., Francis, C. M., Gamba–Rios, M. & Kuczynska, V., (2024). The state of the bats in North America. *Annals of the New York Academy of Sciences*, 1541(1), 115–128.

Albers, J. (1963). *Interaction of color*. Yale University Press.

Anderson, J. R., Hardy, E. E., Roach, J. T., & Witmer, R. E. (1976). *A land use and land cover classification system for use with remote sensor data* (U.S. Geological Survey Professional Paper 964). U.S. Geological Survey. https://doi.org/10.3133/pp964

Annis Water Resources Institute. (2002). *Muskegon River Watershed 319 Management Plan*. 449 pp.

Annis Water Resources Institute. (2014). *Upper Muskegon River Watershed 319 Management Plan*. 190 pp.

Anselin, L. (1995). Local indicators of spatial association—LISA. *Geographical Analysis, 27*(2), 93–115. https://doi.org/10.1111/j.1538-4632.1995.tb00338.x

Anthony, M. A., Bender, S. F., & Van Der Heijden, M. G. (2023). Enumerating soil biodiversity. *Proceedings of the National Academy of Sciences, 120*(33), e2304663120

Araújo, M. B., & Peterson, A. T. (2011). Uses and misuses of bioclimatic envelope modeling. *Ecology, 93*(7), 1527–1539.

Arnheim, R. (1954). *Art and visual perception: A psychology of the creative eye.* University of California Press.

ATTRA (2025). *Intercropping Principles and Production Practices*. https://www.iatp.org/sites/default/files/Intercropping_Principles_and_Production_Practi.htm (accessed Oct. 4, 2025).

Axelsson, Ö., Nilsson, M. E., & Berglund, B. (2010). A principal components model of soundscape perception. *The Journal of the Acoustical Society of America, 128*(5), 2836–2846. https://doi.org/10.1121/1.3493436

Baeckens, S., Herrel, A., Broeckhoven, C., Vasilopoulou-Kampitsi, M., Huyghe, K., Goyens, J., Vanhooydonck, B., Damme, R. V., & Irschick, D. J. (2017). Evolutionary morphology of the lizard chemosensory system. *Scientific Reports, 7*, Article 10141.

Bailey, R. G. (1983). Delineation of ecosystem regions. *Environmental Management, 7*(4), 365–373. https://doi.org/10.1007/BF01866919

Bailey, R. G. (1995). *Description of the ecoregions of the United States* (2nd ed., rev. and expanded). U.S. Department of Agriculture, Forest Service. https://doi.org/10.2737/WO-GTR-139

Bailey, R. G. (1995, rev. 2002). *Description of the ecoregions of the United States* (Misc. Pub. 1391). USDA Forest Service.

Bailey, R. G. (2009). *Ecosystem geography: From ecoregions to sites* (2nd ed.). Springer. https://doi.org/10.1007/978-0-387-89516-1

Barbosa Júnior, M. R., Moreira, B. R. A., Carreira, V. S., Brito Filho, A. L., Trentin, C., Souza, F. L. P., Tedesco, D., Setiyono, T., Flores, J. P., Ampatzidis, Y., Silva, R. P., & Shiratsuchi, L. S. (2024). Precision agriculture in the United States: A comprehensive meta-review inspiring further research, innovation, and adoption. *Computers and Electronics in Agriculture, 221*, 108993. https://doi.org/10.1016/j.compag.2024.108993

Beck, H. E., McVicar, T. R., Vergopolan, N., Berg, A., Lutsko, N. J., Dufour, A., Zeng, Z., Jiang, X., van Dijk, A. I. J. M., & Miralles, D. G. (2023). High-resolution (1 km) Köppen-Geiger maps for 1901–2099 based on constrained CMIP6 projections. *Scientific Data*, 10, Article 724. https://doi.org/10.1038/s41597-023-02549-6 nature.com

Beck, H. E., Zimmerman, N., McVicar, T., Vergopolan, N., Berg, A., & Wood, E. (2018). Data descriptor: Present and future Köppen-Geiger climate classification maps at 1 km resolution. *Scientific Data*, 5, 180214.

Bell, S. (2001). Landscape pattern, perception and visualisation in the visual management of forests. *Landscape and Urban Planning, 54*(1–4), 201–211.

Bell, S. (2012). *Landscape: Pattern, perception and process*. Routledge.

Bett, N. N., & Hinch, S. G. (2016). Olfactory navigation during spawning migrations: A review and introduction of the Hierarchical Navigation Hypothesis. *Biological Reviews, 91*(3), 728–759.

Biehl, L., & Landgrebe, D. (2002). MultiSpec—A tool for multispectral–hyperspectral image data analysis. *Computers & Geosciences, 28*(10), 1153–1159. doi:10.1016/S0098-3004(02)00033-X

Bird, L., Light, A., & Goldsmith, I. (2025). *US clean power development sees record progress, as well as stronger headwinds.* World Resources Institute. https://www.wri.org/insights/clean-energy-progress-united-states (accessed Oct. 4, 2024).

Blackburn, T. M., Pyšek, P., Bacher, S., Carlton, J. T., Duncan, R. P., Jarošík, V., Wilson, J. R. U., & Richardson, D. M. (2011). A proposed unified framework for biological invasions. *Trends in Ecology & Evolution, 26*(7), 333–339. https://doi.org/10.1016/j.tree.2011.03.023

Blais, J. A. R. (1995). Geomatics: A new name for surveying? *Canadian Surveyor, 49*(1), 45–49.

Blann, K. L., Anderson, J. L., Sands, G. R., & Vondracek, B. (2009). Effects of agricultural drainage on aquatic ecosystems: A review. *Critical Reviews in Environmental Science and Technology, 39*(11), 909–1001. https://doi.org/10.1080/10643380801977966

Board, C. (1967). *Maps as models.* In R. J. Chorley & P. Haggett (Eds.), *Models in geography* (pp. 671–725). Methuen.

Boetzl, F. A., Ries, E., Schneider, G., & Krauss, J. (2018). It's a matter of design—how pitfall trap design affects trap samples and possible predictions. *PeerJ, 6*, e5078.

Box, G. E. P. (1976). Science and statistics. *Journal of the American Statistical Association, 71*(356), 791–799.

Bradbury, J. W., & Vehrencamp, S. L. (2011). *Principles of animal communication* (2nd ed.). Sinauer Associates.

Brewer, C. A. (1994). Color use guidelines for mapping and visualization. In A. M. MacEachren & D. R. F. Taylor (Eds.), *Visualization in modern cartography* (pp. 123–147). Pergamon.

Brewer, C. A. (2005). *Designing better maps: A guide for GIS users.* Esri Press.

Brewer, C. A., Hatchard, G. W., & Harrower, M. (2003). ColorBrewer in print: A catalog of color schemes for maps. *Cartography and Geographic Information Science, 30*(1), 5–32.

Brown, J. F., Tollerud, H. J., Barber, C. P., Zhou, Q., Dwyer, J. L., Vogelmann, J. E., Loveland, T. R., Woodcock, C. E., Stehman, S. V., Zhu, Z. & Pengra, B.W., (2020). Lessons learned implementing an operational continuous United States national land change monitoring capability: The Land Change Monitoring, Assessment, and Projection (LCMAP) approach. *Remote sensing of environment*, 238, 111356.

Brown, J. H. (1995). *Macroecology.* University of Chicago Press.

Brown, J. H. (2014). Why are there so many species in the tropics?. *Journal of Biogeography, 41*(1), 8–22.

Brown, J. H., & Lomolino, M. V. (2000). Concluding remarks: Historical perspective and the future of island biogeography theory. *Global Ecology and Biogeography, 9*(1), 87–92.

Bünemann, E. K., Bongiorno, G., Bai, Z., Creamer, R. E., De Deyn, G., de Goede, R., Fleskens, L., Geissen, V., Kuyper, T. W., Mäder, P., Pulleman, M., Sukkel, W., van Groenigen, J. W., & Brussaard, L. (2018). *Soil quality: A critical review. Soil Biology and Biochemistry, 120*, 105–125.

Bünemann, E. K., Schwenke, G. D., & Van Zwieten, L. (2006). Impact of agricultural inputs on soil organisms—a review. *Soil Research, 44*(4), 379–406.

Burrough, P. A. (1986). *Principles of geographical information systems for land resources assessment.* Oxford University Press.

Campbell, J. B., & Wynne, R. H. (2011). *Introduction to remote sensing* (5th ed.). Guilford.

Cardé, R. T. (2021). Navigation along windborne plumes of pheromone and resource-linked odors. *Annual Review of Entomology, 66*, 317–336.

Cardinale, B. J., Duffy, J. E., Gonzalez, A., Hooper, D. U., Perrings, C., Venail, P., . . . Naeem, S. (2012). Biodiversity loss and its impact on humanity. *Nature, 486*(7401), 59–67.

CBE Style Manual Committee. (1994). *Scientific style and format: The CBE manual for authors, editors, and publishers*, ed. E. J. Huth et al. Cambridge University Press.

Ceballos, G., García, A., & Ehrlich, P. R. (2010). The sixth extinction crisis: Loss of animal populations and species. *Journal of Cosmology, 8*(1821), 31.

Chapin, F. S., III, Matson, P. A., & Vitousek, P. M. (2011). *Principles of terrestrial ecosystem ecology* (2nd ed.). Springer. https://doi.org/10.1007/978-1-4419-9504-9

Chave, J., Réjou–Méchain, M., Búrquez, A., Chidumayo, E., Colgan, M. S., Delitti, W. B., Duque, A., Eid, T., Fearnside, P. M., Goodman, R.C. & Henry, M., (2014). Improved allometric models to estimate the aboveground biomass of tropical trees. *Global change biology*, 20(10): 3177–3190.

Chrisman, N., (2006). *Charting the unknown: How computer mapping at Harvard became GIS*. ESRI Press.

Clements, F. E., & Shelford, V. E. (1939). *Bio-ecology*. John Wiley & Sons.

Cliff, A. D., & Ord, J. K. (1973). *Spatial autocorrelation*. Pion.

Cliff, A. D., & Ord, J. K. (1981). *Spatial processes: Models and applications*. Pion.

Comer, P. J., Albert, D. A., Wells, H. L., Hart, B. L., Raab, J. B., Price, D. L., Kashian, D. M., Corner, R. A., & Schuen, D. W. (1993). *Michigan's pre-settlement vegetation, as interpreted from the General Land Office Surveys 1816–1856*. Michigan Natural Features Inventory.

Conchou, L., Lucas, P., Meslin, C., Proffit, M., Staudt, M., & Renou, M. (2019). Insect odorscapes: From plant volatiles to natural olfactory scenes. *Frontiers in Physiology, 10*, 972. https://doi.org/10.3389/fphys.2019.00972

Coppock, J. T., & Rhind, D. W. (1991). The history of GIS. In D. J. Maguire, M. F. Goodchild, & D. W. Rhind (Eds.), *Geographical information systems: Principles and applications* (Vol. 1, pp. 21–43). Longman Scientific & Technical.

Corfield, J. R., Price, K., Iwaniuk, A. N., Gutierrez-Ibanez, C., Birkhead, T., & Wylie, D. R. (2015). Diversity in olfactory bulb size in birds reflects allometry, ecology, and phylogeny. *Frontiers in Neuroanatomy, 9*, Article 102. https://doi.org/10.3389/fnana.2015.00102

Cowardin, L. M., Carter, V., Golet, F. C., & LaRoe, E. T. (1979). *Classification of wetlands and deepwater habitats of the United States* (FWS/OBS-79/31). U.S. Fish and Wildlife Service.

Cressie, N. A. C. (1993). *Statistics for spatial data* (rev. ed.). Wiley.

Croxton, F. E., & Cowden, D. J. (1955). *Applied general statistics* (2nd ed.). Prentice-Hall.

Darwin, C. (1859). *On the Origin of Species*.

Daubenmire, R. (1959). A canopy-coverage method of vegetational analysis. *Northwest Science*, 33, 43–64.

Davis, G. B. (1974). *Management information systems: Conceptual foundations, structure, and development*. McGraw-Hill.

DeMers, M. N. (2009). *Fundamentals of geographic information systems* (4th ed.). Wiley.

Di Gregorio, A., & Jansen, L. J. (1998). A new concept for a land cover classification system. *The Land, 2*(1), 55–65.

Di Marco, M., Watson, J. E. M., Possingham, H. P., & Venter, O. (2017). Limitations and trade-offs in the use of species distribution maps for protected area planning. *Journal of Applied Ecology, 54*(2), 402–411. https://doi.org/10.1111/1365-2664.12771

Edwards, C. B., Zipkin, E. F., Henry, E. H., Haddad, N. M., Forister, M. L., Burls, K. J., Campbell, S. P., Crone, E. E., Diffendorfer, J., Douglas, M. R., Drum, R. G., Fallon, C. E., Glassberg, J., Grames, E. M., Hatfield, R., Hershcovich, S., Hoffman Black, S., Larsen, E. A., Leuenberger, W., . . . Schultz, C. B. (2025). Rapid butterfly declines across the United States during the 21st century. *Science, 387*(6738), 1090–1094. https://doi.org/10.1126/science.adp4671

EEA (2000). *CORINE land cover technical guide—Addendum 2000*. European Environment Agency, Technical Report No. 40.

Elliot, A. J., & Maier, M. A. (2014). Color psychology: Effects of perceiving color on psychological functioning in humans. *Psychological Bulletin, 140*(1), 66–75. https://doi.org/10.1037/a0034857

Ellis, E. C., Klein Goldewijk, K., Siebert, S., Lightman, D., & Ramankutty, N. (2010). Anthropogenic transformation of the biomes, 1700 to 2000. *Global Ecology and Biogeography, 19*(5), 589–606.

Elton, C. S. (1958). *The ecology of invasions by animals and plants*. Methuen.

Emonet, T., & Vergassola, M. (2024). Olfactory cues and memories in animal navigation. *Nature Reviews Physics, 6*(4), 215–216.

Enquist, C. A. F. (2002). Predicted regional impacts of climate change on the geographical distribution and diversity of tropical forests in Costa Rica. *Journal of Biogeography, 29*(4), 519–534.

Erisman, J. W., Sutton, M. A., Galloway, J., Klimont, Z., & Winiwarter, W. (2008). How a century of ammonia synthesis changed the world. *Nature Geoscience, 1*(10), 636–639.

ESA (Ecological Society of America). (2026). Web site describing the R. H. MacArthur award. https://esa.org/about/awards/robert-h-macarthur-award/

Feeney, E., O'Brien, S., Scannell, A., Markey, A., & Gibney, E. R. (2011). Genetic variation in taste perception: Does it have a role in healthy eating? *Proceedings of the Nutrition Society, 70*(1), 135–143. https://doi.org/10.1017/S0029665110003976

Fehrman, K. R., & Fehrman, C. (2004). *Color: The secret influence.* Pearson Prentice Hall.

Fellers, G. M. (1995). *A standardized protocol for surveying aquatic amphibians* (No. 58). National Biological Service, Cooperative Park Studies Unit, University of California, Division of Environmental Studies.

Ferreira, P. M. P., Arcanjo, D. D. R., & Peron, A. P. (2023). Drug development, Brazilian biodiversity and political choices: Where are we heading?. *Journal of Toxicology and Environmental Health, Part B, 26*(5), 257–274.

Field et al. (1998). Primary production of the biosphere: Integrating terrestrial and oceanic components. *Science, 281*(5374), 237–240.

Finnerty, P. B., McArthur, C., Banks, P., Price, C., & Shrader, A. M. (2022). The olfactory landscape concept: A key source of past, present, and future information driving animal movement and decision-making. *BioScience, 72*(8), 745–752. https://doi.org/10.1093/biosci/biac039

Fisher, P. F. (1994). Probable and fuzzy models of the viewshed operation. In M. F. Worboys (Ed.), *Innovations in GIS 1* (pp. 161–175). Taylor & Francis.

Fisher, R. A. (1925). *Statistical methods for research workers.* Oliver & Boyd.

Fleming, T. H., & Eby, P. (2003). *Ecology of bat migration.* In T. H. Kunz & M. B. Fenton (Eds.), *Bat ecology* (pp. 156–208). University of Chicago Press.

Food and Agriculture Organization of the United Nations (FAO). (2002a). *Africover—Multipurpose land cover database for Africa: Concepts and methodology.* FAO, Environment and Natural Resources Service, Working Paper Series No. 8.

Food and Agriculture Organization of the United Nations (FAO). (2002b). *Land tenure studies 1: Land tenure and rural development.*

Food and Agriculture Organization of the United Nations. (2023). *Land use statistics and indicators 2000–2021: Global, regional and country trends* (FAOSTAT Analytical Brief No. 71). Food and Agriculture Organization of the United Nations.

Forister, M. L., Pelton, E. M., & Black, S. H. (2019). Declines in insect abundance and diversity: We know enough to act now. *Conservation Science and Practice, 1*(8), e80.

Francomano, D., Rodríguez González, M. I., Valenzuela, A. E., Ma, Z., Raya Rey, A. N., Anderson, C. B., & Pijanowski, B. C. (2022). Human–nature connection and soundscape perception: Insights from Tierra del Fuego, Argentina. *Journal for Nature Conservation, 65*, 126110.

Frick, W. F., Kingston, T., & Flanders, J. (2020). A review of the major threats and challenges to global bat conservation. *Annals of the New York Academy of Sciences.*

Frick, W. F., Puechmaille, S. J., Hoyt, J. R., et al. (2015). Disease alters macroecological patterns of North American bats. *Global Ecology and Biogeography, 24*(7), 741–749. https://doi.org/10.1111/geb.12290

Friedl, M. A., Sulla-Menashe, D., Tan, B., Schneider, A., Ramankutty, N., Sibley, A., & Huang, X. (2010). MODIS Collection 5 global land cover: Algorithm refinements and characterization of new datasets. *Remote Sensing of Environment, 114*(1), 168–182. https://doi.org/10.1016/j.rse.2009.08.016

Gardner, R. H., Milne, B. T., Turner, M. G., & O'Neill, R. V. (1987). Neutral models for the analysis of broad-scale landscape pattern. *Landscape Ecology, 1*(1), 19–28.

Gardner, R. H., & Urban, D. L. (2007). Neutral models for testing landscape hypotheses. *Landscape Ecology, 22*(1), 15–29.

Gascoyne, K., Marchand, F., Desharnais, J., & Americi, H. (2011). *Tea: History, terroirs, varieties.* Firefly Books.

Gatti, R. C., Reich, P. B., Gamarra, J. G., Crowther, T., Hui, C., Morera, A., . . . Park, M. (2022). The number of tree species on Earth. *Proceedings of the National Academy of Sciences of the United States of America, 119*(6), e2115329119.

Getis, A., & Ord, J. K. (1992). The analysis of spatial association by use of distance statistics. *Geographical Analysis, 24*(3), 189–206. https://doi.org/10.1111/j.1538-4632.1992.tb00261.x

Goodchild, M. F. (1992). Geographical information science. *International Journal of Geographical Information Systems, 6*(1), 31–45.

Goodchild, M. F. (2007, April). Geography prospers from GIS. *ArcWatch* (Esri online magazine).

Goodchild, M. F. (2018). Reimagining the history of GIS. *Annals of GIS, 24*(1), 1–8.

Granö, J. G. (1929). *Reine Geographie: Eine methodologische Studie, beleuchtet mit Beispielen aus Finnland und Estland.* Tilgmann.

Granö, J. G. (1930). *Puhdas maantiede: Tutkimusesimerkeillä Suomesta ja Virosta valaistu metodologinen selvitys*. WSOY.
Granö, J. G. (1997). *Pure geography* (O. Granö & A. Paasi, Eds.; M. Hicks, Trans.). Johns Hopkins University Press.
Groom, M. J., Meffe, G. K., & Carroll, C. R. (2005). *Principles of conservation biology* (3rd ed.). Sinauer Associates.
Hagen-Zanker, A., & Lajoie, G. (2008). Neutral models of landscape change as benchmarks in the assessment of model performance. *Landscape and Urban Planning, 86*(3–4), 284–296.
Hallmann, C. A., Sorg, M., Jongejans, E., Siepel, H., Hofland, N., Schwan, H., . . . de Kroon, H. (2017). More than 75 percent decline over 27 years in total flying insect biomass in protected areas. *PLOS ONE, 12*(10), e0185809.
Hansen, M. C., Potapov, P. V., Moore, R., Hancher, M., Turubanova, S. A., Tyukavina, A., Thau, D., Stehman, S. V., Goetz, S. J., Loveland, T. R., Kommareddy, A., Egorov, A., Chini, L., Justice, C. O., & Townshend, J. R. G. (2013). High-resolution global maps of 21st-century forest cover change. *Science, 342*(6160), 850–853.
Harrower, M., & Brewer, C. A. (2003). ColorBrewer.org: An online tool for selecting colour schemes for maps. *The Cartographic Journal, 40*(1), 27–37.
Hartman, E. J., & Abrahams, M. V. (2000). Sensory compensation and the detection of predators: The interaction between chemical and visual information. *Proceedings of the Royal Society B: Biological Sciences, 267*(1443), 571–575. https://doi.org/10.1098/rspb.2000.1039
Hesselbarth, M. H. K., Sciaini, M., With, K. A., Wiegand, K., & Nowosad, J. (2019). Landscapemetrics: An open-source R tool to calculate landscape metrics. *Ecography*, 42, 1648–1657.
Holdridge, L. R. (1947). Determination of world plant formations from simple climatic data. *Science, 105*(2727), 367–368.
Holdridge, L. R. (1967). *Life zone ecology* (rev. ed.). San José: Tropical Science Center.
Holt, B. G., Lessard, J.-P., Borregaard, M. K., Fritz, S. A., Araújo, M. B., Dimitrov, D., Fabre, P.-H., Graham, C. H., Graves, G. R., Jønsson, K. A., Nogués-Bravo, D., Wang, Z., Whittaker, R. J., Fjeldså, J., & Rahbek, C. (2013). An update of Wallace's zoogeographic regions of the world. *Science, 339*(6115), 74–78.
Homer, C. G., Dewitz, J. A., Yang, L., Jin, S., Danielson, P., et al. (2020). Completion of the 2016 National Land Cover Database for the Conterminous United States—Representing a decade of land cover change information. *Remote Sensing of Environment*, 237, 111485.
Homer, C., Huang, C., Yang, L., Wylie, B., & Coan, M. (2004). Development of a 2001 National Land-Cover Database for the United States. *Photogrammetric Engineering & Remote Sensing*, 70(7), 829–840.
Hrozencik, R. A., & Aillery, M. (2021). *Trends in U.S. irrigated agriculture: Increasing resilience under water supply scarcity* (Economic Information Bulletin No. 229). U.S. Department of Agriculture, Economic Research Service.
Humboldt, A. von, & Bonpland, A. (1807). *Essai sur la géographie des plantes, accompagné d'un tableau physique des régions équinoxiales*. Paris: Levrault & Schoell; Tübingen: J. G. Cotta.
Huntley, E. R., & Rosenblum, M. (2020). The omega affair: Discontinuing the University of Michigan department of geography (1975–1982). *Annals of the American Association of Geographers, 110*(2), 364–384.
Huss, C. P., Holmes, K. D., & Blubaugh, C. K. (2022). Benefits and risks of intercropping for crop resilience and pest management. *Journal of Economic Entomology, 115*(5), 1350–1362.
Huston, M. A. (1994). *Biological diversity: The coexistence of species on changing landscapes*. Cambridge University Press.
International Cartographic Association (ICA). (1966). *Definitions*. In *Proceedings of the 6th International Cartographic Conference*, Amsterdam.
International Union for Conservation of Nature (IUCN). (2025, April 16). *The benefits of bats: Celebrating the critical contributions of bats in International Bat Appreciation Day*. IUCN.
IPCC. (2021). Annex VII: Glossary (J. B. R. Matthews, V. Möller, R. van Diemen, J. S. Fuglestvedt, V. Masson-Delmotte, C. Méndez, S. Semenov, & A. Reisinger, Eds.). In V. Masson-Delmotte et al. (Eds.), *Climate change 2021: The physical science basis* (pp. 2215–2256). Cambridge University Press. https://doi.org/10.1017/9781009157896.022.
Isaaks, E. H., & Srivastava, R. M. (1989). *Applied geostatistics*. Oxford University Press.
Itten, J. (1961). *The art of color*. Van Nostrand Reinhold.
Itten, J. (1970). *The elements of color: A treatise on the color system of Johannes Itten based on his book The art of color* (E. van Hagen, Ed.; E. L. van Hagen, Trans.). Van Nostrand Reinhold.
IUCN. (2019). Mapping standards and data quality for the IUCN red list spatial data (Version 1.18). IUCN.

IUCN Standards and Petitions Committee. (2022). *Guidelines for using the IUCN red list categories and criteria version 14* (Version 14). IUCN. https://www.iucnredlist.org/resources/redlistguidelines

IUCN Standards and Petitions Committee. (2024, March). *Guidelines for using the IUCN red list categories and criteria* (Version 16). IUCN.

IUCN. (2025). *Background and history*. https://www.iucnredlist.org/about/background-history

Jackson, M. J., Bell, S. B. M., & Diaz, B. M. (1983). Database developments in the experimental cartography unit. *Cartographica: The International Journal for Geographic Information and Geovisualization*, *20*(3), 55–68.

Jasinski, M. F., Stutzer, D. A., & Siqueira, P. (2023). Remote sensing of fractional vegetation and canopy structure: Review and outlook for next-generation observations. *Remote Sensing of Environment*, 295, 113641.

Jebb, D., Huang, Z., Pippel, M., Hughes, G. M., Lavrichenko, K., Devanna, P., Winkler, S., Jermiin, L. S., Skirmuntt, E. C., Katzourakis, A., Burkitt-Gray, L., Ray, D. A., Sullivan, K. A. M., Roscito, J. G., Kirilenko, B. M., Dávalos, L. M., Corthals, A. P., Power, M. L., Jones, G., . . . Teeling, E. C. (2020). Six reference-quality genomes reveal evolution of bat adaptations. *Nature*, *583*(7817), 578–584. https://doi.org/10.1038/s41586-020-2486-3

Jones, K. A., Niknami, L. S., Buto, S. G., & Decker, D. (2022). *Federal standards and procedures for the National Watershed Boundary Dataset (WBD)* (5th ed.; Techniques and methods 11–A3). US Geological Survey. https://doi.org/10.3133/tm11A3

Joppa, L. N., Butchart, S. H. M., Hoffmann, M., Bachman, S. P., Akçakaya, H. R., Moat, J. F., Böhm, M., Holland, R. A., Newton, A., Polidoro, B., & Hughes, A. (2016). Impact of alternative metrics on estimates of extent of occurrence for extinction risk assessment. *Conservation Biology*, *30*(2), 362–370. https://doi.org/10.1111/cobi.12591

Kahl, S., Wood, C. M., Eibl, M., & Klinck, H. (2021). BirdNET: A deep learning solution for avian diversity monitoring. *Ecological Informatics, 61*, 101236. https://doi.org/10.1016/j.ecoinf.2021.101236

Kaiser, C. (2024). *Largest wind farms in the United States*. LandApp. https://www.landapp.com/post/largest-wind-farms-in-the-us?utm_source=chatgpt.com (accessed Oct. 4, 2025).

Kareiva, P., & Marvier, M. (2003). Conserving biodiversity coldspots: Recent calls to direct conservation funding to the world's biodiversity hotspots may be bad investment advice. *American Scientist, 91*(4), 344–351.

Kareiva, P., & Marvier, M. (2012). What is conservation science?. *BioScience, 62*(11), 962–969.

Keane, R. E., Loehman, R. A., & Holsinger, L. M. (2019). Selecting a landscape model for natural resource management applications. *Current Landscape Ecology Reports*, 4, 31–40. https://doi.org/10.1007/s40823-019-00036-6

Kingston, T., Florens, F. B. V., & Vincenot, C. E. (2023). Large Old World fruit bats on the brink of extinction: Causes and consequences. *Annual Review of Ecology, Evolution, and Systematics, 54*, 237–257. https://doi.org/10.1146/annurev-ecolsys-110321-055122

Klein Goldewijk, K., Beusen, A., van Drecht, G., & de Vos, M. (2011). The HYDE 3.1 spatially explicit database of human-induced global land-use change over the past 12,000 years. *Global Ecology and Biogeography*, *20*(1), 73–86. https://doi.org/10.1111/j.1466-8238.2010.00587.x

Köppen, W. (1884). Die Wärmezonen der Erde, nach der Dauer der heißen, mäßigen und kalten Zeit und nach der Wirkung der Wärme auf die organische Welt betrachtet.

Köppen, W., & Geiger, R. (1936). *Handbuch der Klimatologie*.

Kunz, T. H., de Torrez, E. B., Bauer, D., Lobova, T., & Fleming, T. H. (2011). Ecosystem services provided by bats. *Annals of the New York Academy of Sciences, 1223*(1), 1–38.

Lambin, E. F., Geist, H. J., & Lepers, E. (2003). Dynamics of land-use and land-cover change in tropical regions. *Annual Review of Environment and Resources*, *28*(1), 205–241.

Lambin, E. F., Turner, B. L., Geist, H. J., Agbola, S. B., Angelsen, A., Bruce, J. W., . . . Xu, J. (2001). The causes of land-use and land-cover change: Moving beyond the myths. *Global Environmental Change*, *11*(4), 261–269.

Lan, Y., Zhang, K., Han, X., Chen, Z., Ling, M., You, H., & Chen, J. (2024). The spatiotemporal variation in biodiversity and its response to different future development scenarios: A case study of Guilin as an internationally renowned tourist destination in China. *Applied Sciences*, *14*(5), 2101. https://doi.org/10.3390/app14052101

Langefors, B. (1966). *Theoretical analysis of information systems*. Studentlitteratur.

Lark, T. J., Salmon, J. M., & Gibbs, H. K. (2015). Cropland expansion outpaces agricultural and biofuel policies in the United States. *Environmental Research Letters*, *10*(4), 044003.

Latham, J., Cumani, R., Rosati, I., & Bloise, M. (2014). *Global land cover SHARE (GLC-SHARE) database beta-release version 1.0–2014*. Food and Agriculture Organization of the United Nations.

Laudon, K. C., & Laudon, J. P. (2018). *Management information systems: Managing the digital firm* (15th ed.). Pearson.

Leal, W. S. (2013). Odorant reception in insects: Roles of receptors, binding proteins, and degrading enzymes. *Annual Review of Entomology, 58*, 373–391.

Lee, T., Quinn, M. S., & Duke, D. (2006). Citizen, science, highways, and wildlife: Using a web-based GIS to engage citizens in collecting wildlife information. *Ecology and Society, 11*(1).

Leemans, R. (1990). *Possible changes in natural vegetation patterns due to global warming* (IIASA Working Paper WP-90-008). International Institute for Applied Systems Analysis.

Lehmann, J., Bossio, D. A., Kögel-Knabner, I., & Rillig, M. C. (2020). The concept and future prospects of soil health. *Nature Reviews Earth & Environment, 1*(10), 544–553.

Leuschner, W. A., Young, J. A., Waldon, S. A., & Ravlin, F. W. (1996). Potential benefits of slowing the gypsy moth's spread. *Southern Journal of Applied Forestry, 20*(2), 65–73.

Li, X., Tian, H., Lu, C., & Pan, S. (2023). Four-century history of land transformation by humans in the United States (1630–2020): Annual and 1 km grid data for the HIStory of LAND changes (HISLAND-US). *Earth Syst. Sci. Data*, 15, 1005–1035. https://doi.org/10.5194/essd-15-1005-2023

Liebhold, A. M., Halverson, J. A., & Elmes, G. A. (1992). Gypsy moth invasion in North America: A quantitative analysis. *Journal of Biogeography*, 513–520.

Locey, K. J., & Lennon, J. T. (2016). Scaling laws predict global microbial diversity. *Proceedings of the National Academy of Sciences*. https://doi.org/10.1073/pnas.1521291113

Lomolino, M. V., Riddle, B. R., & Whittaker, R. J. (2017). *Biogeography* (5th ed.). Sinauer Associates.

López-Vicente, M., Pérez-Bielsa, C., López-Montero, T., Lambán, L. J., & Navas, A. (2014). Runoff simulation with eight different flow accumulation algorithms: Recommendations using a spatially distributed and open-source model. *Environmental Modelling & Software*, 62, 11–21.

Losos, J. B., & Ricklefs, R. E. (Eds.). (2009). *The theory of island biogeography revisited*. Princeton University Press.

Lugo, A. E., Brown, S. L., Dodson, R., Smith, T. S., & Shugart, H. H. (1999). The Holdridge life zones of the conterminous United States in relation to ecosystem mapping. *Journal of Biogeography, 26*(5), 1025–1039.

MacArthur, R. H. (1958). Population ecology of some warblers of northeastern coniferous forests. *Ecology, 39*(4), 599–619.

MacArthur, R. H., & Wilson, E. O. (1963). An equilibrium theory of insular zoogeography. *Evolution, 17*(4), 373–387.

MacArthur, R. H., & Wilson, E. O. (1967). *The theory of island biogeography*. Princeton University Press.

Máchová, R., & Lněnička, M. (2017). Evaluating the quality of open data portals on the national level. *Journal of Theoretical and Applied Electronic Commerce Research, 12*(1), 21–41. https://doi.org/10.4067/S0718-18762017000100003

Maguire, D. J. (1991). An overview and definition of GIS. In D. J. Maguire, M. F. Goodchild, & D. W. Rhind (Eds.), *Geographical information systems: Principles and applications* (Vol. 1, pp. 9–20). Longman/John Wiley & Sons.

Maguire, K., Tanner, S., Winikoff, J. B., & Williams, R. (2024). *Utility-scale solar and wind development in rural areas: Land cover change (2009–20)*. US Department of Agriculture, Economic Research Service. ERR-330.

Maidment, D. R. (Ed.). (1993). *Handbook of hydrology*. McGraw-Hill.

Maidment, D. R. (Ed.). (2002). *Arc hydro: GIS for water resources*. Esri Press.

Malonza, P. K. (2015). Patterns of reptile and amphibian species richness along an elevational gradient in Mt Kenya (central Kenya). *PLoS ONE*.

Marschner, F. J. (1930, rev. 1959). *The original vegetation of Minnesota*. US Department of Agriculture.

Marshall, J. (1957). *The birds of the pine bark woodland in southern Arizona and adjacent Mexico*. Pacific Coast, Avifauna no. 32.

Masser, I. (2005). *GIS worlds: Creating spatial data infrastructures*. ESRI Press.

Matheron, G. (1963). Principles of geostatistics. *Economic Geology, 58*(8), 1246–1266.

Mayo, J. H., Straka, T. J., & Leonard, D. S. (2003). The cost of slowing the spread of the gypsy moth (*Lepidoptera: Lymantriidae*). *Journal of Economic Entomology, 96*(5), 1448–1454.

Mayr, E. (1933). Die Vogelwelt Polynesiens. *Mitteilungen aus dem Zoologischen Museum in Berlin, 19*, 306–323.

Mayr, E. (1940). The origin and history of the bird fauna of Polynesia. In *Proceedings of the Sixth Pacific Science Congress* (Vol. 4, pp. 197–216).

Mayr, E. (1943). The zoogeographic position of the Hawaiian Islands. *The Condor*, *45*(2), 45–48. https://doi.org/10.2307/1364376

Mayr, E. (1944). Wallace's line in the light of recent zoogeographic studies. *The Quarterly Review of Biology*, *19*(1), 1–14. https://doi.org/10.1086/394684

McGarigal, K., Cushman, S. A., & Ene, E. (2012). *FRAGSTATS v4: Spatial pattern analysis program for categorical and continuous maps* [Computer software]. University of Massachusetts, Amherst.

McGarigal, K., & Marks, B. J. (1995). *FRAGSTATS: Spatial pattern analysis program for quantifying landscape structure* (Gen. Tech. Rep. PNW-GTR-351). US Department of Agriculture, Forest Service, Pacific Northwest Research Station.

McGarigal, K., Tagil, S., & Cushman, S. A. (2009). Surface metrics: An alternative to patch metrics for the quantification of landscape structure. *Landscape Ecology*, *24*(3), 433–450.

MDNR (Michigan Department of Natural Resources). (1981). *Michigan resource information system (MiRIS): Land use/land cover mapping program—Technical documentation.* Land Resource Programs Division, Michigan DNR.

Mittelbach, G. G., et al. (2007). Evolution and the latitudinal diversity gradient: Speciation, extinction and biogeography. *Ecology Letters*, *10*(4), 315–331.

Monke, J. (2024). *Farm bill primer: Background and status* (CRS Report IF12047). Congressional Research Service.

Monsi, M., & Saeki, T. (1953). Über den Lichtfaktor in den Pflanzengesellschaften und seine Bedeutung für die Stoffproduktion. *Japanese Journal of Botany*, 14, 22–52.

Montgomery, S. L. (2017). *The Chicago guide to communicating science*. University of Chicago Press.

Mora, C., Tittensor, D. P., Adl, S., Simpson, A. G. B., & Worm, B. (2011). How many species are there on earth and in the ocean? *PLoS Biol*, *9*(8), e1001127. https://doi.org/10.1371/journal.pbio.1001127

Moran, P. A. P. (1950). Notes on continuous stochastic phenomena. *Biometrika*, *37*(1–2), 17–23.

Morgan, R. P. C. (2009). *Soil erosion and conservation*. John Wiley & Sons.

Myers, D. E. (1994). Spatial interpolation: An overview. *Geoderma*, *62*(1–3), 17–28.

Naiman, R. J., & Rogers, K. H. (1997). Large animals and system-level characteristics in river corridors: Implications for river management. *BioScience*, *47*(8), 521–529. https://doi.org/10.2307/1313120

Nair, P. R., & Nair, V. D. (2014). *Scientific writing and communication in agriculture and natural resources* (No. 15530). Springer International Publishing.

Nassauer, J. I. (1995). Messy ecosystems, orderly frames. *Landscape Journal*, *14*(2), 161–170.

Nearing, M. A. (2013). Soil erosion and conservation. In J. Wainwright & M. Mulligan (Eds.), *Environmental modelling: Finding simplicity in complexity* (2nd ed.). John Wiley & Sons.

Notter, B., MacMillan, L., Viviroli, D., Weingartner, R., & Liniger, H. P. (2007). Impacts of environmental change on water resources in the Mt. Kenya region. *Journal of Hydrology*, *343*(3–4), 266–278.

O'Callaghan, J. F., & Mark, D. M. (1984). The extraction of drainage networks from digital elevation data. *Computer Vision, Graphics, and Image Processing*, *28*(3), 323–344. https://doi.org/10.1016/S0734-189X(84)80011-0

O'Donoghue et al. (2017). Saving the mountain bongo (*Tragelaphus eurycerus isaaci*): Assessment of the genetic status of captive bongos as a source for genetic reinforcement of wild populations. *Journal of Zoo & Aquarium Research*, *5*(3), 123–132.

Olson, D. M., Dinerstein, E., Wikramanayake, E. D., Burgess, N. D., Powell, G. V. N., Underwood, E. C., D'Amico, J. A., Itoua, I., Strand, H. E., Morrison, J. C., Loucks, C. J., Allnutt, T. F., Ricketts, T. H., Kura, Y., Lamoreux, J. F., Wettengel, W. W., Hedao, P., & Kassem, K. R. (2001). Terrestrial ecoregions of the world: A new map of life on Earth. *BioScience*, *51*(11), 933–938. https://doi.org/10.1641/0006-3568(2001)051[0933:TEOTWA]2.0.CO;2

Olson, J. M., Alagarswamy, G., Andresen, J. A., Campbell, D. J., Davis, A. Y., Ge, J., Huebner, M., Lofgren, B. M., Lusch, D. P., Moore, N. J., & Pijanowski, B. C. (2008). Integrating diverse methods to understand climate–land interactions in East Africa. *Geoforum*, *39*(2), 898–911.

Olson, J. M., Butt, B., Atieno, F., Maitima, J., Smucker, T. A., Muchungu, E., . . . Xu, H. (2004). *Multi-scale analysis of land use and management change on the eastern slopes of Mt. Kenya*. LUCID Project, International Livestock Research Institute.

Omernik, J. M. (1987). Ecoregions of the conterminous United States. *Annals of the Association of American Geographers, 77*(1), 118–125. https://doi.org/10.1111/j.1467-8306.1987.tb00149.x

Omernik, J. M., & Griffith, G. E. (2014). Ecoregions of the conterminous United States: Evolution of a hierarchical spatial framework. *Environmental Management, 54*(6), 1249–1266. https://doi.org/10.1007/s00267-014-0364-1

O'Neill, R. V., Gardner, R., & Turner, M. G. (1992). A hierarchical neutral model for landscape analysis. *Landscape Ecology, 7*(1), 55–61.

Ord, J. K., & Getis, A. (1995). Local spatial autocorrelation statistics: Distributional issues and an application. *Geographical Analysis, 27*(4), 286–306. https://doi.org/10.1111/j.1538-4632.1995.tb00912.x

O'Sullivan, D., & Perry, G. L. W. (2013). Spatial simulation: Exploring pattern and process. John Wiley & Sons.

Pauley, M., & Hill, A. (2025). Sources of variance in the volatile contribution of juniper to gin. *Journal of the Institute of Brewing, 131*(2).

Peat, J., Elliott, E., Baur, L., & Keena, V. (2013). *Scientific writing: Easy when you know how*. John Wiley & Sons.

Peuquet, D. J. (1984, repr. 2011). A conceptual framework and comparison of spatial data models. *Cartographica: The International Journal for Geographic Information and Geovisualization, 21*(4), 66–113.

Pidgeon, A. M., Radeloff, V. C., Flather, C. H., Lepczyk, C. A., Clayton, M. K., Hawbaker, T. J., & Hammer, R. B. (2007). Associations of forest bird species richness with housing and landscape patterns across the USA. *Ecological Applications, 17*(7), 1989–2010. https://doi.org/10.1890/06-1489.1

Pijanowski, B. C. (2024). *Principles of soundscape ecology: Discovering our sonic world*. University of Chicago Press. https://doi.org/10.7208/chicago/9780226824284.001.0001

Pijanowski, B. C., Farina, A., Gage, S. H., Dumyahn, S. L., & Krause, B. L. (2011). What is soundscape ecology? An introduction and overview of an emerging new science. *Landscape Ecology, 26*(9), 1213–1232. https://doi.org/10.1007/s10980-011-9600-8

Pijanowski, B. C., & Rivas Fuenzalida, F. (2024). Soundscapes and vibroscapes. In S. M. Scheiner (Ed.), *Encyclopedia of biodiversity* (3rd ed., pp. 408–431). Elsevier.

Pijanowski, B. C., Rivas Fuenzalida, F., Banerjee, S., Minghim, R., Lima, S. L., Bowers-Sword, R., Ruiz Guzman, S., Revuelta-Acosta, J., Adeniji, A. E., Grimes, S. E., Sarker, S. K., Hossain, M. R., Tasnim Anika, T., & Savage, D. (2024). Soundscape analytics: A new frontier of knowledge discovery in soundscape data. *Current Landscape Ecology Reports, 9*, 88–107. https://doi.org/10.1007/s40823-024-00101-9

Pijanowski, B. C., Rivas Fuenzalida, F., Kang, J., Werner, C., Francis, A., McPherson, M., Heinz, M., Grimes, S., Post, J., Rhodes, O. E., Mediastika, C. E., Sors, T. G., & Park, T. H. (2025). A sonic One Health framework: Characterizing health of the environment, animals, and humans using a soundscape perspective. *BioScience*. Advance online publication. https://doi.org/10.1093/biosci/biaf109

Pijanowski, B. C., & Robinson, K. D. (2011). Rates and patterns of land use change in the Upper Great Lakes States, USA: A framework for spatial temporal analysis. *Landscape and Urban Planning, 102*(2), 102–116.

Pijanowski, B. C., Villanueva-Rivera, L. J., Dumyahn, S. L., Farina, A., Krause, B. L., Napoletano, B. M., Gage, S. H., & Pieretti, N. (2011). Soundscape ecology: The science of sound in the landscape. *BioScience, 61*(3), 203–216. https://doi.org/10.1525/bio.2011.61.3.6

Pimentel, D., & Kounang, N. (1998). Ecology of soil erosion in ecosystems. *Ecosystems, 1*(5), 416–426.

Plourde, J. D., Pijanowski, B. C., & Pekin, B. K. (2013). Evidence for increased monoculture cropping in the central United States. *Agriculture, Ecosystems & Environment, 165*, 50–59.

Pogue, M. G. (2009). Biodiversity of *Lepidoptera*. In R. G. Foottit & P. H. Adler (Eds.), *Insect biodiversity: Science and society* (pp. 325–355). Wiley-Blackwell.

Pontius Jr., R. G., Shusas, E., & McEachern, M. (2004). Detecting important categorical land changes while accounting for persistence. *Agriculture, ecosystems & environment, 101*(2–3), 251–268.

Porteous, J. D. (1985). Smellscape. *Progress in Human Geography, 9*(3), 356–378.

Potapov, P., Li, X., Hernandez-Serna, A., Tyukavina, A., Hansen, M., Kommareddy, A., Pickens, A., Turubanova, A., Tang, H., Silva, C .E., Armston, J., Dubayah, R., Blair, J. B., Hofton, M. (2020) Mapping global forest canopy height through integration of GEDI and Landsat data. Remote Sensing of Environment, 112165.

Powell, A. (2007, October 11). Biologists remember landmark theory. *Harvard Gazette*.

Purrington, C., (2025). *Poster design recommendations*. https://colinpurrington.com/tips/poster-design (accessed December 12, 2025).

Rabinowitz, D. (1981). Seven forms of rarity. In H. Synge (Ed.), *The biological aspects of rare plant conservation* (pp. 205–217). John Wiley & Sons.

Raisz, E. (1948). *General cartography*. McGraw-Hill.

Ramsar Convention Secretariat. (2016). *An introduction to the Ramsar Convention on wetlands* (7th ed.). Gland, Switzerland: Ramsar Convention Secretariat. (Original work published 1971). Retrieved from https://www.ramsar.org/sites/default/files/documents/library/handbook1_5ed_introductiontoconvention_final_e.pdf

RESOLVE Biodiversity & Wildlife Solutions, World Wildlife Fund (WWF), & partners. (2017). *Ecoregions 2017* [Data set; global terrestrial ecoregions grouped into 14 biomes and 8 realms]. Google Earth Engine/Ecoregions2017 (CC-BY 4.0).

Rew, R., & Davis, G. (1990). NetCDF: An interface for scientific data access. *IEEE Computer Graphics and Applications, 10*(4), 76–82.

Rhind, D. W. (2008). *Encyclopedia of geographic information science* (K. Kemp, ed.). Sage.

Rich, P. M. (1990). Characterizing plant canopies with hemispherical photographs. *Remote Sensing Reviews*, *5*(1), 13–29.

Riitters, K. H., O'Neill, R. V., Hunsaker, C. T., Wickham, J. D., Yankee, D. H., Timmins, S. P., Jones, K. B., & Jackson, B. L. (1995). A factor analysis of landscape pattern and structure metrics. *Landscape Ecology, 10*(1), 23–39. https://doi.org/10.1007/BF00158551

Ripple, W. J., Wolf, C., Newsome, T. M., Galetti, M., Alamgir, M., Crist, E., . . . & 15,364 Scientist Signatories from 184 Countries. (2017). World scientists' warning to humanity: A second notice. *BioScience*, *67*(12), 1026–1028.

Rosenberg, K. V., Dokter, A. M., Blancher, P. J., Sauer, J. R., Smith, A. C., Smith, P. A., . . . Marra, P. P. (2019). Decline of the North American avifauna. *Science*, *366*(6461), 120–124.

Sánchez-Bayo, F., & Wyckhuys, K. A. G. (2019). Worldwide decline of the entomofauna: A review of its drivers. *Biological Conservation, 232*, 8–27.

Sangermano, F. (2022). Acoustic diversity of forested landscapes: Relationships to habitat structure and anthropogenic pressure. *Landscape and Urban Planning, 226*, 104508. https://doi.org/10.1016/j.landurbplan.2022.104508

Schaible, G. (2017). Use of pressure-sprinkler systems for irrigation has increased in the western United States. ERS-USDA. https://ers.usda.gov/data-products/charts-of-note/chart-detail?chartId=84617#:~:text=There%20are%20two%20main%20types,%2C%20%22Understanding%20Irrigated%20Agriculture.%22 (accessed October 4, 2025).

Seaber, P. R., Kapinos, F. P., & Knapp, G. L. (1987). *Hydrologic unit maps* (U.S. Geological Survey Water-Supply Paper 2294). US Government Printing Office. https://doi.org/10.3133/wsp2294

Seibold, S., Gossner, M. M., Simons, N. K., Blüthgen, N., Müller, J., Ambarlı, D., . . . Weisser, W. W. (2019). Arthropod decline in grasslands and forests is associated with landscape-level drivers. *Nature*, *574*, 671–674.

Sharov, A. A., Leonard, D., Liebhold, A. M., Roberts, E. A., & Dickerson, W. (2002). "Slow the spread": A national program to contain the gypsy moth. *Journal of Forestry*, *100*(5), 30–36.

Shelford, V. E. (1963). *The ecology of North America*. University of Illinois Press.

Simberloff, D. (2022). Edward O. Wilson (1929–2021): It all started with ants. *The American Naturalist*, 200(5), 627–633. https://doi.org/10.1086/721257

Simunovic, M. P. (2010). Colour vision deficiency. *Eye*, *24*(5), 747–755. https://doi.org/10.1038/eye.2009.251

Skalski, J. R., & Robson, D. S. (2012). *Techniques in wildlife investigations: Design and analysis of capture data*. Academic Press.

Skellam, J. G. (1951). Random dispersal in theoretical populations. *Biometrika*, *38*(1–2), 196–218.

Smith, N. (1987). Academic war over the field of geography: The elimination of geography at Harvard, 1947–1951. *Annals of the Association of American Geographers*, *77*(2), 155–172.

Sollaci, L. B., & Pereira, M. G. (2004). The introduction, methods, results, and discussion (IMRAD) structure: A fifty-year survey. *Journal of the Medical Library Association*, *92*(3), 364.

Soulé, M. E. (1985). What is conservation biology? *BioScience*, *35*(11), 727–734.

Soulé, M. E., & Simberloff, D. (1986). What do genetics and ecology tell us about the design of nature reserves? *Biological Conservation*, *35*(1), 19–40.

Souza, C. M., Shimbo, J. Z., Rosa, M. R., et al. (2020). Reconstructing three decades of land use and land cover changes in Brazil using Landsat imagery: The MapBiomas project. *Remote Sensing*, *12*(17), 2735.

Spalding, M. D., Fox, H. E., Allen, G. R., Davidson, N., Ferdaña, Z. A., Finlayson, M. A. X., Halpern, B. S., Jorge, M. A., Lombana, A. L., Lourie, S. A. & Martin, K. D., (2007). Marine ecoregions of the world: a bioregionalization of coastal and shelf areas. *BioScience*, 57(7): 573–583.

Stoddard, J. L., Herlihy, A. T., Peck, D. V., Hughes, R. M., Whittier, T. R., & Tarquinio, E. (2008). A process for creating multimetric indices for large-scale aquatic surveys. *Journal of the North American Benthological Society*, *27*(4), 878–891. https://doi.org/10.1899/08-053.1

Stork, N. E. (2018). How many species of insects and other terrestrial arthropods are there on Earth? *Annual Review of Entomology*, *63*, 31–45.

Strahler, A. N. (1952). Hypsometric (area-altitude) analysis of erosional topography. *Geological Society of America Bulletin*, *63*(11), 1117–1142.

Sueur, J., Pavoine, S., Hamerlynck, O., & Duvail, S. (2008). Rapid acoustic survey for biodiversity appraisal. *PLoS ONE*, *3*(12), e4065. https://doi.org/10.1371/journal.pone.0004065

Sundaram, M., Slik, J. W. F., Sørensen, M. M., Faurby, S., & Svenning, J.-C. (2019). Accumulation over evolutionary time as a major cause of current richness patterns in the world's woody flora. *Proceedings of the Royal Society B: Biological Sciences*, *286*(1914), 20191887. https://doi.org/10.1098/rspb.2019.1887

Tang, Z., Engel, B. A., Pijanowski, B. C., & Lim, K. J. (2005). Forecasting land use change and its environmental impact at a watershed scale. *Journal of Environmental Management*, *76*(1), 35–45.

Taufer, F., Pálsdóttir, A. M., & Hedblom, M. (2025). Psychological and physiological responses to smells from nature—Potential health benefits for urban dwellers. *NPJ Urban Sustainability*, *5*, 80. https://doi.org/10.1038/s42949-025-00274-0

ter Steege, H., Pitman, N. C. A., Sabatier, D., Baraloto, C., Salomão, R. P., . . . Silman, M. R. (2013). Hyperdominance in the Amazonian tree flora. *Science*, *342*(6156), Article 1243092. https://doi.org/10.1126/science.1243092

Thorsby, J. (2017). Understanding the content and features of open data portals in American cities. *Government Information Quarterly*, *34*(1), 53–61. https://doi.org/10.1016/j.giq.2016.12.002

Tihelka, E. (2021). The evolution of insect biodiversity. *Current Biology*, *31*(19), R1290–R1295.

Tilman, D., & Kareiva, P. (Eds.). (1997). *Spatial ecology: The role of space in population dynamics and interspecific interactions* (Vol. 30). Princeton University Press.

Tobin, P. C., & Blackburn, L. M. (Eds.). (2007). *Slow the spread: A national program to manage the gypsy moth* (General Technical Report NRS-6). U.S. Department of Agriculture, Forest Service, Northern Research Station. https://doi.org/10.2737/NRS-GTR-6

Tobin, P. C., & Whitmire, S. L. (2005). Spread of gypsy moth (*Lepidoptera: Lymantriidae*) and its relationship to defoliation. *Environmental Entomology*, *34*(6), 1448–1455.

Tomlinson, R. F. (1968). A geographic information system for regional planning. In G. A. Stewart (Ed.), *Land evaluation* (pp. 200–210). Macmillan.

Torge, W. (1980, revised 2001). *Geodesy* (3rd ed.). De Gruyter.

Torge, W., & Müller, J. (2012). *Geodesy* (4th ed.). De Gruyter.

Trubek, A. B. (2008). *The taste of place: A cultural journey into terroir*. University of California Press.

Tucker, C. J. (1979). Red and photographic infrared linear combinations for monitoring vegetation. *Remote Sensing of Environment*, *8*(2), 127–150.

Turner, M. G., Gardner, R. H., & O'Neill, R. V. (2001). *Landscape ecology in theory and practice: Pattern and process*. Springer-Verlag. https://doi.org/10.1007/b97434

UN DESA. (2026). *Handbook on the definition of key terms*. https://unstats.un.org/capacity-development/handbook/html/topic.htm (accessed Dec 5, 2025).

United States Department of Agriculture, Animal and Plant Health Inspection Service. (2024). *Spongy moth program manual.* https://www.aphis.usda.gov/sites/default/files/spongy-moth-program-manual.pdf

United States Department of Agriculture Forest Service. (1995). *Landscape aesthetics: A handbook for scenery management* (Agricultural Handbook 701). USDA Forest Service.

United States Department of Agriculture, National Agricultural Statistics Service. (n.d.). *Major cropping regions of the United States* [Map]. https://www.nass.usda.gov/Publications/AgCensus/2022/Online_Resources/Ag_Census_Web_Maps/Overview/

US Environmental Protection Agency. (2013). *Level III and IV ecoregions of the conterminous United States* (EPA/ORD). US EPA, Office of Research and Development.

US Environmental Protection Agency. (2023, July 25). *Nonpoint source pollution: The nation's largest water quality problem.* US Environmental Protection Agency. https://www.epa.gov/nps/nonpoint-source-pollution

USGS. (2017). *Irrigation methods: A quick look.* https://www.usgs.gov/water-science-school/science/irrigation-methods-a-quick-look#overview (accessed Oct. 4, 2025).

USGS. (2025). *Pesticide national synthesis project.* https://water.usgs.gov/nawqa/pnsp/usage/maps/compound_listing.php

van Everdingen, R. O. (Ed.). (1998, revised 2005). *Multi-language glossary of permafrost and related ground-ice terms.* International Permafrost Association/National Snow and Ice Data Center.

van Klink, R., Bowler, D. E., Gongalsky, K. B., Swengel, A. B., Gentile, A., & Chase, J. M. (2020). Meta-analysis reveals declines in terrestrial but increases in freshwater insect abundances. *Science, 368*(6489), 417–420.

Van Leeuwen, C., Roby, J.-P., & de Rességuier, L. (2018). Soil-related terroir factors: A review. *OENO One, 52*(2), 173–188. https://doi.org/10.20870/oeno-one.2018.52.2.2208

Verdin, K. L., & Verdin, J. P. (1999). A topological system for delineation and codification of the Earth's river basins. *Journal of Hydrology, 218*(1–2), 1–12. https://doi.org/10.1016/S0022-1694(99)00011-6

Vincent, H., Hole, D. G., Butchart, S. H. M., Willis, S. G., Martínez-Meyer, E., Rodríguez, J. P., & Seddon, A. W. R. (2022). Congruence between global crop wild relative hotspots, biodiversity hotspots, and protected areas. *Biological Conservation, 265*, 109402. https://doi.org/10.1016/j.biocon.2021.109402

Wagner, D. L. (2021). Insect declines in the Anthropocene: Death by a thousand cuts. *Proceedings of the National Academy of Sciences, 118*(2), e2023989118.

Wallace, A. R. (1876). *The geographical distribution of animals: With a study of the relations of living and extinct faunas as elucidating the past changes of the Earth's surface* (Vols. 1–2). London: Macmillan.

Walter, H. (1979). *Vegetation of the Earth and ecological systems of the geo-biosphere.* New York: Springer.

Wan, L., Kendall, A. D., Rapp, J., & Hyndman, D. W. (2024). Mapping agricultural tile drainage in the US Midwest using explainable random forest machine learning and satellite imagery. *Science of the Total Environment, 950*, 175283.

Ware, C. (2013). *Information visualization: Perception for design* (3rd ed.). Morgan Kaufmann.

Warshall, P. (1995). *Biodiversity and management of the Madrean Archipelago: The Sky Islands of southwestern United States and northwestern Mexico.* USDA Forest Service, General Technical Report RM-GTR-264.

Wells, M. C., & Lehner, P. N. (1978). The relative importance of the distance senses in coyote predatory behavior. *Animal Behaviour, 26*(1), 251–258. https://doi.org/10.1016/0003-3472(78)90025-8

Whitmire, S. L., & Tobin, P. C. (2006). Persistence of invading gypsy moth populations in the United States. *Oecologia, 147*(2), 230–237.

Whittaker, R. H. (1960). Vegetation of the Siskiyou Mountains, Oregon and California. *Ecological Monographs, 30*(3), 279–338.

Whittaker, R. H. (1972). Evolution and measurement of species diversity. *Taxon, 21*(2–3), 213–251.

Whittaker, R. H. (1975). *Communities and ecosystems* (2nd ed.). Macmillan.

Wiens, J. J. (2023). How many species are there on Earth? Progress and problems. *PLoS Biology, 21*(11), e3002388.

Williams, S. D., Barkla, B. J., Rose, T. J., & Liu, L. (2022). Does coffee have terroir and how should it be assessed? *Foods, 11*(13), 1907. https://doi.org/10.3390/foods11131907

Wilson, E. O. (1971). *The insect societies*. Belknap Press of Harvard University Press.

Wilson, E. O. (1975). *Sociobiology: The new synthesis*. Harvard University Press.

Wilson, E. O. (1994). *Naturalist*. Island Press.

Wilson, E. O. (2013). *Letters to a young scientist*. Liveright.

Wilson, E. O. (2016). *Half-Earth: Our planet's fight for life*. Liveright.

Wilson, J. P., & Gallant, J. C. (Eds.). (2000). *Terrain analysis: Principles and applications*. Wiley.

Winters-Michaud, C. P., & Hrozencik, R. A. (2024, October 23). *2022 Census of Agriculture: Irrigated cropland reaches a 30-year low in 2022*. U.S. Department of Agriculture, Economic Research Service.

With, K. A. (2002). The landscape ecology of invasive spread. *Conservation Biology*, *16*(5), 1192–1203.

With, K. A., & King, A. W. (1997). The use and misuse of neutral landscape models in ecology. *Oikos*, 219–229.

Wright, D. H. (1983). Species–energy theory. *Oikos*, 41, 496–506.

Wright, R. G., Murray, M. P., & Merrill, T. (1998). Ecoregions as a level of ecological analysis. *Biological Conservation*, *86*(2), 207–213. https://doi.org/10.1016/S0006-3207(98)00002-0

Wu, J., & Vankat, J. L. (1995). *Island biogeography: Theory and applications*. In W. A. Nierenberg (Ed.), *Encyclopedia of environmental biology* (Vol. 2, pp. 371–379). Academic Press.

Xian, G., & Homer, C. (2010). Updating the National Urban Impervious Surface Map to reflect 2001–2006 urban growth. *Photogrammetric Engineering & Remote Sensing*, *76*(7), 758–765.

Yang, D., Pijanowski, B. C., and Gage, S. H. (1998). Analysis of gypsy moth (*Lepidoptera*: *Lymantriidae*) population dynamics in Michigan using geographic information systems. *Environmental Entomology*, *27*(4), 842–852.

Yang, L., Huang, C., Homer, C., Wylie, B., & Coan, M. (2003). An approach for mapping large-area impervious surfaces: synergistic use of Landsat-7 ETM+ and high spatial resolution imagery. *Canadian Journal of Remote Sensing*, *29*(2).

Zanaga, D., Van De Kerchove, R., Daems, D., De Keersmaecker, W., Brockmann, C., Kirches, G., Wevers, J., Cartus, O., Santoro, M., Fritz, S., Lesiv, M., Herold, M., Tsendbazar, N.-E., Xu, P., Ramoino, F., & Arino, O. (2022). *ESA WorldCover 10 m 2021 v200*. https://pure.iiasa.ac.at/id/eprint/18478/.

Zhang, M., & Mei, J. (2008, December). Design and implementation for mobile-based GIS urban information system. In 2008 *International Symposium on Information Science and Engineering* (Vol. 1, pp. 464–467). IEEE.

Zhang, X., et al. (2021). A 30 m Global Land Cover Dataset for 2015: First integration of Landsat and MODIS data. *Earth System Science Data*, *13*, 2753–2772. https://doi.org/10.5194/essd-13-2753-2021

Zhang, X., Zhao, T., Xu, H., Liu, W., Wang, J., Chen, X., & Liu, L. (2024). GLC_FCS30D: The first global 30 m land-cover dynamics monitoring product with a fine classification system for the period from 1985 to 2022 generated using dense-time-series Landsat imagery and the continuous change-detection method. *Earth System Science Data*, *16*, 1353–1381. https://doi.org/10.5194/essd-16-1353-2024

Zhou et al. (2022) showed that global urban expansion between 2003 and 2018 increased mean annual LST by ~1.5 °C in urban cores. *Nature Communications*, 13, 6075.

Zulauf, C., Schnitkey, G., Coppess, J., & Paulson, N. (2024, September 18). Loss of US farmland in the 21st century: The national perspective from the census of agriculture. *Farmdoc Daily*, (14), 169. Department of Agricultural and Consumer Economics, University of Illinois at Urbana-Champaign.

INDEX

Page numbers in italics indicate Figures.

ABOUT THE AUTHOR

BRYAN C. PIJANOWSKI is a professor in the Department of Forestry and Natural Resources at Purdue University. He is also the director of the Center for Global Soundscapes and a recognized leader in soundscape ecology, a field of science and natural resource management that uses sound as a measure of environmental and cultural health. Pijanowski has published more than 190 peer-reviewed papers. He is an American Association for the Advancement of Science Fellow; a designated Distinguished Landscape Ecologist by the International Association of Landscape Ecology-North America (IALE-NA); the society's highest honor; a recipient of Purdue's Lowell S. Hardin Award for Excellence in International Agriculture; and serves as the executive producer of the giant-screen, big-screen, and domed-theater interactive educational experience *Global Soundscapes! A Mission to Record the Earth*. He is also the author of *Principles of Soundscape Ecology: Discovering Our Sonic World* and *Listening to Place: Participatory Soundscape Planning for Communities*.

www.ingramcontent.com/pod-product-compliance
Lightning Source LLC
Chambersburg PA
CBHW082007220726
48566CB00036B/163